# MICROBIOLOGY PERSPECTIVES

*A Photographic Survey of the Microbial World*

REPOBLIKAN'I MADAGASIKARA
PAOSITRA 1996
LOUIS PASTEUR
1822 - 1895
1750 FMG
ARIARY 350

GUINÉ-BISSAU
1988
correios
Amanita muscaria
reprodução assexual
780.POO

REPUBLIQUE FRANÇAISE
0,45
POSTES
1973
HANSEN
CENTENAIRE DE LA DÉCOUVERTE
DU BACILLE DE LA LEPRE

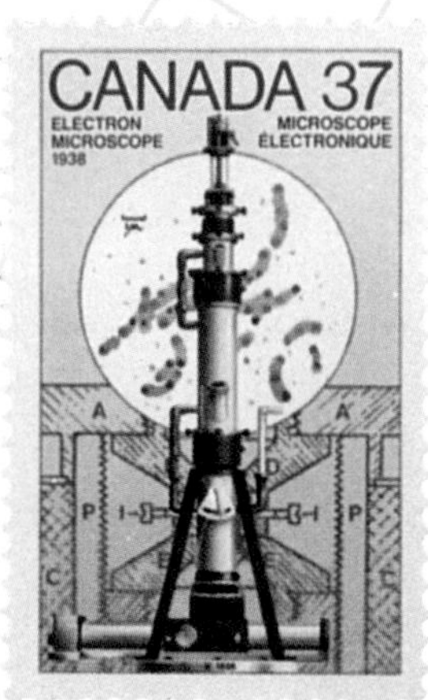
CANADA 37
ELECTRON
MICROSCOPE
1938
MICROSCOPE
ÉLECTRONIQUE

마라리아박멸운동 기념우표
1962년 4월7일 발행
대한민국우표
REPUBLIC OF KOREA
40
마라리아박멸운동기념
대한민국 체신부

MONACO
LUTTE CONTRE LA TUBERCULOSE
1996
8,00
DECOUVERTE DU VACCIN B.C.G. (1921)
CROIX-ROUGE MONÉGASQUE
C. Guérin & A. Calmette
P. LAMBERT
ITVF

45
AIDS
SIDA
One World. One Hope.
Unis dans l'espoir.
Canada

# MICROBIOLOGY PERSPECTIVES

## *A Photographic Survey of the Microbial World*

GEORGE WISTREICH
East Los Angeles College

PRENTICE HALL
Upper Saddle River, NJ 07458

**Library of Congress Cataloging-in-Publication Data**

Wistreich, George A.
Microbiology perspectives: a photographic survey of the microbial world / George A. Wistreich.
p. cm.
Includes index.
ISBN 0-13-856824-3 (papercover)
1. Microbiology--Atlases. I. Title.
QR54.W57 1998
579'.022'2--dc21 98-22035
CIP

Executive Editor: David Kendric Brake
Acquisitions Editor: Linda Schreiber
Assistant Vice President and Director of Production: David W. Riccardi
Special Projects Manager: Barbara A. Murray
Production Editor: Dawn Blayer
Cover Design: Joseph Sengotta

Cover Photos: **FRONT COVER**

**1st row**: Guiné-Bissau, 1988 issue of Amanita muscaria;

**2nd row**, *left*: Bahamas, 1982 issue celebrating the centenary of *Mycobacterium tuberculosis* discovery; *right*: Republique Francaise, 1973 issue showing Dr. Armauer G. Hansen and honoring the discovery of *Mycobacterium leprae*.

**3rd row**: *left*: Canada, 1996 issue promoting AIDS awareness; *center*: Canada, 1988 issue honoring the development of the electron microscope at the University of Toronto by James Hiller and Albert Prebus under the direction of Eli Burton; *right*: Republique de Guinee, 1970 issue to emphasize a campaign against smallpox and measles.

**4th row**: *left*: Korea, 1962 issue promoting the WHO drive to eradicate malaria; *center and right*: Republique de Guinee, 1970 issues showing a woman being immunized and Edward Jenner, respectively. (Stamps are part of a stamp set emphasizing a campaign to elimate smallpox and measles.)

**BACK COVER**

**Top**: Repoblikan '1 Madagasikara, 1995 issue honoring the achievements of Louis Pasteur.

**Middle**: Nigeria, 1978 issue promoting the global eradication of smallpox.

**Bottom**: Monaco, 1996 issue honoring the development of BCG vaccine by Camille Guérin and Albert Leon Calmette



Printed in the United States of America

10 9 8 7 6 5

ISBN 0-13-856824-3

Prentice-Hall International (UK) Limited, *London*
Prentice-Hall of Australia Pty. Limited, *Sydney*
Prentice-Hall Canada, Inc., *Toronto*
Prentice-Hall Hispanoamericana, S.A., *Mexico*
Prentice-Hall of India Private Limited, *New Delhi*
Prentice-Hall of Japan, Inc., *Tokyo*
Pearson Education Asia Pte. Ltd., *Singapore*
Editoria Prentice-Hall do Brasil, Ltda., *Rio de Janeiro*

*To Renée, my wife, and to my sons Eddie and Phillip, whose love and support have been such an important part of my personal and professional life, and to my colleagues, former instructors, and students, who have provided the motivation and inspiration to devolop my microbiology perspectives.*

# Contents

# Preface

*Microbiology Perspectives: A Survey of the Microbial World* provides a visual guide to many of the specific properties, activities, and procedures associated with microorganisms and helminths (worms). It also discusses selected features of a number of infectious diseases. In short, this book is intended to present microorganisms to anyone who is interested and curious about the microbial world. Individuals, either familiar or not familiar with this microscopic world of life, can gain some perspective as to the many types of microorganisms and their activities that exist in the world around them. For individuals not directly involved in industrial, laboratory, or medically related activities, this orientation and exposure provides a broad perspective of how microorganisms and their various activities influence everyday situations.

*Microbiology Perspectives* is not intended to replace currently available texts and manuals; rather, it is designed to show the reader several timely and relevant properties of microorganisms and the results of their actions, many of which either are not included or are not adequately presented in texts or manuals. Toward this end, a limited amount of explanatory material accompanies the many color photographs. This information also is included so that individual sections may serve as self-contained study guides and as aids in laboratory situations.

Perhaps more than most science specialties, microbiology depends heavily on color and visual representation of microbial reactions and activities. In the selection of illustrations for this publication, care was taken to include microorganisms and activities commonly encountered in the laboratory, as well as microorganism activities and situations that are not included in texts but are nevertheless significant to microbial or disease agent identification.

The color plates are used to illustrate the microbial colonial and microscopic features, selected biochemical characteristics of the most frequently encountered microorganisms, various disease states associated with microorganisms, and clinical and diagnostic features of selected infectious diseases. The illustrations demonstrate most effectively the way in which microorganisms and helminths appear in clinical laboratories and the various ways in which preliminary or definitive identifications are made. It should be noted, however, that no standard magnification or enlargement has been applied throughout. The sizes of organisms are not necessarily proportionate from one illustration to another. Magnifications and enlargements have been selected so as to provide the greatest clarity possible.

It is hoped that *Microbiology Perspectives* will be a useful learning tool and functional reference, not only for undergraduate students taking an introductory biological science or microbiology course but also for other students taking more advanced programs in the biological sciences, medical technology, medicine, biotechnology, dentistry, and veterinary medicine.

## Acknowledgment

As with any book, numerous people have been involved and have helped in its production, and I would like to take this opportunity to acknowledge them. I am particularly grateful to my fellow microbiologists and investigators from around the world who generously contributed their marvelous photographs of clinical states, microorganisms, and related subjects; to Difco Laboratories, Becton Dickinson and Company, and Becton Dickinson Microbiology Systems for providing samples of their products; to Sylvia Files for the effort and care she exercised in typing portions of the manuscript; to Dawn Blayer who masterminded the operation with enthusiasm and efficiency; to Barbara Murray for her invaluable expertise in guiding various aspects of publication; and to Linda Schreiber and David Brake for their generous support and interest in the development of true microbiological perspectives.

# Pronunciation Guide

In *Microbiology Perspectives* you will find phonetic pronunciations for many terms and specific microorganisms. Taking time to sound out new terms and to say them aloud once or twice will help you master one of the tasks in microbiology-related activities—learning its specialized vocabulary. The following key explains the system used for the pronunciations.

1. The strongest accented syllable appears in capital letters: e.g., microbe (MĪ-krōb) and microscope (MĪ-krō-skōp). A syllable that has a secondary accent is followed by a single prime ('): microaerophilic (mī-krō-Ā-er-o-fil-ī'k) and micrococcus (mī'-krō-KOK-us).
2. Vowels pronounced with long sounds are indicated by a line above the vowel and are pronounced as in the following words.
   ā as in māke
   ē as in bē
   ī as in ivy
   ō as in pōle
   ū as in ūnit
3. Vowels not marked for long sounds are pronounced with the short sound, as in the following words.
   a as in above
   e as in bet
   i as in sip
   o as in not
   u as in bud
4. Other phonetic symbols are used to indicate the following sounds.
   soo as in sue
   kyoo as cute
   oy as in oil

A list of most of the microorganisms described in *Microbiology Perspectives* and their phonetic pronunciations can be found on the front and back inside covers.

# About the Author

George Wistreich is Professor of Microbiology and Chair of the Department of Life Sciences, East Los Angeles College, Monterey Park, California. He has taught for over 30 years in the various areas of microbiology, general biology, and electron microscopy. He received his B.A. in Bacteriology and his M.S. in Infectious Diseases from the University of California, Los Angeles. After receiving his Ph.D., he completed a National Institutes of Health postdoctoral fellowship in the Department of Medical Microbiology and Immunology at the School of Medicine, University of Southern California. He has also completed several programs, seminars, workshops, and courses in molecular biology, genetic engineering, development in infectious diseases, and curriculum design. Dr. Wistreich belongs to a number of scientific and professional organizations; in addition, he has been elected as Fellow to the American Academy of Microbiology, the Linnean Society of London, the American Institute of Chemists, and the Royal Society of Health. He has served on the Board of Education, American Society of Microbiology for 12 years; nine of these years were in the capacity of chair of the Precollege Education Committee.

Among the awards he has received during his years of teaching are the Chancellor's award for student success and the outstanding educator award from the Chicanos for Creative Medicine. Dr. Wistreich is the author or co-author of over 20 scientific articles and over 60 laboratory manuals, workbooks, and textbooks.

gwrndx.exe

# SECTION 1

# Introduction

*It is characteristic of Science and Progress that they continually open new fields to our vision.*

*—Louis Pasteur*

Microbiology as a science includes the study of bacteria **(bacteriology)**, fungi **(mycology)**, viruses **(virology)**, protozoa **(protozoology)**, and certain algae (**algology** or **phycology**). The study of worms, or helminths **(helminthology)**, is frequently also incorporated into microbiology courses because of the ability of some of these forms of life to cause disease. In addition, the study of the immune system **(immunology)**, which includes approaches to disease prevention and diagnosis, is considered a subdivision of microbiology.

Microbiology had its beginnings with the discovery of microscopic forms of life by Anton van Leeuwenhoek in 1685. The major developments in microbiology did not begin, however, until in the 1870s, Louis Pasteur and others (Figure 1) conclusively demonstrated that microorganisms are living entities (organisms) capable of self-multiplication and are not spontaneously generated from nonliving organic matter. This section provides an introduction to the world of microorganisms.

## A Brief View of Classification

All forms of life having a cellular form of organization are grouped into one of five biological categories called kingdoms: **Prokaryotae** (the bacteria), **Protista** (protozoa and certain algae), **Fungi, Plantae**, and **Animalia**. Most **organisms** (living forms) are placed into these kingdoms according to properties that can be seen and measured. These include cell type, the number of cells in each organism, and the method of acquiring energy.

## A Word about Microscopes

Observing cells, the basic units of life, in action is made difficult by the fact that they are, in general, very small and transparent to visible light and thus are invisible to the unaided eye. The approaches used to overcome these limitations include **microscopy** and the application of dyes that stain different cellular parts. Staining techniques used in microbiology are discussed in a later section.

Since the early 1950s, the technical advances in microscopy and associated techniques have steadily provided scientists with new approaches to probe more deeply into the structure, organization, and functions of microorganisms and other forms of life (Figures 2 and 3).

The objectives of microscopy include (1) magnifying the observed image, (2) maximizing the resolving power (the ability to see detail), and (3) distinguishing the various elements in the material being viewed. The third objective involves providing **contrast** and is best achieved with the use of stains.

Historically, a significant amount of information about cells has been provided by light microscopy. Light microscopes can magnify about 2,000 times and show details as fine as 0.25 μm (micrometer). A micrometer is one-millionth of a meter (about 1/25,000th of an inch). With electron microscopy (a more specialized system for examining materials), a concentrated beam of electrons (negatively charged particles) can be used to examine cellular features. Two general types of electron microscopes are currently in use. The **transmission electron microscope (TEM)** provides flat, two-dimensional views (Figures 3 and 4), and the **scanning electron microscope (SEM)** adds depth to a view (Figure 2). A combined version of these two instruments, the **scanning transmission electron microscope (STEM)**, also exists. Table 1 compares selected microscopes used in microbiology and related sciences. A photograph of an image formed with a microscope is called a **micrograph**.

**Figure 1**
Photograph, taken in Paris in 1894, showing several outstanding French physicians and microbiologists of the time. Front row, left to right: A. Calmette, L. Martine, E. Roux, L. Pasteur, E. Nocard, H. Pottevin, and F. Mesnill, Back row, left to right: E. Viala, Rebound, L. Merieux, A. Fernbach, R. Chaillou, A. Borrel, H. Marnier, A. Marie, A. Veillon, and E. Fernback. (Courtesy of the Pasteur Institute.)

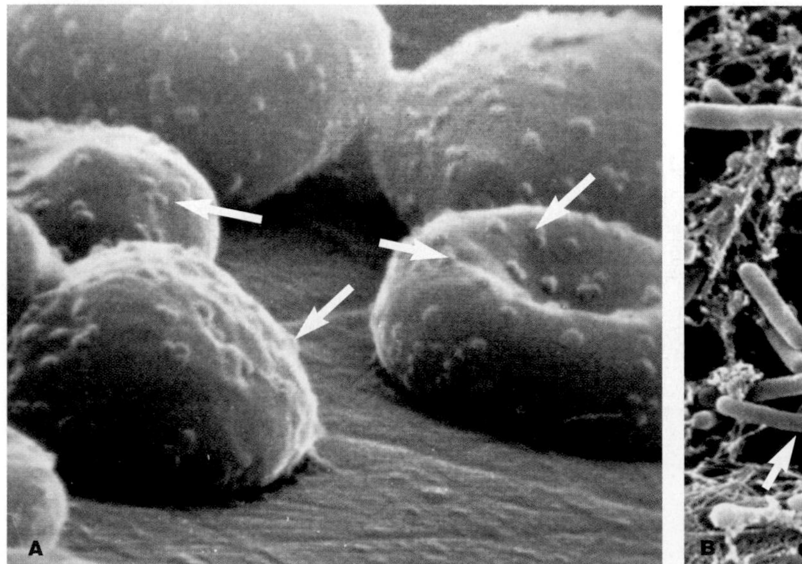


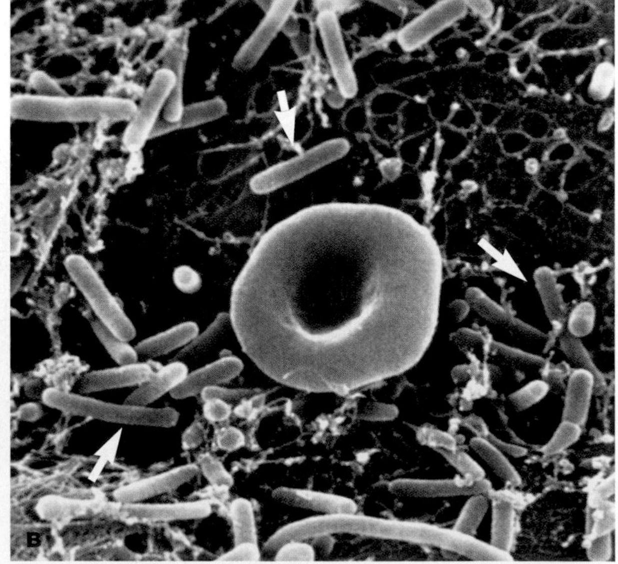


**Figure 2**
Size relationships and surface features of microorganisms as shown by scanning electron microscopy. (a) Influenza viruses (arrows) attached to the surfaces of red blood cells. (b) Bacteria (arrows) surrounding a single much larger human red blood cell. Refer to Figure 5 for actual sizes.

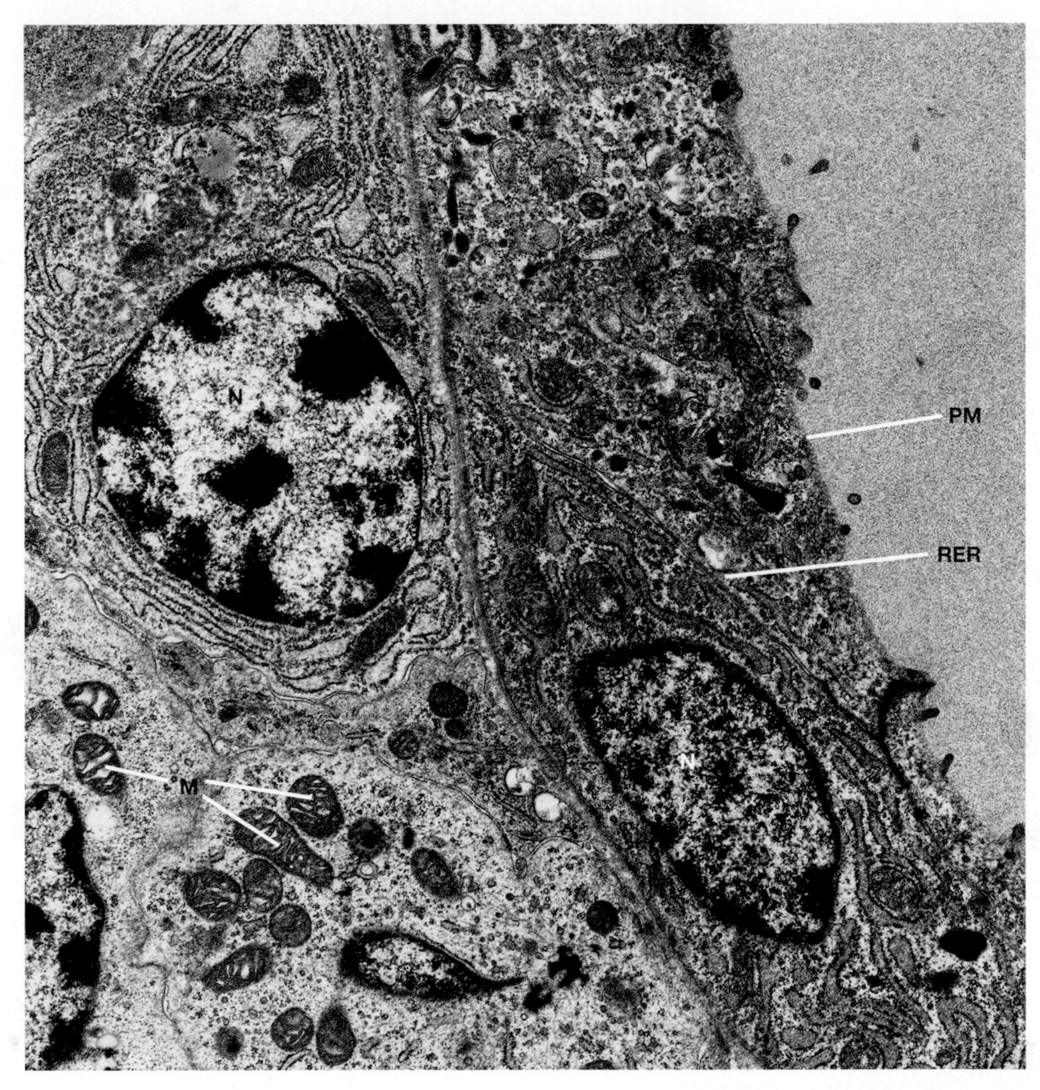


**Figure 3**
A transmission electron micrograph of two eukaryotic animal cells. A number of typical organelles can be seen. N, nucleus; M, mitochondria; RER, rough endoplasmic reticulum; PM, plasma membrane.

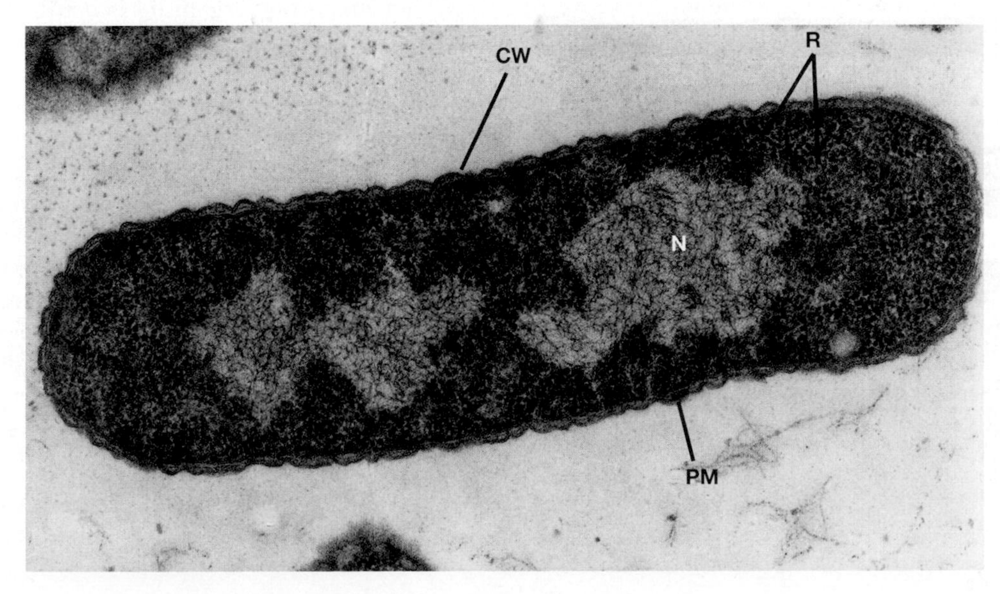


**Figure 4**
A transmission electron micrograph of a prokaryotic cell. A number of typical parts are shown. CW, cell wall; N, nucleoid; R, ribosomes; PM, plasma membrane.

**Table 1** Comparisons of Microscopes

| Microscope | Type of Illumination | Magnification | Types of Specimen; Preparation Techniques |
|---|---|---|---|
| Light | Visible Light | 1,000–2,000X | Living, dead; stained, unstained |
| Fluorescent | Ultraviolet light | 1,500X + | Living; stained, immunofluorescence (fluorescent-antibody technique)[a] |
| Transmission electron microscope (TEM) | Electrons | 100,000–1 millionX | Dead; stained, coated with thin layers of heavy metals, and ultrathin slicing |
| Scanning electron microscope (SEM) | Electrons | 100,000–500,000X | Dead; coated with thin layers of heavy metals |
| Scanning transmission electron microscope (STEM) | Electrons | 100,000–500,000X + | Similar to those for TEM and SEM |

[a] This diagnostic technique uses natural defense molecules; or *antibodies* produced by humans and many animals in response to foreign substances known as *antigens.* Such antibody molecules are coated (tagged) with fluorescent dye molecules and are used to detect microorganisms (antigens) and other foreign substances.

## Eukaryotes and Prokaryotes

Two fundamentally different kinds of cells are recognized. These are **prokaryotic** and **eukaryotic**. The term *karyotic* refers to the nucleus, a membrane-enclosed, or membrane-bound, cellular structure containing genetic material; the prefix *eu* means "true" in Greek. Thus, eukaryotic cells are typically recognized by the presence of a membrane-enclosed nucleus (Figure 3). In addition, eukaryotic cells are larger than prokaryotic cells and contain a variety of other **organelles** (specialized cellular structures enclosed by membranes). Organelles perform specific functions such as the formation of organic molecules and the production of energy. Table 2 summarizes the typical components of eukaryotic and prokaryotic cells.

Prokaryotic cells do not have a well-defined nucleus (Figure 4). Their genetic material is found in a general cytoplasmic area known as the **nucleoid.** Moreover, prokaryotic cells are much smaller than eukaryotic cells (Figure 5), and they lack membrane-enclosed or -bound organelles. All organisms in the kingdom Prokaryotae have a prokaryotic type of cellular organization, whereas the cells of organisms in the other four kingdoms are eukaryotic.

Organisms of the kingdoms Prokaryotae and Protista are typically unicellular (single-celled), although some appear in strands or in thick cellular mats but have little communication or organization among the cells. Fungi, plants, and animals (which include helminths) are multicellular, and the activities of both individual cells and entire organ systems are completely dependent on one another.

All organisms need energy to perform their various activities and life processes (reproduction, growth, metabolism, motility, storage, and transport). Organisms that are photosynthetic capture energy from sunlight and store it in molecules such as sugars and oils. These forms of life, which include plants, some prokaryotes, and algae (protist) are called **autotrophs,** meaning "self-feeders." Organisms incapable of carrying out photosynthesis must obtain their energy prepackaged in the molecules manufactured by other forms of life. These organisms are called **heterotrophs**, meaning "other feeders." Most prokaryotes and protozoa and all fungi and animals are heterotrophs. **Pathogens** (disease-causing organisms) are heterotrophic.

## Viruses

Viruses are given a position independent of prokaryotic and eukaryotic organisms because of several significant differences. Clearly, viruses are not cells (Figure 6). They have no membranes of their own and none of the other structures found among prokaryotes or eukaryotes. Viruses cannot be seen with an ordinary light microscope. Special instruments, electron microscopes, must be used to observe these submicroscopic forms. In addition, viruses cannot move or grow, make their own source of energy, or replicate (reproduce) outside of a living cell. Viruses lack the enzymatic machinery to totally replicate themselves and to perform essential metabolic processes. These submicroscopic forms are totally dependent on living cells for their survival. They are obligate intercellular parasites. The section titled "Virology" discusses viruses more fully.

**Table 2** Summary of Typical Eukaryotic and Prokaryotic Cell Components and a Comparison with Viruses

| Cell Part and/or Related Structures | Functions | Eukaryotic | | | | Prokaryotic Bacteria, Cyanobacteria, etc. | Viruses |
|---|---|---|---|---|---|---|---|
| | | *Animal* | *Plant* | *Protist* | *Fungus* | | |
| Cell wall | Protection, structural support | None | X | X[a] | X | X | None |
| Plasma membrane | Control of substances moving into and out of cell | X | X | X | X | X | None |
| Nucleus | Physical separation and organization of DNA | X | X | X | X | X | One |
| DNA | Encoding of hereditary information | X | X | X | X | X | Either DNA or RNA, never both |
| RNA | Transcription translation of DNA messages into specific proteins | X | X | X | X | X | Either DNA or RNA, never both |
| Nucleolus | Assembly of ribosomal subunits | X | X | X | X | None | None |
| Ribosome | Protein synthesis | X | X | X | X | X | None |
| Endoplasmic reticulum (ER) | Initial modification of many newly forming proteins | X | X | X | X | None | None |
| Golgi body | Final modification of proteins and lipids, sorting and packaging them for use inside cell or for export; lipid synthesis | X | X | X | X | None | None |
| Lysosome | Intracellular digestion | X | X | X | X | None | None |
| Mitochondrion | (ATP) formation | X | X | X | X | None | None |
| Photosynthetic pigments | Light-energy conversion | None | X | X | None | X | None |
| Chloroplast | Photosynthesis, some starch storage | None | X | X | None | None[b] | None |
| Central vacuole | Increasing cell surface area, storage | None | X | None | X | None | None |
| Cytoskeleton | Cell shape, internal organization, basis of cell motion | X | X | X | X | None | None |
| 9+2 Flagellum | Movement | X | X | X | X | None[c] | None |
| 9+2 Cilium | Movement | X | X | X | X | None | None |
| Pilus | Attachment or transfer of genetic material | None | None | None | None | X | None |
| Spore | Reproduction | None | X | X[d] | X | | None |

[a]Found only with algae.
[b]Photosynthesizing prokaryotes have structures called chromatophores.
[c]Flagella are not showing the 9+2 arrangement are found among prokaryotes.
[d] Found among algae.

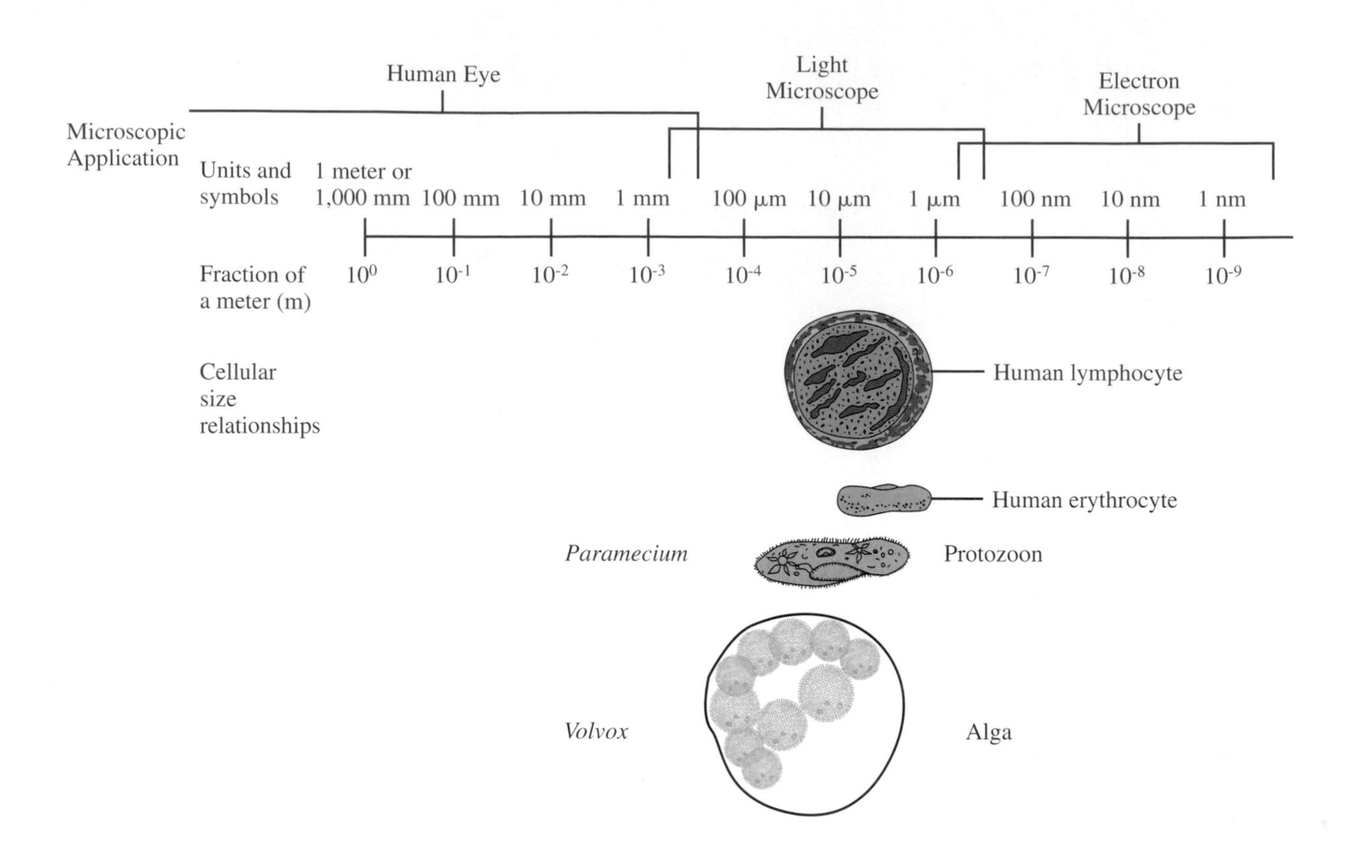


**Figure 5**
The size relationships of selected microorganisms and types of cells and the respective magnification ranges of specific microscopes. The metric units of measure used in the biological sciences include: 1 meter = 1,000 millimeters (mm); 1 mm = 1,000 micrometers (µm); 1 µm = 1,000 nanometers (nm).

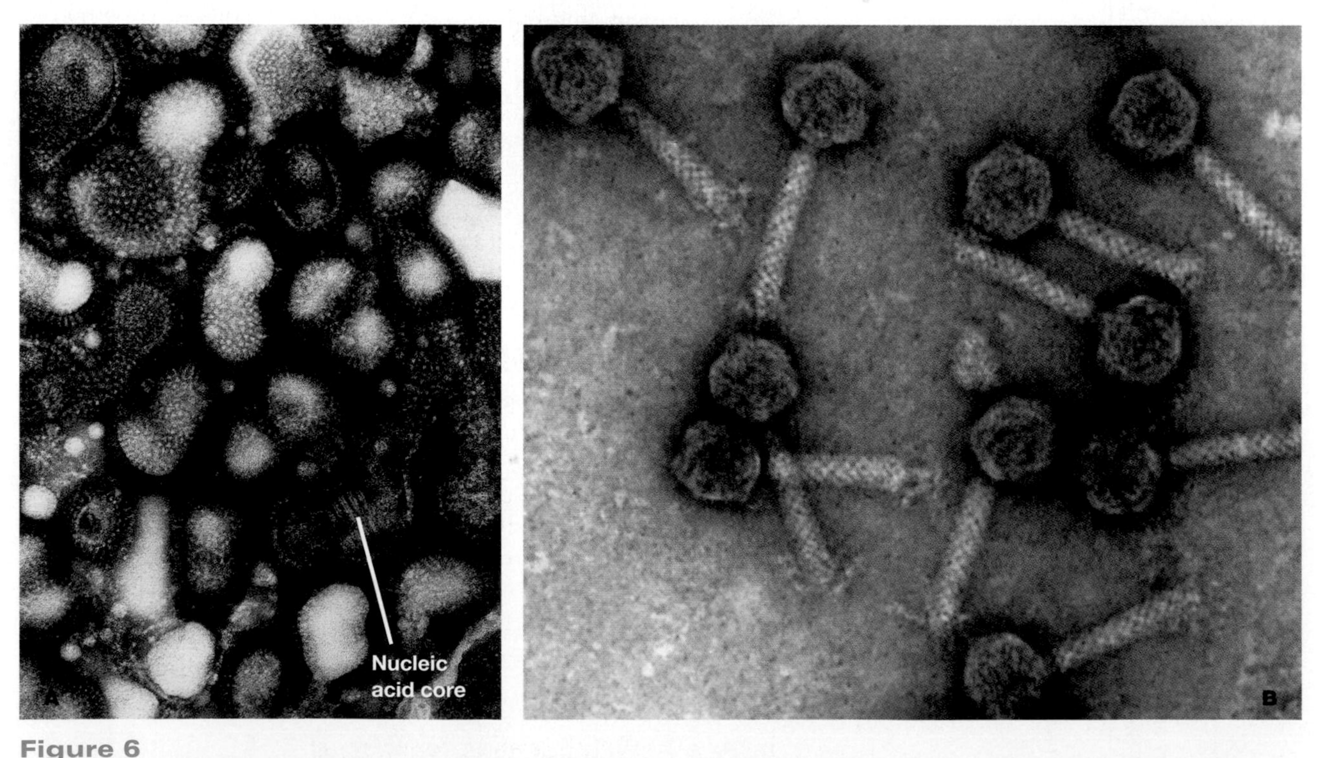


**Figure 6**
Transmission electron micrographs of two different viruses both of which clearly lack the structures found in eukaryotic and prokaryotic cells. (a) Influenza virus particles. The coiled nucleic acid component of some individual viruses can be seen (195,000X). (b) Bacterial viruses. Note the presence of heads and tails.

# SECTION 2

# Modern Imaging Techniques

*One enlarges science in two ways; by adding new facts and by simplifying what already exists.*

*—Claude Bernard*

For most infectious diseases, microbiological isolation and identification techniques are the fastest and most specific approaches to the determination of the causative agent. However, there are also a number of clinical conditions for which a reliable method of locating the site of infection would be extremely useful to diagnosis and determining the effectiveness of treatment. Such conditions include bacterial and fungal infections involving the entire body (systemic infections), situations in which no specific diagnosis is made, and situations in which fever of an unknown origin exists.

In the last 25 years, techniques for the diagnosis and treatment for a wide range of diseases and previously unsuspected disorders have advanced as a consequence of the improved capability and availability of medical imaging. **Radiography** (RĀ-dē-ō'-graf-ē), or the conventional x-ray (Figure 7), is well known for its diagnostic value and has been in use since the late 1940s. This imaging technique involves the passage of a single barrage of x-rays through the body to produce a two-dimensional image of the internal parts of the body.

Newer techniques, which include **computerized tomography (CT) scanning**; **magnetic resonance imaging,** or MRI; **ultrasonography**; and **scintigraphic techniques** provide sharply focused objective information helpful in precisely locating sites of infections and inflammation. These techniques also contribute to the understanding of body functions and disease processes. Brief explanations of the techniques are provided here because several of them are associated with the detection or diagnosis of infectious diseases.

## Computerized Tomography (CT) Scanning

After the injection of material to provide contrast of body structures, an x-ray beam is moved in an arc around the body. A series of cross-sectional images (slices) of an individual's body is produced by a scanning device that is processed by a computer and displayed on a video monitor. Three-dimensional views of body parts are constructed by stacking a series of images taken at different levels through an organ, one on top of another. These CT scans are widely used for observing evidence of disease (such as lesions) or changes in an organ (Figure 8).

## Magnetic Resonance Imaging (MRI)

Magnetic resonance imaging uses radio-frequency radiation in the presence of a carefully controlled magnetic field. It measures the response of positive subatomic particles (protons) to a pulse of radio waves while they are subjected to a strong magnetic field. The response produces high-quality cross-sectional images of the body (Figure 9). Magnetic resonance imaging is valuable in providing images of the brain, heart, large blood vessels, and soft body tissues, and in finding certain disease agents. The injection of contrast material is not required.

## Ultrasonography

Ultrasonography (ul-tra-son-OG-ra-fē) is a technique that uses high-frequency sound waves that bounce off tissues and are recorded by a scanning device as it is passed over the body. Signals from the scanner are transmitted to form an image called a **sonogram** (SO-no-gram) on a video monitor (Figure 10). The technique is used to detect cardiovascular disorders, abnormal masses in certain body organs, and multiple pregnancies and to find gallstones and related disorders.

## Scintigraphic Techniques

Scintigraphy involves the injection of a chemical to which some radioactive material has been attached.

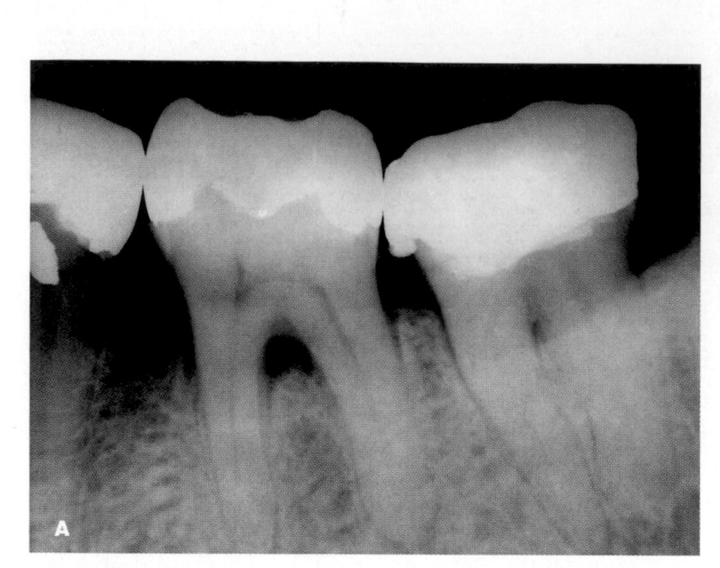


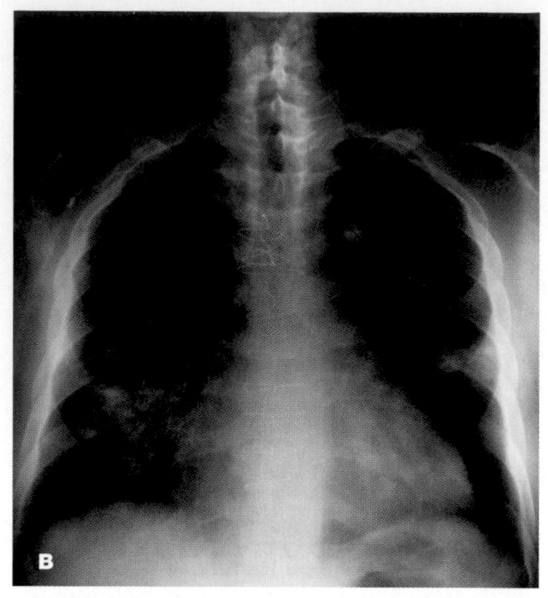


**Figure 7**
Two examples of x-rays or radiographs. (a) Results of a dental x-ray examination showing crowns (top) on molars and the beginning of gum disease (dense areas). (b) A chest x-ray showing extensive disease in the lungs. This imaging technique is used as a major screening approach for the detection of a variety of respiratory diseases and abnormalities.

As this radiopharmaceutical (rā-dē-ō-far-ma-SOO-ti-kal) circulates, radiation detectors determine its uptake or distribution within the body. The detector is a chemical substance that scintillates (sin-ti-LĀTES), or gives off energy in the form of a flash of light. The energy is picked up by a camera, and the information is processed by a computer to produce an image (Figure 11). Scintigraphic techniques have major roles in the detection of various types of cancer and in locating sites of infection and inflammation.

## Modern Imaging Techniques

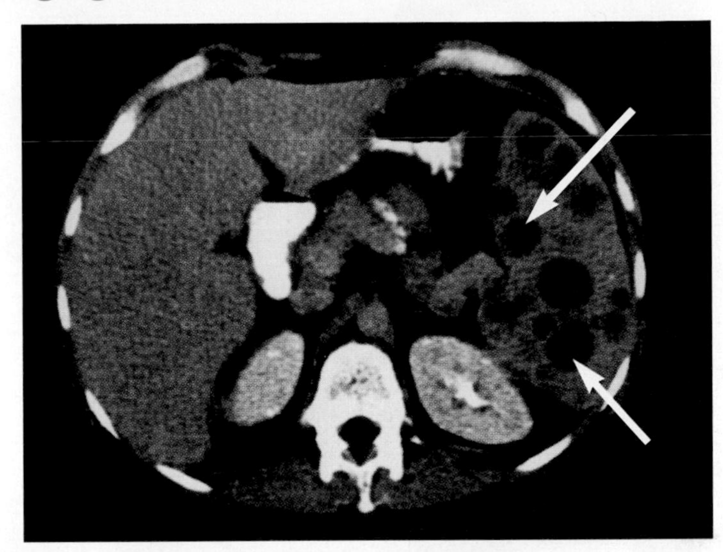

**Figure 8**
A CT scan showing the liver injury (dark circular areas) caused by a microbial infection (arrows).

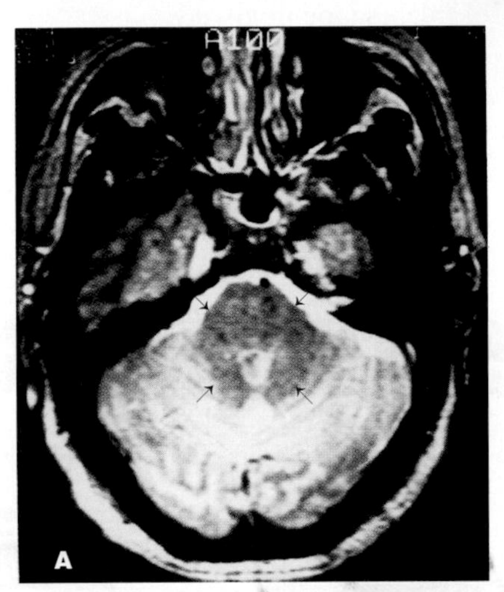


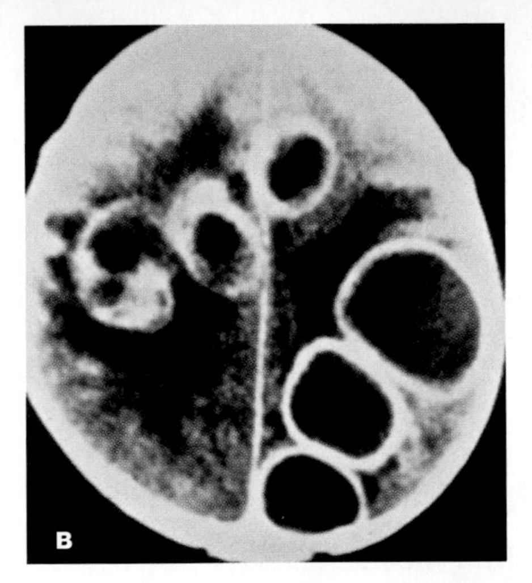


**Figure 9**
Two examples of magnetic resonance imaging (MRI). (a) The damaging effects of neurosyphilis (arrows). (From Tuite, M., L. Ketonen, K. Kieburtz, and B. Handy. *Amer. J. Neuroradiology.* 14(1993):257–63.) (b) The extensive destruction (large holes) caused by a microbial infection of the brain.

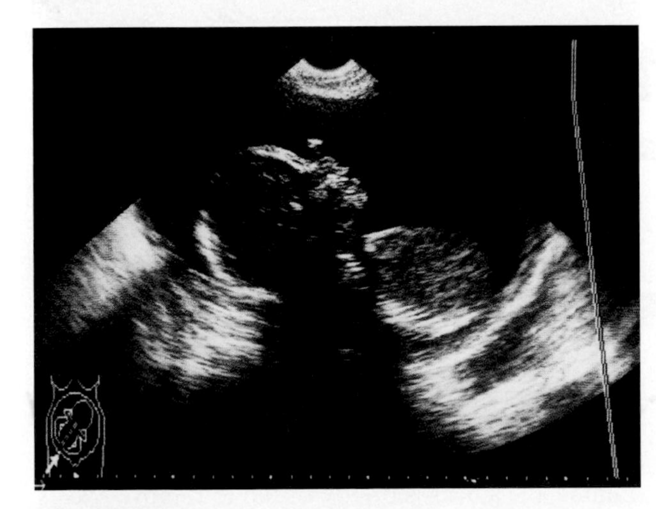

**Figure 10**
A sonogram of a human fetus. The general outline of the body is evident.

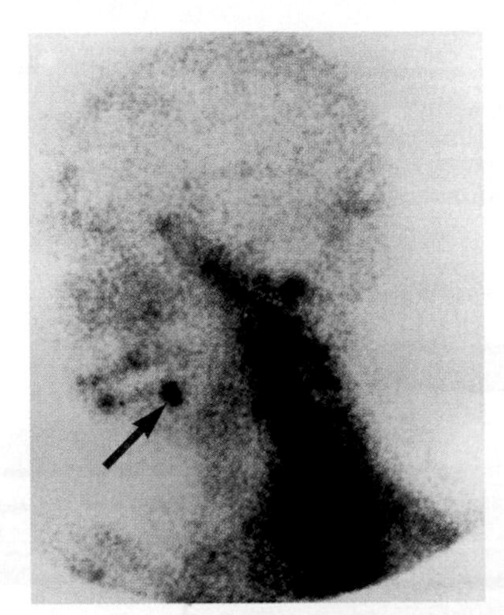

**Figure 11**
A scintigram showing the signal produced by an increased uptake of radioactive material and the location of a dental bacterial infection. (From K. Siminoski. *Clin. Inf. Dis.* 16(1993): 550–554.)

# SECTION 3 Bacteriology

*. . . it is of the highest importance, therefore not to have useless facts elbowing out the useful ones.*

*—Sir Arthur Conan Doyle*

Bacteria, fungi, protozoa, algae, and viruses constitute a group of biological forms that differ in a variety of properties. However, they do resemble one another in their small size and relative simplicity of structure and organization. The remaining portions of this section present various laboratory approaches, techniques, and materials used not only to study bacteria but also to identify them. In addition, a number of the properties of bacteria commonly studied in the laboratory as well as pathogens and representative disease states are considered.

## Bacterial Morphology, Morphological Arrangements, and Staining Techniques

The first application of the microscope to the study of microorganisms occurred only in the late 1600's, when a highly curious Dutch merchant, Anton van Leeuwenhoek applied his talents to lens grinding and the construction of simple, single-lens microscopes. His clear, accurate descriptions of the "little animalcules" he saw in specimens taken from teeth, the throat, pond water, rain barrels, and other sources defined the major microscopic types of microorganisms known to this day.

The objectives of microscope usage today are not appreciably different from those of van Leeuwenhoek. These include magnifying the image, maximizing the detail **(resolution)** of cells and their parts, and achieving sufficient contrast with which to distinguish microorganisms and cellular parts from the background of a microscopic viewing area.

The type of specimen preparation used is determined by the condition of the specimen (living or dead), the purpose of the viewing or examination, and the type of microscope involved.

### Living Preparations versus Simple Staining

The direct examination of live microorganisms can be extremely useful in determining size and shape relationships, movement (motility), and reactions to various chemicals and other substances. Nevertheless, live and unstained organisms are difficult to see and to distinguish from the fluid in which they are suspended (Figure 12). Reducing the intensity of the light source, however, can increase contrast, and thereby one can see the outline and arrangement of cells with less difficulty.

From a practical standpoint, contrast is increased by using dyes, which are either natural or artificial organic compounds. Such dyes are used in the laboratory for the direct staining of specimens to make cells or their parts more visible than they would be unstained (Figure 13). The surfaces of microorganisms are negatively charged. Therefore, positively charged (basic) dyes or their components will be naturally attracted to those negatively charged surfaces. Dyes commonly used include crystal violet, safranin, carbol fuchsin, and methylene blue.

Before they are stained, bacteria are usually suspended in water or some other liquid on a clean microscope slide and are then spread in a thin, even film. The film is allowed to dry in air, and the organisms are "fixed" (attached) to the slide by gentle heating. The preparation is known as a **fixed smear** and is ready for staining.

Simple staining procedures involve the application of a single stain to the fixed smear. The time required for staining varies with the dye being used. After the staining, the smear is briefly rinsed to remove excess stain. The slide is then dried and examined with the appropriate microscope.

Simple stained smears are used to detect the morphology (shape), cellular arrangement, and relative size of bacteria. They can also be used to detect the presence of certain structures such as spores (Figure 13). Such preparations do not, however, reveal internal structural details.

## Bacterial Morphology and Morphological Arrangements

The general shapes of bacteria—**rod, coccus** (spherical), and **spiral**—were described by Anton van Leeuwenhoek in the late 1600s. A new morphological type of prokaryote, the square, was described in 1980 by A. E. Walsby. These cells appear as flat, rectangular boxes with perfectly straight edges. In addition to these differences in cellular shape, definite patterns in numbers and arrangements of cells are known to exist among different bacterial species (see Figure 17). In the case of spherical, or coccus, forms five patterns can be found (Figure 14). These are pairs of cells **(diplococci)**, chains of four or more cells **(streptococci)**, four cells in a square arrangement **(tetrads)**, irregular groups of cells resembling grape clusters **(staphylococci)**, and eight cells grouped into a cuboidal packet **(sarcinae)**.

Rod-shaped bacteria (Figure 15), also referred to as **bacilli** (singular, *bacillus*), can be found occasionally in pairs **(diplobacilli)** and in chains (**streptobacilli**). However, these morphological patterns of rod-shaped forms are not as constant as in the case of the cocci and therefore should not necessarily be considered as characteristic for particular types of bacteria (species). In certain cases, because of the limitations of a microscope, it is difficult to distinguish the shape of extremely small rods, which appear to be almost cocci. Such cocci-like organisms are referred to as **coccobacilli** or **coccoid** in shape (Figure 15c).

Two groups of coiled or spiral-shaped bacteria are known. One group, the spirochetes (Figure 16a), consists of flexible, waving forms with several coils. The second group, the spirilla (Figure 16b), are rigid bacteria possessing one or several curves. Spiral forms that are short and do not form complete coils are called vibrios (Figure 16c).

Figure 17 summarizes the characteristic cell shapes and morphological arrangements found among bacteria (see page 13).

Variations in the general shapes and sizes of bacteria are frequently seen and can be explained in terms of environmental factors. **Pleomorphism** is the term used to denote these modifications when they occur under favorable conditions. Under unfavorable conditions, these variations are called **involution forms.**

## Differential Staining Techniques

As indicated earlier, for most cases of infectious diseases microbial isolation and identification techniques offer the fastest and most specific determinations of causative agents. On the other hand, finding disease agents in the course of preparing cells, tissue sections (slices), or both often adds important information and in some situations is crucial to a timely diagnosis. Several examples of microorganisms in tissues are shown in later sections.

New dimensions in techniques for staining and identifying microorganisms have been provided by biotechnological advances. Some of these are briefly discussed in this section.

Differential staining procedures distinguish structures within a cell or distinguish one type of cell from another. In the laboratory, all stains and related chemicals or processes can functionally be divided into one of four categories: **primary stain, mordant, decolorizer,** and **secondary stain, or counterstain.**

A primary stain is generally a basic (alkaline), positively charged dye. Examples include crystal violet, safranin, carbol fuchsin, and methylene blue. Mordants usually are chemicals that fix (firmly attach) the primary stain to the target microorganisms. A common example is Gram's iodine used in the Gram-staining procedure (Table 3). A decolorizer is a chemical that removes unattached stain from target microorganisms. After decolorizer has been applied, the target microorganisms are the color of the primary stain, and nontarget organisms and the surrounding background should be colorless. Acetone-alcohol, used in the Gram stain, and acid-alcohol (a combination of 3% hydrochloric acid and alcohol), used in the acid-fast staining procedure, are examples of common decolorizers. Most but not all differential staining procedures require the application of a counterstain to color and show the presence of nontarget organisms (Table 3). Safranin used in the Gram-staining procedure and methylene blue in the acid-fast are counterstains.

Differential stains are widely used for the detection and identification of bacteria and other microorganisms. Table 3 summarizes the features of differential and related staining procedures, and Figures 18 through 26 show the results of these procedures.

Variations of differential staining procedures include the applications of fluorescence microscopy. Certain dyes, called **fluors** or **fluorochromes,** have the property of becoming excited (raised to a higher energy level) after absorbing ultraviolet (UV) light (light of short wavelength). As the excited molecules return to their normal state, they release the excess energy in the form of visible light or a longer wavelength than that which first excited them. This property of be-

## Bacterial Morphology, Morphological Arrangements, and Staining Techniques

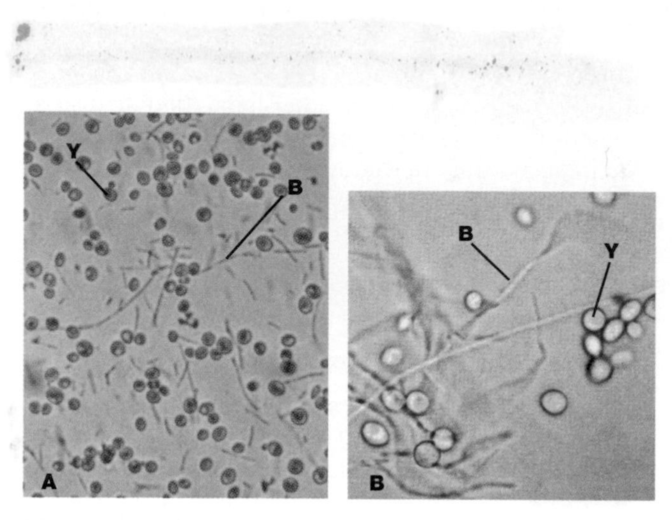


**Figure 12**
A preparation of living bacteria (B) and yeast (Y). (a) A low-power view (100X). (b) A higher magnification (450X).

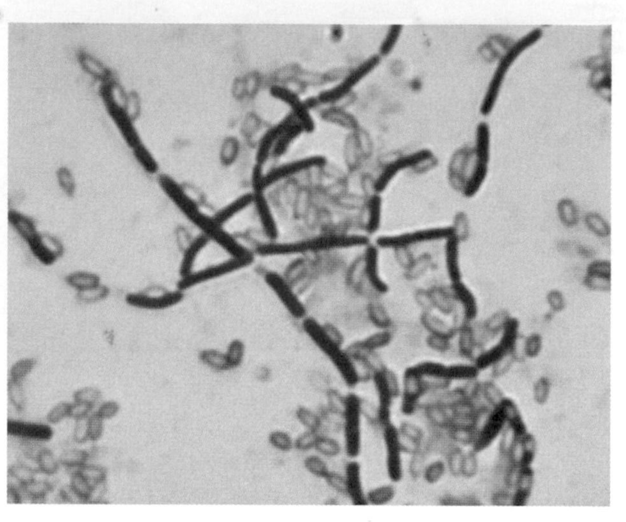

**Figure 13**
The bacterial rod, *Bacillus subtilis*, simple stained with crystal violet. This bacterium can form heat-resistant structures known as spores. In simple stained preparations such as this one, spores do not stain but appear as clear zones within the bacterial cells in which they were formed.

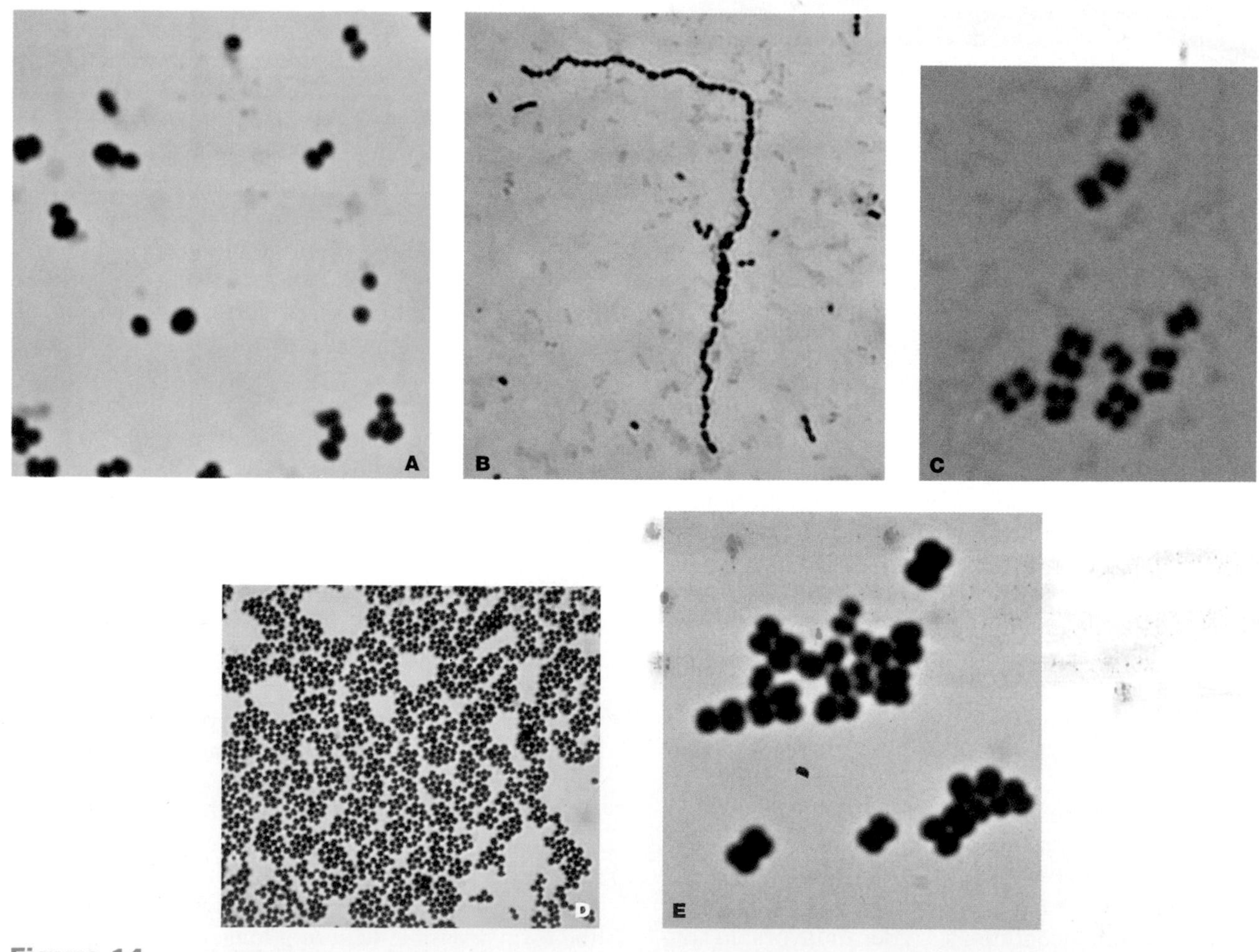


**Figure 14**
The morphological arrangement (pattern) of cocci. (a) Diplococcus (two cells). (b) Streptococcus (chains). (c) Tetrads (four cells). (d) Staphylococcus (grapelike clusters of cells). (e) Sarcinae (cuboidal packets of eight of cells). Note the thickened appearance of the cocci (arrows) exhibiting this pattern.

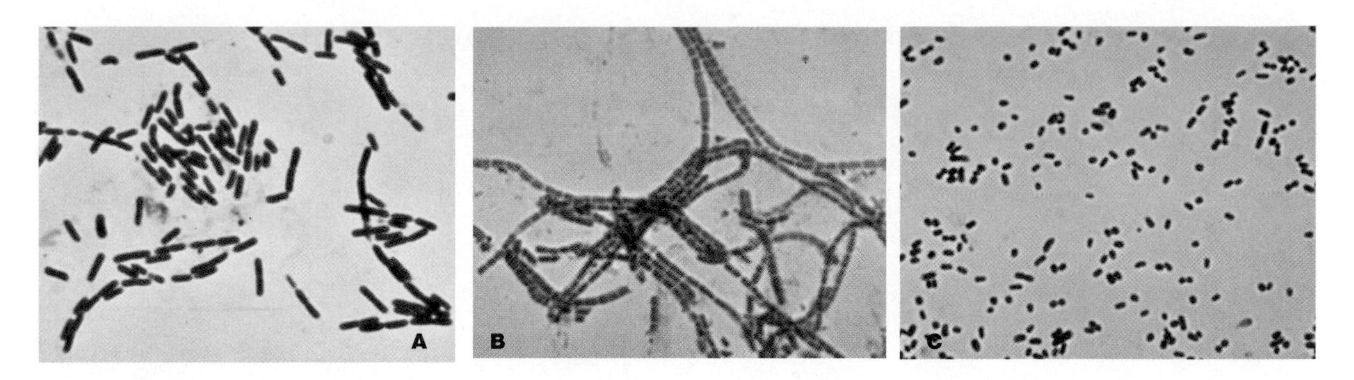


**Figure 15**
Patterns of rod-shaped bacteria. (a) Diplobacillus (single cells also are present). (b) Streptobacillus (chains). (c) The cocco-bacillus (coccoid form).

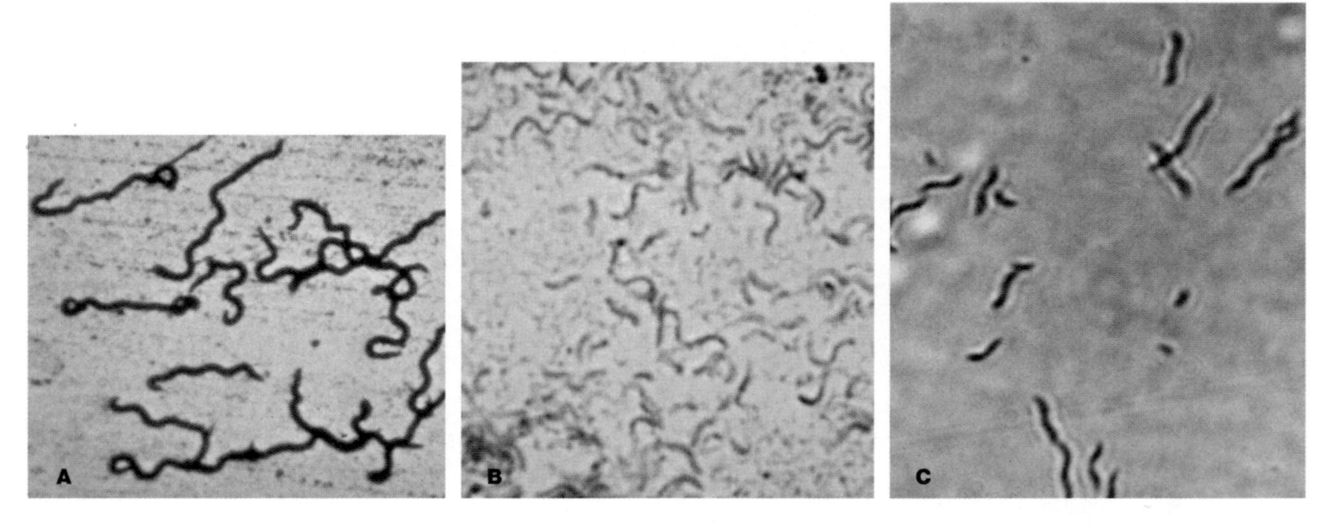


**Figure 16**
Patterns of spiral and related forms of bacteria. (a) Spirochetes. (b) Spirilla. (c) Vibrios.

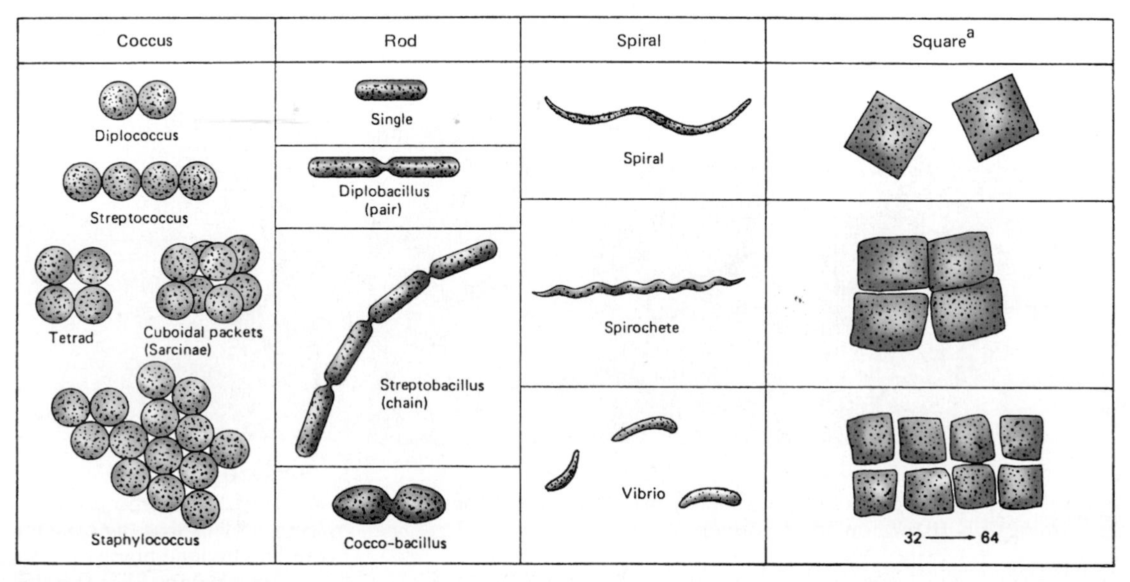


**Figure 17**
A summary of bacterial cell shapes and arrangements.

## Differential Staining Techniques

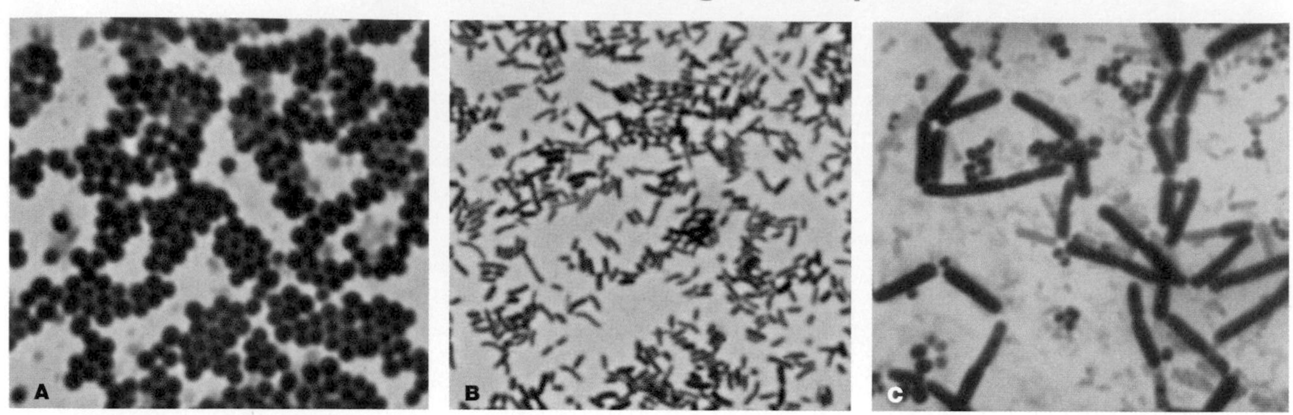

**Figure 18**
Gram reactions. (a) Gram-positive cocci. (b) Gram-negative rods. (c) A mixed smear showing gram-positive cocci and rods and gram-negative rods.

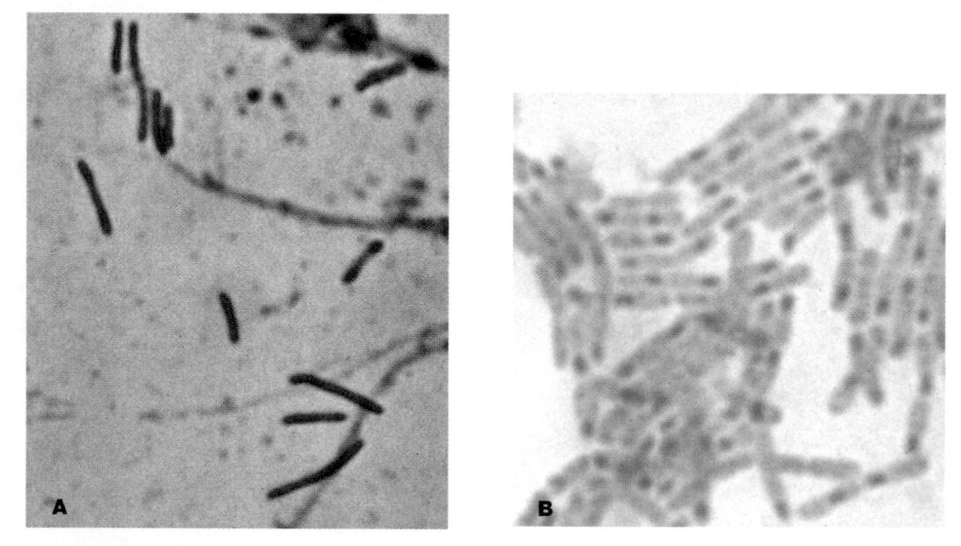

**Figure 19**
The acid-fast reaction. (a) Red, acid-fast cells. (b) A non-acid-fast reaction showing blue cells.

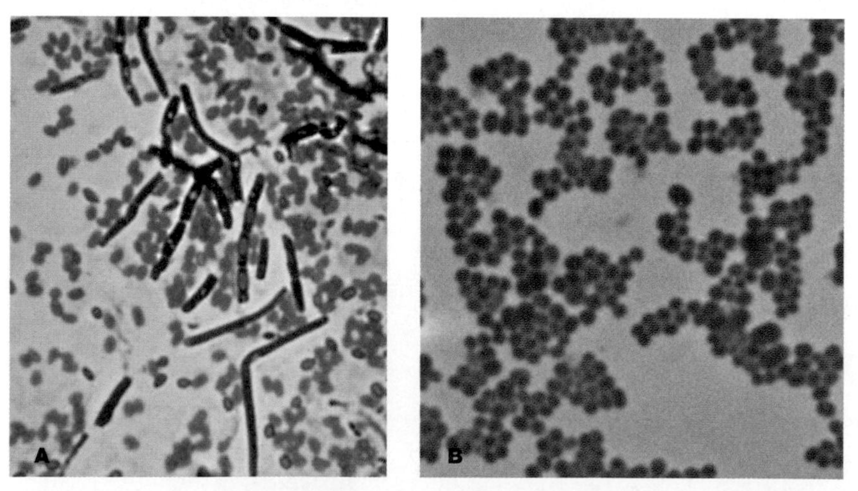

**Figure 20**
The Schaeffer-Fulton spore stain. (a) Green spores inside and outside of red non-spore forming cells. (b) A non-spore-former.

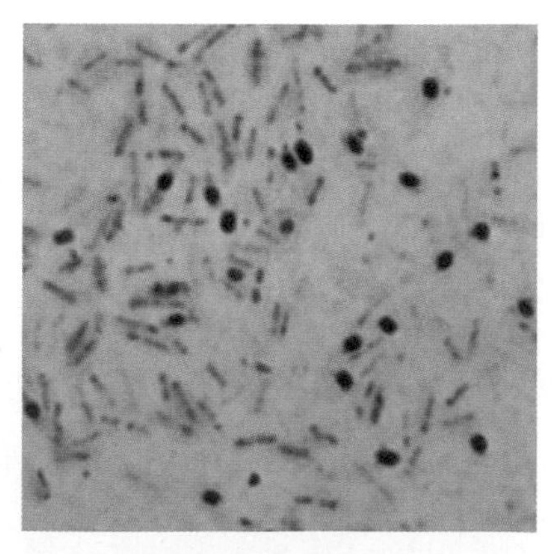

**Figure 21**
The results of a spore stain using carbol fuchsin as the primary stain and methylene blue as the secondary or counterstain. Spores stain red.

**Table 3** Summary of Differential and Related Staining Reactions

| Staining Procedure | Description | Reaction(s) and/or Cellular Appearance | Application(s) and/or Examples |
|---|---|---|---|
| Gram stain | *Primary stain:* crystal violet<br>*Mordant:* Gram's iodine<br>*Decolorizing agent:* acetone-alcohol or 95% ethanol<br>*Counterstain:* safranin | Gram-positive: purple cells.<br>Gram-negative: pink cells. | Divides bacteria into one of two groups:<br>Gram-positives: *Staphylococcus, Bacillus.*<br>Gram-negatives: *Escherichia, Pseudomonas, Salmonella, etc.* |
| Acid-fast stain | *Primary stain:* carbol fuchsin<br>*Decolorizing agent:* acid-alcohol<br>*Counterstain:* methylene blue | Non-acid fast: blue cells.<br>Acid-fast: red cells. | Divides bacteria into one of two groups:<br>Acid-fast: *Mycobacterium, Nocardia.*<br>Non-acid fast: most other bacteria. |
| Spore stain (Schaeffer-Fulton) | *Primary stain:* malachite green<br>*Counterstain:* safranin<br>*Decolorizing agent:* none | Spores present: green oval structures inside and outside of red cells.<br>Spores absent: red cells without any green oval structures. | Separates spore-formers from non-spore-formers. |
| Spore stain | *Primary stain:* carbolfuchsin<br>*Counterstain:* methylene blue<br>*Decolorizing agent:* none | Spores present: red oval structure inside and outside of blue cells.<br>Spores absent: blue cells without any red oval structures. | Spores: *Bacillus, Clostridium.*<br>Nonspores: most other bacteria. |
| Capsule stain | Culture is mixed with 1 drop of Indian ink. After air-drying, safranin is applied. | Clear (halo-like) area surrounding pink cells.[a]<br>No clear area surrounding pink cells. | Demonstrates the presence of capsule-formers.<br>Capsule present: *Klebsiella, Streptococcus pneumoniae.*<br>Capsule absent: *Escherichia, Mycobacterium.* |
| Flagella stain | A combination of mordant and stain is applied to the smear. The mordant acts to deposit more stain onto the flagella. | Flagella present: long wavy structures extending from cells in various patterns.<br>Flagella absent: no wavy structures extending from cells. | Shows presence and arrangement of flagella.<br>Flagella are found with several organisms, including *Salmonella, Escherichia, and Bacillus.* |
| Fluorescent-antibody staining | Specific fluorescent-dye-tagged (attached) antibody is applied to smear containing unknown bacteria or other types of cells. | Cells fluoresce: the color(s) exhibited by cells are determined by the fluorescent dye molecules used; colors can include red, green, yellow, and orange. Cells do not fluoresce in negative tests. | Identifies specific organisms; can be used for diagnosis of diseases. Only microorganisms and other cells containing specific antigen fluoresce. |

[a] The capsule itself does not stain.

coming self-luminous is called **fluorescence.** Fluorescing objects appear brightly lit against a dark background (Figures 24–26), with the color depending on the fluorochrome being used.

The principal use of fluorescence microscopy is a diagnostic technique called **immunofluorescence**, or the **fluorescent-antibody technique. Antibodies** are natural defense protein molecules produced by humans and many other animals in response to substances recognized by their respective immune systems as being foreign. These foreign substances or cells are known as **antigens.**

Fluorescent antibodies for a particular antigen are obtained when a laboratory animal is injected with a specific antigen, such as a bacterium. The animal then begins forming specific antibodies against the injected antigen. After a sufficient amount of antibodies have been produced, they are removed from the blood of the animal and chemically combined with fluorochrome molecules. The resulting fluorescent antibodies can then be used to detect and identify the same antigen that was injected into the laboratory animal, cells, tissues or other types of clinical specimens. Fluorescent antibodies are applied to specimens on slides containing an unknown organism. If the unknown organism is the same bacterium that was injected into the animal, the fluorescent antibodies bind to the surface of the bacteria causing them to fluoresce, or glow.

Modern technology has provided alternatives to fluorescent antibodies for identifying bacteria. These include the attachment of fluorochromes to nucleic acid components to detect highly specific ribonucleic acid parts of bacterial ribosomes (Figure 25).

An alternative technique to the Gram stain has been developed that can be applied to living bacteria only (Figure 26). The **LIVE Bac Light** procedure uses a mixture of nucleic acid–fluorescent dye stains that can differentiate between living gram-positive and gram-negative organisms. When stained with the unique nucleic acid–fluorescent dye combinations, gram-positive cells fluoresce yellow-orange, and gram negative organisms fluoresce bright green.

## Basic Cultivation Techniques

The cultivation of microorganisms requires the use of nutrient preparations called **culture media** (singular, *medium*). Natural media such as milk, vegetable slices, and certain meat infusions contain soluble organic and inorganic substances that are the necessary factors for growth. However, the exact chemical compositions of natural media are unknown and quite variable. On the other hand—in the case of **chemically defined,** or **synthetic, media**—the kinds and exact amounts of all ingredients are known. Media of this type can be duplicated to specification and used to study the effects of specific compounds on microbial growth.

There are three general forms of culture media: **solid, semisolid**, and **broth** (liquid). Each type of medium has particular properties that make it more or less suitable for certain growth situations (Figure 27). Most solid and semisolid media contain **agar** as a solidifying agent. It is a complex polysaccharide extracted commercially from certain species of red marine algae such as *Gelidium, Gracilaria,* and *Rhodophyta.*

### Isolation Techniques

Certain procedures have become indispensable to bacteriologists. Among them are the standard **pour plate** (Figure 28) and **streak plate** (Figure 29) **techniques**. These methods can be effective in both the detection and the determination of numbers and kinds of different microorganisms present in specimens. Microorganisms such as bacteria grow as visible accumulations of identical cells. Forming **colonies** on the surfaces of agar media contained in Petri dishes. These dishes are constructed to allow air but not dust to pass through.

The pour plate technique consists of: (1) cooling a melted agar-containing medium (1.5% agar) to approximately 42° to 45°C and (2) inoculating the medium with a specimen just before pouring it into a sterile Petri dish. Thus, bacteria are distributed throughout the agar and trapped in position as the medium hardens. Although the solidified medium restricts bacterial movement from one area to another, it is of a soft enough consistency to permit growth. Growth occurs both on the surface and within the inoculated medium. Unfortunately, there are several disadvantages to this technique, including the following: (1) colonies of several species may present a similar appearance in the agar environment; (2) certain species of bacteria may not grow in this environment; and (3) difficulty may be encountered in removing (picking) colonies for further study. Figure 28 shows the features of the pour plate.

The streak plate procedure is another example of a dilution technique. It was originally developed by two bacteriologists, Friederich Löeffler and Georg Gaffky, in the laboratory of Robert Koch. The preparation of a streak plate involves the spreading of a single loopful of material **(inoculum)** containing microorganisms over the surface of an agar medium that has been allowed to solidify. Numerous streaks are made to allow for a greater separation of the organisms in the inoculum. Figure 29 shows the results obtainied with the streak plate technique.

After inoculation by any method, plates are incubated at desired temperatures. Water of condensation may form in Petri dishes as a consequence of the high concentration of water in agar. Prepared plates are in-

## Differential Staining Techniques *(Continued)*

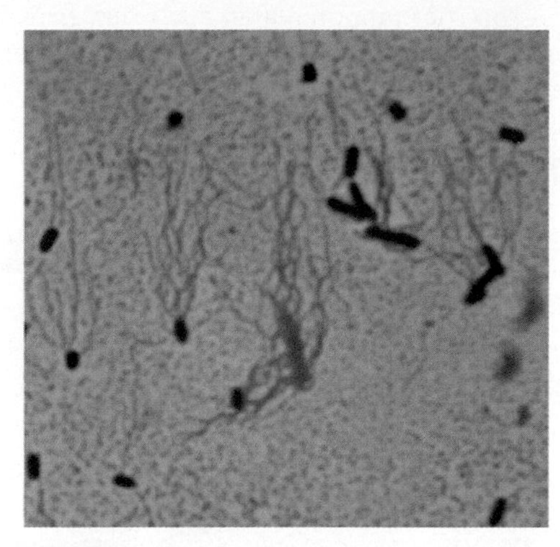

**Figure 22**
Flagella at the ends of the bacterium *Salmonella typhi,* the cause of typhoid fever.

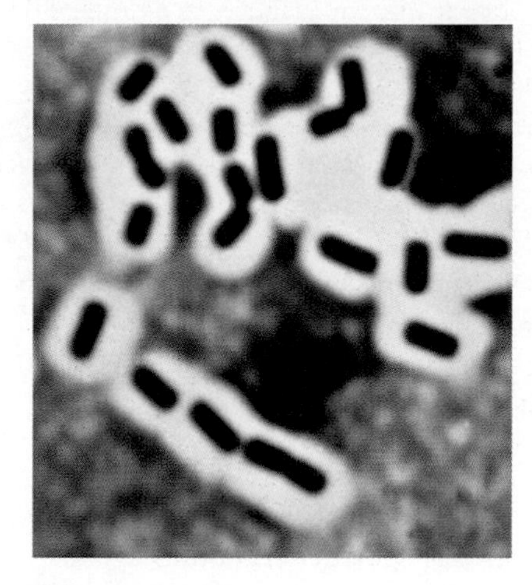

**Figure 23**
The capsule stain. Cells are surrounded by clear areas formed by capsules.

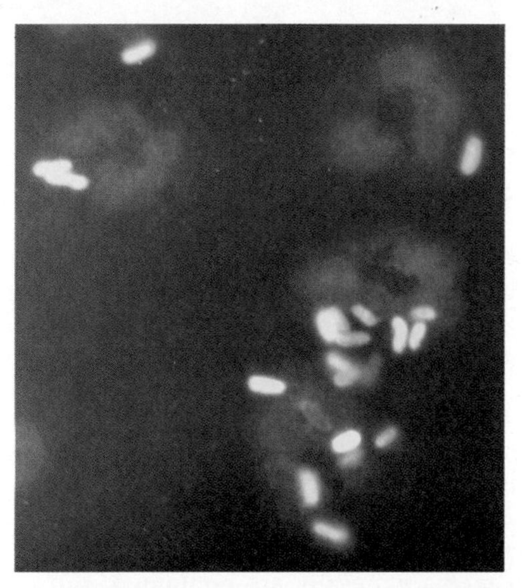

**Figure 24**
Immunofluorescence (fluorescent-antibody staining). Specific antibody molecules to which fluorescent dye molecules are attached can differentiate among cells. Bacteria appear as greenish yellow cells.

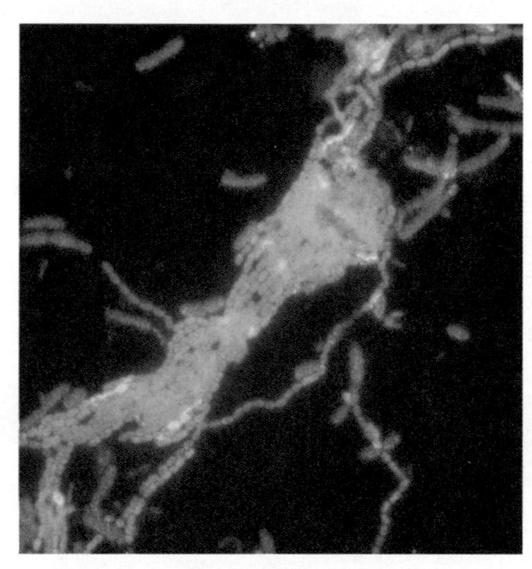

**Figure 25**
The use of fluorescent dye–labeled nucleic acid probes to detect and identify bacteria. (From M. Wagner, R. Amann, H. Lemmer, and K.-H., Schleifer. *Appl. Environ. Microbiol.* 59(1993):1520–1525.)

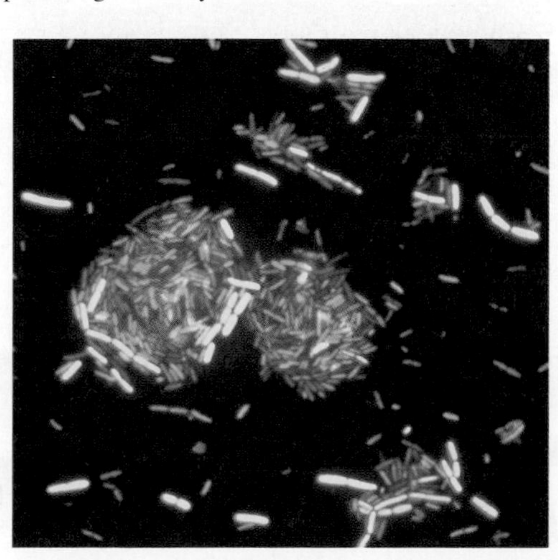

**Figure 26**
A variation of the Gram stain using LIVE Bac Light. Gram stains with living gram-positive *Bacillus cereus* (yellow-orange) and gram-negative *Pseudomonas aeruginosa* (green). (Courtesy of Molecular Probes, Inc.)

**Figure 27**
An assortment of different media that may be used in bacterial detection, isolation, and identification. Plate, broth, and agar slant media are shown. (Courtesy of Becton Dickinson Microbiology Systems.)

cubated in an inverted position to prevent water from forming on media surfaces and causing bacterial colonies to run together.

## Agar Plate, Broth, and Agar Slant Characteristics

The pigmentation, size, shape, and overall general appearance of bacterial colonies growing in or on agar plates (Figure 30) can serve as identifying features when viewed under both reflected and transmitted light. Pigmentation and several other cultural characteristics are influenced by incubation temperature. Colonies may appear to be transparent, opaque, or translucent. Some species produce a distinguishing fluorescent pigment that is evident when cultures are examined under ultraviolet light. Other colonies exhibit specific surface properties and may appear as dry (powdery), contoured, rough, smooth (glistening), or rugose (wrinkled), or they may have concentric rings or ridges. Some organisms also produce characteristic odors. Such odors may be sweet, foul, soil-like, or musty.

In addition to the agar plates, solid medium preparations can be made into slants. Such preparations are made by pouring a melted agar medium into test tubes and allowing the material to solidify at an angle. Agar slants are valuable in studying growth characteristics and in maintaining pure cultures. The patterns of growth on agar slants are important characteristics used in bacterial identification (Figure 31). Liquid or broth media generally are contained in test tubes closed by nonabsorbent plugs or special closure caps to prevent the entering of unwanted microorganisms. After the inoculation of broth media, bacteria may exhibit a particular form of growth. These include clouding of the medium **(turbidity)**, accumulations of cells at the tube bottom **(sediment),** and the formation of a thin surface film **(pellicle)**. Broth media can also be used to perform various biochemical tests. Several of these are presented later. Figure 31 and 32 show selected properties of agar slant and broth cultures, respectively.

## Anaerobic Cultivation

Anaerobic bacteria are unable to grow on or near the surfaces of semisolid or solid media in air at atmospheric pressure. The early bacteriologists believed that successful cultivation of strict anaerobes (organisms that can survive and grow in environments complete-

## Isolation Techniques

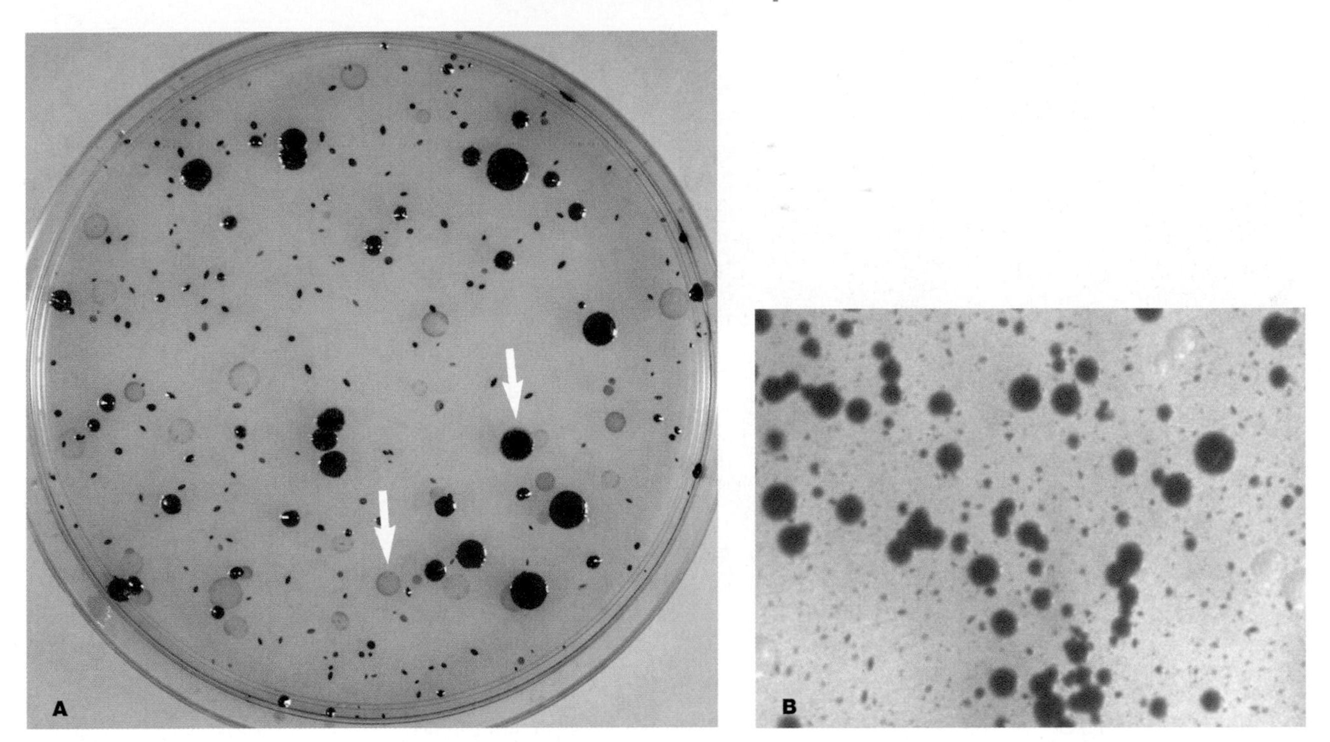


**Figure 28**
The pour plate technique. (a) A plate of *Micrococcus luteus* and *Serratia marcescens.* Note the distribution of differently pigmented colonies. (b) Higher magnification of this preparation showing the various shapes of bacterial colonies.

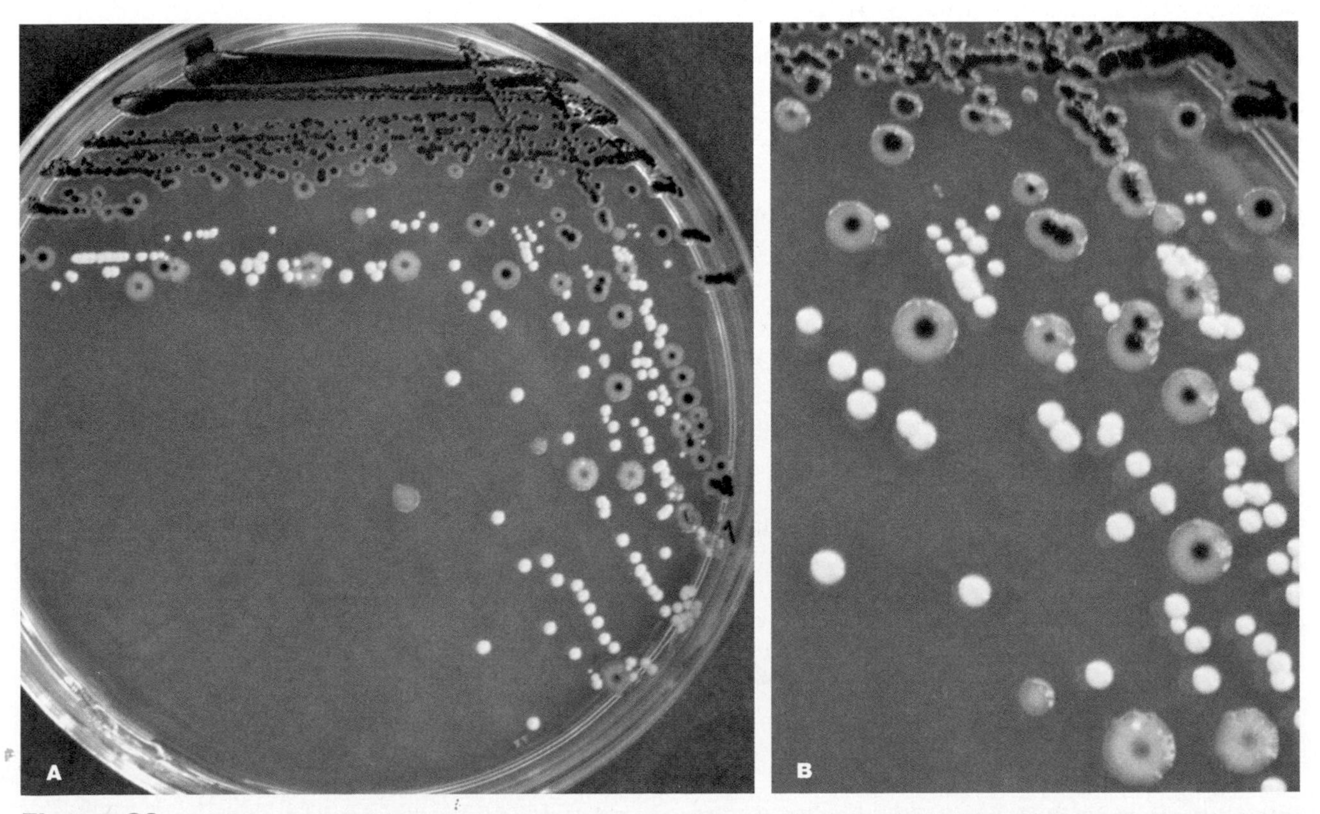


**Figure 29**
The streak plate technique. (a) A plate with a mixed culture containing *Micrococcus luteus* and *Serratia marcescens.* (b) A close-up of the agar surface. Note the well-separated colonies.

## Agar Plate, Agar Slant, and Broth Characteristics

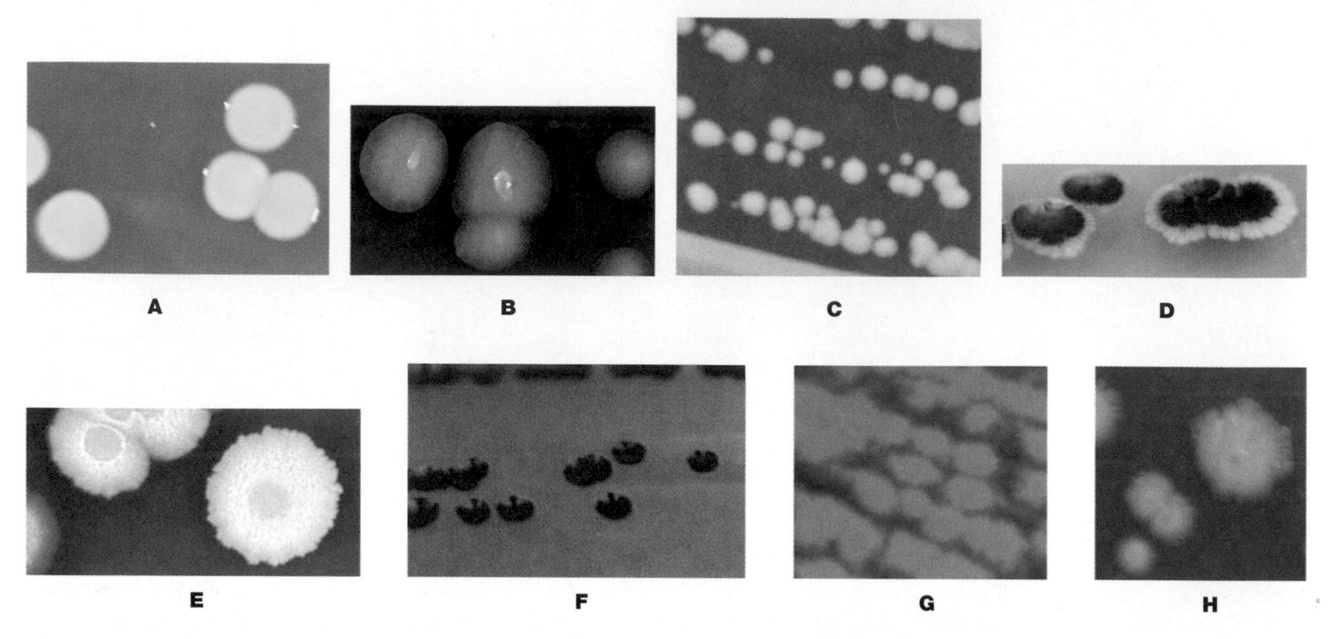


**Figure 30**
Examples of bacterial colony characteristics. (a) Circular, entire, smooth, opaque, yellow-pigmented colonies. (b) Irregular, undulate, umbonate, opaque, cream-colored colonies. (c) Punctiform (less than 1 mm in diameter), smooth, entire, opaque colonies among larger circular ones. (d) Unusual red-and-white–pigmented colonies that also exhibit irregular, lobate, and opaque properties. (e) Irregular concentric, undulate, opaque, somewhat raised colonies. (f) Red-pigmented, circular, smooth, entire, opaque, convex colonies. (g) Filamentous, opaque, umbonate colonies. (h) Irregular, opaque, umbonate colonies.

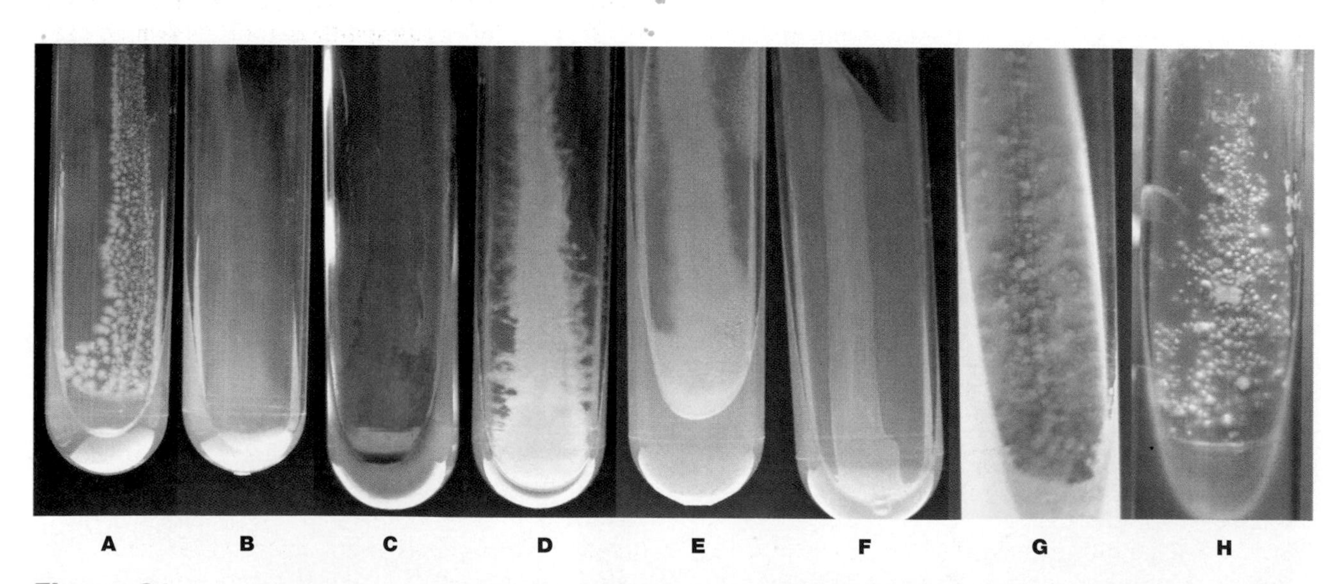


**Figure 31**
Selected agar slant patterns. (a) Beaded form of growth. (b) Spreading growth. (c) Flat, red, and filiform. (d) Arborescent (branched). (e) Rhizoid growth. (f) Filiform, yellow growth. (g) Orange growth of *Mycobacterium phlei.* (h) Nonpigmented growth of a *Mycobacterium* species. Both species of *Mycobacterium* are growing on a form of Löwenstein-Jensen medium.

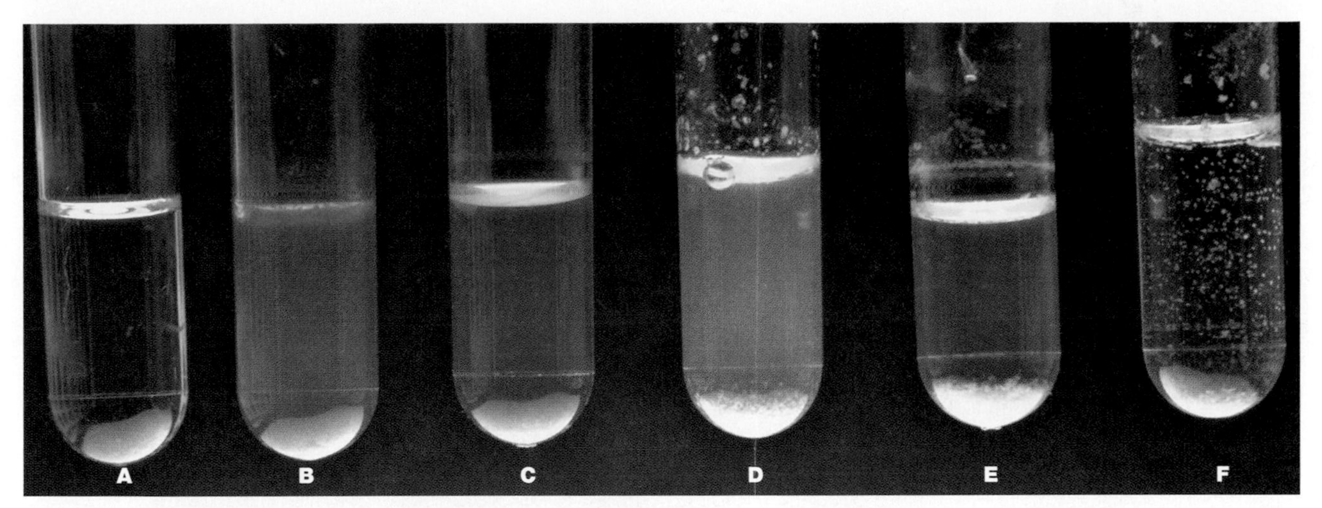


**Figure 32**
Selected broth culture patterns. (a) Clear, uninoculated broth. (b) Turbid (cloudy) growth and red pigmentation. (c) Turbid and green pigmentation in the upper portion of the broth. (d) Turbid growth. (e) Slightly turbid and ring formation (1 ml of the broth was removed to demonstrate the presence of the ring. (f) Somewhat clear broth with flocculent growth.

ly without free oxygen) could be achieved only through methods that excluded all free oxygen. Providing anaerobic conditions requires only the removal of free oxygen from the immediate environment of the microorganisms or simply the maintenance of a low oxidation-reduction (redox) potential, or $E_h$, in the media. The $E_h$ is a measure of the tendency of a preparation to be oxidized or reduced. In routinely used laboratory media, oxygen is primarily responsible for the increasing $E_h$. Various redox dyes can be used to estimate the $E_h$ of a medium or culture. Useful dyes are reversibly oxidized or reduced and are colored in the oxidized state and colorless in the reduced state (see Figure 35).

Procedures for cultivating and identifying anaerobic bacteria are much like those for aerobes. The difference lies in the incubation atmosphere. Reducing agents, which are nontoxic, are added to most anaerobic media to depress and maintain the redox potential at low levels. Examples of such agents are sodium thioglycollate and cysteine hydrochloride.

Various types of containers are used to provide an anaerobic environment for organisms. Among the most widely used are the Brewer anaerobic jar with a heat-activated catalyst and the GasPak Anaerobic System (Figure 33). With the Brewer apparatus, hydrogen or a mixture of gases is introduced into the anaerobic jar after it has been sealed. Electrical heat activation of a platinum catalyst present in the lid creates anaerobic conditions.

The GasPak jar differs from all other anaerobic jars in that it has no external connections and it utilizes a room-temperature catalyst system, thus rendering electrical connections or other means of heating the catalyst unnecessary. This anaerobic jar is specifically designed to be used with a disposable hydrogen and carbon dioxine–generating system. When water is added to a GasPak envelope, hydrogen gas is released. This gas, in turn, reacts with oxygen in the presence of a catalyst to produce anaerobic conditions. Carbon dioxide is also produced in quantities sufficient to support the growth of anaerobes that require it. A disposable anaerobic indicator containing methylene blue is used to determine whether an anaerobic condition exists within the system.

Procedures that do not involve complicated pieces of equipment or elaborate techniques are also commonly used for the cultivation of some anaerobic, **facultative** organisms (which can grow with or without oxygen) and **microaerophilic** organisms (which prefer low oxygen tension). These include the paraffin-plug technique (Figure 34) and the use of media containing a reducing compound, such as sodium thioglycollate (Figure 35) to reduce the oxygen content. In the paraffin-plug technique, the tube of medium is heated for several minutes, cooled quickly, and then inoculated. A layer of melted paraffin approximately one-quarter of an inch thick is poured onto the top of the medium. The culture is then ready for incubation.

Thioglycollate media are used in the cultivation of anaerobic, microaerophilic, and aerobic bacteria and for the detection of bacteria in normally sterile materials. The ingredients of these media support the growth of a wide variety of microorganisms having a broad range of growth requirements. The presence of sodium thioglycollate lowers the oxidation-reduction potential of the media, while resazurin serves as the oxidation-reduction (OR) indicator and indicates the status of oxidation.

## Anaerobe Cultivation

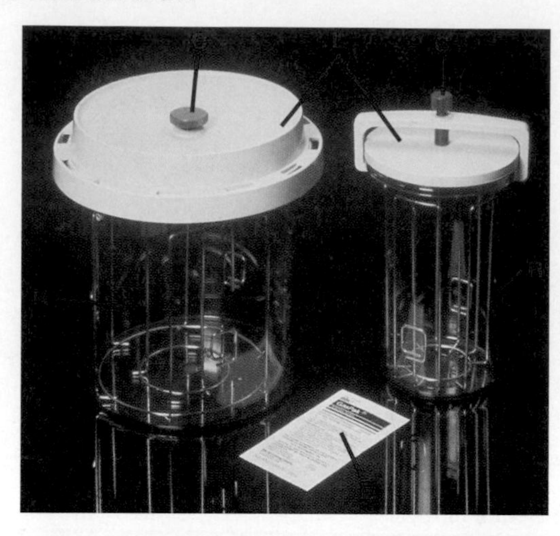

**Figure 33**
A commerical anaerobic cultivation device. (a) The components of the GasPak Anaerobic System. GasPak anaerobic jar (J), lid need film (L), clamp screw (C), charged catalyst reaction chamber (R), transparency from disposable hydrogen + GasPak carbon dioxide–generator Lab Manual envelope (E), and GasPak disposable anaerobic indicator (I).

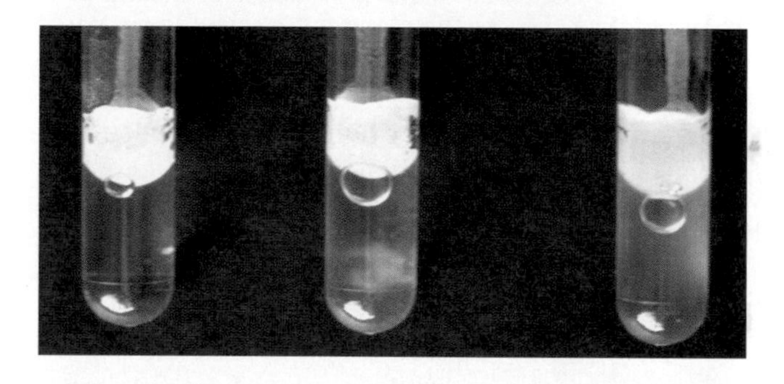

**Figure 34**
The paraffin-plug technique. (Left tube) Uninoculated with a paraffin plug. (Center tube) Turbidity (cloudiness) indicating growth. (Right tube) Growth under anaerobic conditions.

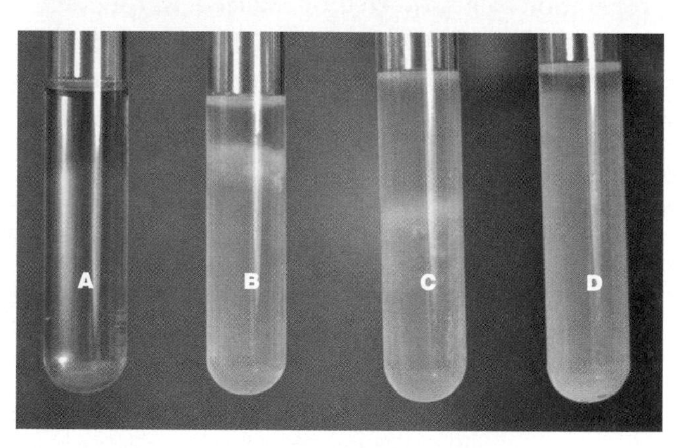


**Figure 35**
Representative growth characteristics of bacteria in sodium thioglycollate broth. (a) Uninoculated. Note the presence of the upper pink zone (for aerobes) and the lower yellow zone (for anaerobes). (b) Shows the growth of a typical anaerobic organism. (c, d) Show facultative (adjustable) organisms.

# Biochemical Activities of Microorganisms

## Differential, Selective, and Selective and Differential Media

A large number of media have been developed to aid in the isolation, differentiation, and identification of microorganisms such as bacteria and fungi. Differential media, which do not prevent the growth of organisms, contain one or more substances **(substrates)** that can be enzymatically attacked. On the basis of the type of reaction produced, colonies of one organism can be distinguished from others growing on the same plate. Blood agar (which contains whole blood) often is used to isolate and distinguish organisms that can enzymatically attack hemoglobin differently (Figure 36). A selective medium is defined as one that permits the growth of certain organisms while preventing or retarding the growth of others. Selection, in general, can be carried out through (1) control of ingredients of the medium, (2) alteration of atmospheric components, or (3) adjustment of incubation temperature.

Several media in use today incorporate both selective and differential substances (Table 4). One example of such preparations is Bacto Brilliant Green agar (Figure 37). This medium is a highly selective preparation used for the isolation of *Salmonella* species other than *S. typhi,* the causative agent of typhoid

## Differential, Selective, and Selective an Differential Media

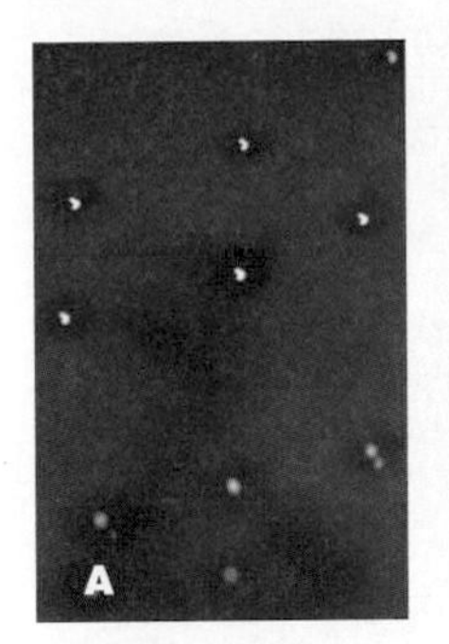

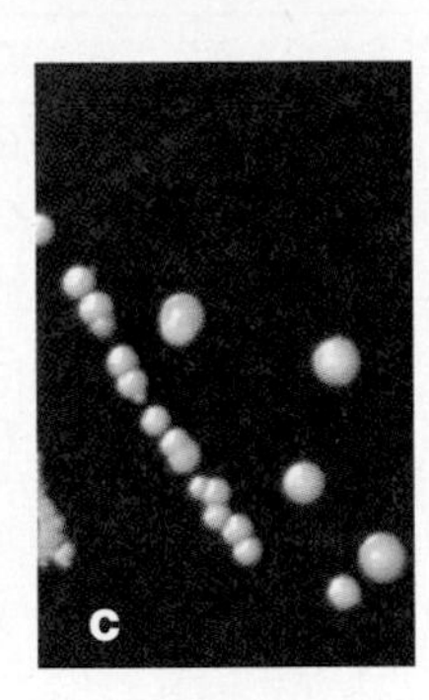


**Figure 36**
Representative hemolytic reactions on the differential medium blood agar. (a) Alpha hemolysis. Greenish discoloration of the medium usually surrounds bacterial colonies. (b) Beta hemolysis. Here, clear zones surround the individual bacterial colonies. (c) Gamma hemolysis, or nonhemolytic reactions. Discolorations, or clear zones, are not present.

**Figure 37**
Bacto Brilliant Green agar, a useful medium for the isolation and identification of *Salmonella* species. Typical non-lactose-fermenting *Salmonella* colonies appear as slightly pink-white opaque colonies surrounded by a brilliant red medium. Colonies of lactose-fermenting organisms form yellow-green colonies surrounded by intense yellow-green zones on the medium. An uninoculated plate also is shown as a control.

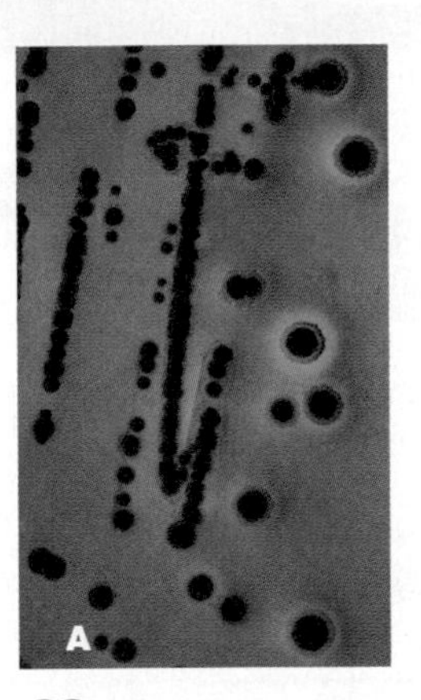

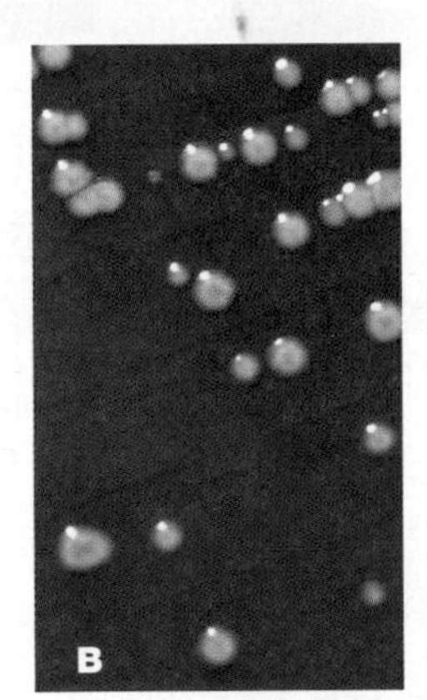

**Figure 38**
Eosin-methylene blue agar reactions. (a) Colonies of a lactose-fermenter. (b) Colonies of a non-lactose-fermenter. (c) A metallic sheen can form on the colony surfaces of certain lactose-fermenters.

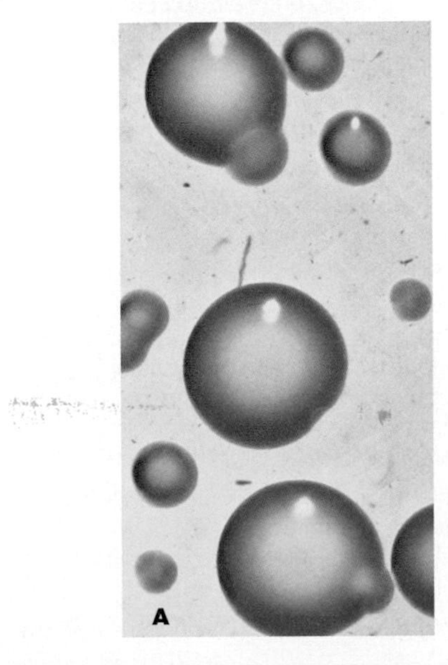

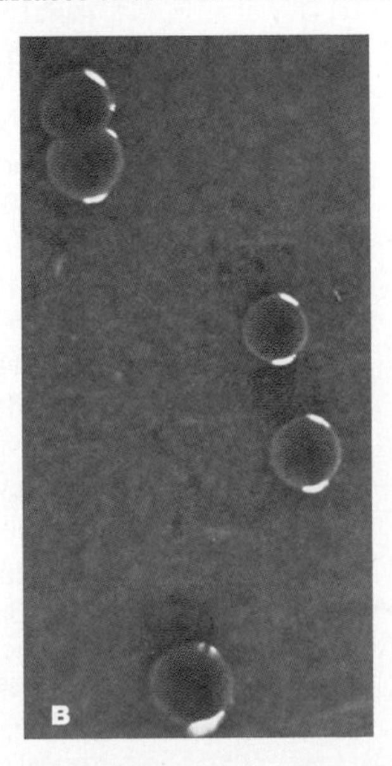

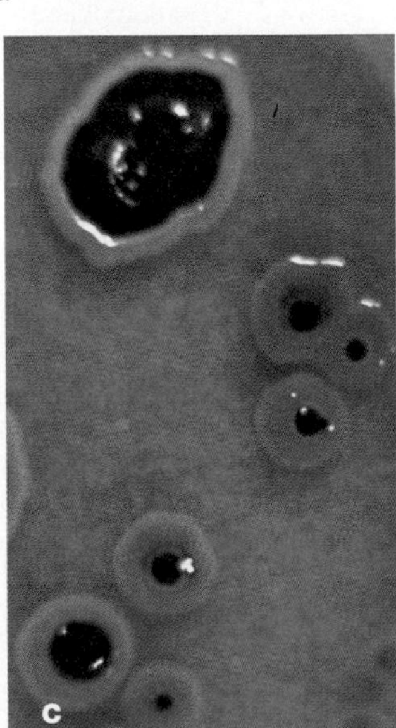


**Figure 39**
Hektoen enteric agar reactions. Three major types of reactions are shown. (a) Rapid lactose-fermenters (yellow to salmon-pink colonies). (b) Non-lactose-fermenters (green colonies). (c) Hydrogen sulfide ($H_2S$) producers (black-centered colonies). *Escherichia coli* is a rapid lactose fermenter; *Proteus vulgaris* and related species are well-known $H_2S$ producers.

**Table 4** Selective and Differential Media Properties

| Medium | Selective Agent(s) | Organisms Encouraged to Grow |
|---|---|---|
| Brilliant green agar | Brilliant green | Gram-negative rods [a] |
| Erosin-methylene blue agar | Eosin Y, methylene blue | Gram-negative rods |
| Hektoen enteric agar | Bile salts | Gram-negative rods |
| MacConkey agar | Bile salts, crystal violet | Gram-negative rods |
| Mannitol-salt agar | Sodium chloride | *Staphylococcus aureus* |

[a] This medium is not used for the isolation of *Salmonella typhi.*

fever, from stools or other specimens suspected of containing the organisms. The growth of other bacteria is almost completely inhibited by the presence of brilliant green dye. This medium also contains lactose and sucrose as substrates for enzymatic action. Typical *Salmonella* form lightly pink-white opaque colonies surrounded by brilliant red areas. The few lactose- or sucrose-fermenting organisms that can grow on the medium are easily differentiated from *Salmonella* by the formation of yellow-green colonies surrounded by intense yellow-green zones. Table 5 summarizes a few selective and differential media, their substrates, and associated reactions and Figures 37 through 41 show some examples. The distinctive reactions produced by pathogenic bacteria on other types of media are described later.

## Extracellular Degradation of Polysaccharides, Proteins, Lipids, and DNA

Metabolism is the sum total of the biochemical reactions required to maintain adequate nutritional levels and functional cellular activities. These reactions include the degradative, or breakdown, process called **catabolism** and the synthetic process known as **anabolism.** The two processes occur simultaneously and complement each other to provide for the essential needs of the cell. All metabolic reactions occur as a series of steps leading from one compound to another and are the bases of **metabolic pathways.** Such biochemical reactions are of value in bacterial identification.

Several microorganisms form and then release enzymes into their environment, such as a culture medi-

**Table 5** Summary of Reactions Associated with Selected Differential (D) and Selective Differential (SD) Media Used for Isolation and/or Identification

| Medium | Substrate(s) | Type of Medium | Reaction and Descriptions |
|---|---|---|---|
| Blood agar | Hemoglobin | D | 1. Alpha hemolysis (green zones around colonies)<br>2. Beta hemolysis (clear zones around colonies)<br>3. Gamma hemolysis (no zone around colonies) |
| Brilliant green agar | Lactose, sucrose | SD | 1. Lactose-fermenter (yellow-green colonies)<br>2. Non-lactose fermenter (pink to white colonies surrounded by brilliant red zones) |
| Eosin-methylene blue agar | Lactose, sucrose | SD | 1. Lactose-fermenter (dark purple colonies or colonies with dark centers and transparent colorless borders)<br>2. Non-lactose or non-sucrose fermenters (colorless colonies) |
| Hektoen enteric agar | Lactose, sucrose, salicin, and amino acids containing sulfur | SD | 1. Lactose-fermenter (salmon-pink colonies)<br>2. Non-lactose-fermenters (green, most colonies)<br>3. Salicin-fermenters (pink zones around colonies)<br>4. Non-salicin-fermenters (no change)<br>5. $H_2S$ producers (colonies with black centers) |
| MacConkey agar | Lactose | SD | 1. Lactose-fermenter (pink-red colonies surrounded by pink zones due to precipitated bile)<br>2. Non-lactose fermenter (colorless and translucent colonies) |
| Mannitol-salt agar | Mannitol | SD | 1. Mannitol-fermenter (colonies surrounded by yellow zones)<br>2. Non-mannitol fermenter (small colonies with no color yellow change) |

um. Some of these microbial extracellular enzymes can degrade, or break down, large molecules in the environment surrounding the cells. Most microbial extracellular enzymes are referred to as being **hydrolytic**. These enzymes are so named because they break large molecules into small ones and are generally classified according to the large types of molecules they can degrade. Such enzymes include **esterases,** which decompose fats and lipids, **glycosidases,** which break apart polysaccharides, and **proteinases,** which break down proteins.

Most of the carbohydrates available to microorganisms are in the form of polysaccharides. Two common examples are cellulose and starch, both of which are composed of smaller units **(polymers)** of glucose. The basic differences between the chemical and physical characteristics of these substances depend upon the structural arrangement of their glucose units. If this were not the case, then cellulase, the enzyme that breaks down cellulose into simple sugar units, would also degrade starch. The enzyme responsible for hydrolysis of starch is called amylase. Cellulase and amylase are examples of extracellular enzymes **(exoenzymes);** they can be secreted through the cell wall in order to degrade complex substances into units that can readily enter the cell. An example of this type of metabolic pattern is the action of amylase on starch (Figure 42) yielding maltose (a disaccharide composed of two glucose molecules).

The degradative action of an organism on an intact protein is analogous to such action on carbohydrates. In the case of starch, the enzyme amylase degrades the polysaccharide into units that can readily be absorbed by the cell. With a protein, such as casein (milk protein), the enzyme caseinase accomplishes much the same result in yielding polypeptides (Figure 43).

Many microorganisms are able to degrade fats and oils (in the process called **lipolysis**) and thus obtain acetate for carbohydrate metabolism and amino acid synthesis. The presence of a lipase in the enzymatic functions of a microorganism can be considered a potential indication of its invasiveness because animal cell membranes are largely composed of lipid. This degradative ability is yet another characteristic of microorganisms that can be useful in classification. The lipolytic activity of microorganisms appears in the area of lipolysis (Figure 44).

The production of deoxyribonuclease (DNAse), another extracellular enzyme, is useful for the isolation and differentiation of several bacterial species. DNAse agar is used to demonstrate the breakdown of DNA. Flooding inoculated plates after incubation with 0.1 N hydrochloric acid will show clear areas around the colonies that degraded the DNA in a medium (Figure 45).

Because microbial species vary in the kinds of extracellular enzymes they possess, demonstrating the presence or absence of a particular enzyme is of value in the identification of unknown organisms. Table 6 summarizes the features of tests used to demonstrate the presence or extracellular enzymes with specific organic macromolecules of importance in metabolism.

## Carbohydrate Metabolism: An Introduction to Intracellular Metabolism

Carbohydrates are the prime sources of energy and carbon for the synthesis of cellular substance and carbon skeletons. Simple sugars such as glucose and galactose have been described as the initiators of metabolic reactions. Most of the carbohydrates are available to microorganisms in the form of polysaccharides. Intracellular enzymes are used by cells to further the metabolic breakdown of carbohydrates produced by extracellular enzymes or to synthesize more complex cellular molecules. Two examples of intracellular enzymes are maltase and lactase. The former acts upon maltose to yield two molecules of glucose; the latter decomposes the carbohydrate lactose and yields one molecule of glucose and one of galactose.

The ability of an organism to attack and break down various carbohydrates can be determined easily by the use of a suitable nutrient medium containing the carbohydrate and a pH (acid-base) indicator. A pH indicator commonly used is phenol red, which is yellow at pH 6.9 (acid) and red at pH 8.5 (alkaline). Gas production in a carbohydrate medium may detected through the use of an inverted (Durham) tube that serves to trap any gas formed. The formation of acid and gas is an indication that the carbohydrate is enzymatically attacked (Figure 46).

Oxidative and fermentative production of acid may often be distinguished in a carbohydrate medium made semisolid by the addition of 0.3% agar. The preparation also contains a pH indicator. In this determination, two tubes of medium are inoculated by stabbing (inserting the inoculating tool down to the bottom of the tube), and the medium in one tube is layered with sterile mineral oil or similar material to exclude oxygen. Fermentative organisms produce acid throughout the medium in both tubes, whereas oxidative organisms produce acid only in the tube without mineral oil (Figure 47). Strict anaerobes produce acid only in the tube with mineral oil. Facultative anaerobes, organisms that can metabolize under aerobic conditions, produce acid in both tubes. Table 7 summarizes the results obtainable with a glucose oxidative/fermentative (O/F) agar medium.

Other carbohydrate tests used to differentiate among bacterial species are the methyl red (MR) and Voges-Proskauer (VP) reactions. In the methyl red test, organisms first metabolize glucose aerobically, thus

## Differential, Selective, and Selective and Differential Media *(continued)*

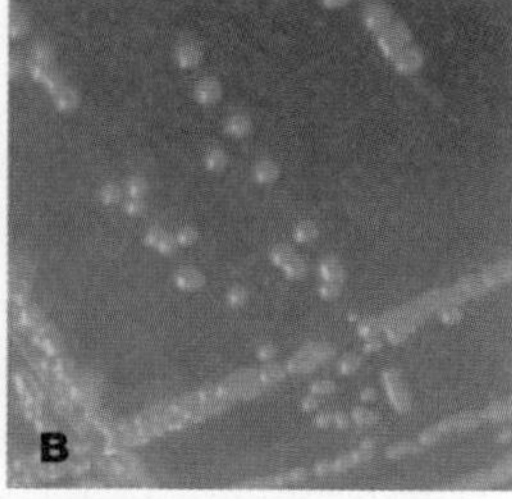


**Figure 40**
MacConkey agar reactions. (a) Colonies of a lactose-fermenter. (b) Colonies of a non-lactose-fermenter.

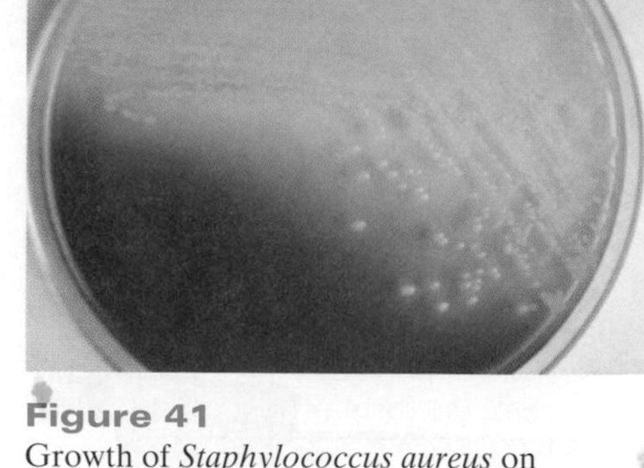

**Figure 41**
Growth of *Staphylococcus aureus* on mannitol-salt agar. The presence of yellow zones around the bacterial growth indicates acid formation from mannitol.

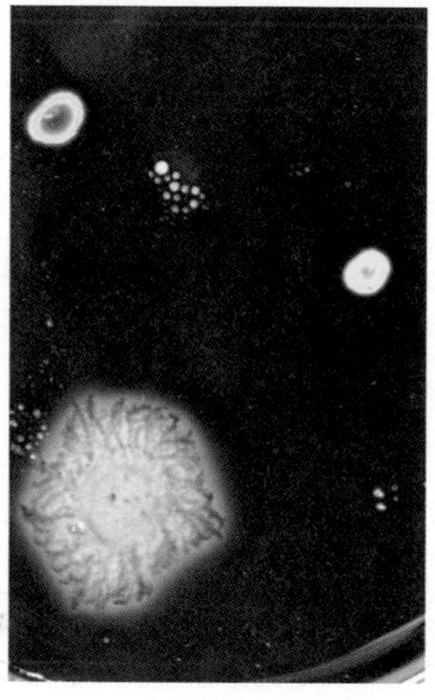

**Figure 42**
Starch hydrolysis (amylase production). A typical starch hydrolytic reaction as produced by the bacterium *Bacillus subtilis.* The complete absence of starch (hydrolysis) upon the addition of an iodine reagent is indicated by the yellow background surrounding the bacterial colonies. A negative reaction is indicated by a dark purple background.

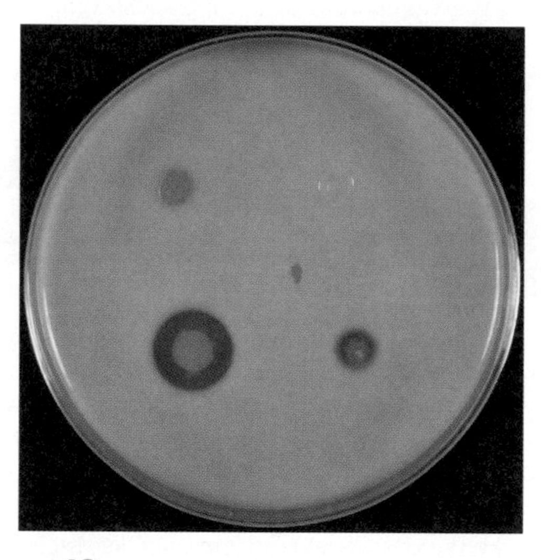

**Figure 43**
Caseinase activity. Casein digestion can be seen as clear zones around bacterial growths. One clearly negative result also is shown.

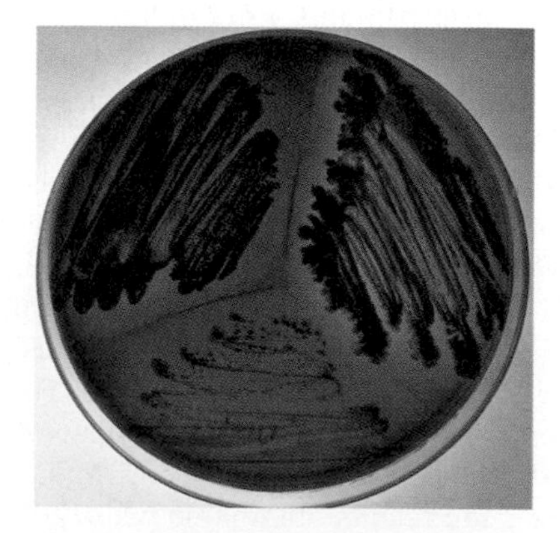

**Figure 44**
The demonstration of lipid hydrolysis using Bacto-spirit blue agar containing Bacto-lipase reagent. The lipolytic activity of a bacterial species is recognized by the deep blue color or clearing that develops in the medium surrounding the test organism. A comparable color change is not observed with nonlipolytic microorganisms.

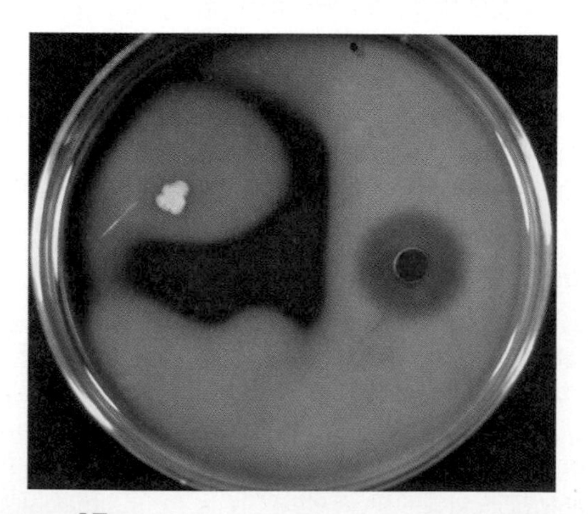

**Figure 45**
DNAse reactions. The presence of the enzyme is indicated by a clear zone around the bacterial growth. The cloudy area (negative reaction) is caused by the addition of hydrochloric acid, which precipitates the DNA.

**Table 6** Summary of Extracellular Enzymes

| Test | Substrate | Reagent(s) and Incubation Times | Positive Reaction | Negative Reaction |
|---|---|---|---|---|
| Casein degradation | Casein (milk protein) | No reagent; 24–48 hours | Clear zones around bacterial colonies | No clear zones around colonies |
| DNAse production | DNA | 0.1 *N* hydrochloric acid (HCl); 24–48 hours | Clear zones around colonies after the addition of HCl | Cloudy zones around colonies after addition of HCl |
| Lipid hydrolysis | Lipid | Spirit blue dye; 24–48 hours | Clear and/or dark areas surrounding growth | No clear or dark area around colonies |
| Starch hydrolysis | Starch | Lugol's or Gram's iodine; 24–48 hours | Yellow zones around colonies after the addition of iodine | Purple or dark zones around colonies after the addition of iodine |

**Table 7** Oxidation/Fermentation Reactions

| Type of Metabolism | Aerobic Conditions (No Mineral Oil Layer) | Anaerobic Conditions (Mineral Oil Layer) |
|---|---|---|
| Oxidative | Acid (yellow) | Alkaline (green) |
| Fermentative | Acid (yellow) | Acid (yellow) |
| Nonsaccharolytic | Alkaline (green) | Alkaline (green) |

exhausting all of the available oxygen by means of respiratory metabolism. This is then followed by one of two types of glucose fermentation, **mixed-acid** or **butylene glycol.** The type of fermentation and the end products formed are of value in species identification. Enteric bacterial species that perform the mixed-acid fermentation of glucose excrete large quantities of **acetic, formic, lactic,** and **succinic acids** and **ethanol.** These excreted metabolic products lower pH significantly to approximately 4.2, which is detectable by the methyl red indicator (Figure 48).

The Voges-Proskauer (VP) test is used to detect **acetoin** (a-SET-ō-in), also known as **acetylmethyl carbinol.** This is an intermediate compound produced by organisms carrying out the butylene glycol type of glucose fermentation. A positive reaction is indicated by the formation of a pink or cherry-red color upon the addition of alpha-naphthol and potassium hydroxide (KOH) plus creatine solutions (Figure 49).

## Nitrogen Metabolism

Many different nitrogen-containing compounds are involved in the metabolism of a living cell. Proteins are involved in enzymatic and structural activities; purines and pyrimidines of nucleic acids are concerned with genetic mechanisms and certain syntheses; and inorganic compounds act as electron donors or acceptors.

Several differential media and tests are routinely used to detect enzymatic reactions involving nitrogen-containing compounds. These reactions include protein and protein-like compound degradation (gelatin, tryptophan, and phenylalanine), urease activity, and nitrate reduction (Table 8).

The indole test is used to determine the ability of an organism to cleave the amino acid tryptophan into **indole, ammonia,** and **pyruvic acid** (Figure 50). During the metabolic activities of microorganisms, amino acids enter cells and are subsequently degraded by specific intracellular enzymes, the **decarboxylases** and **deaminases.** The former remove the carboxyl group (-COOH) from an amino acid, and the latter remove the amino group (-NH) from an amino acid. If the organism being studied has the enzyme **tryptophanase,** the amino group will be removed and indole will be formed from tryptophan. Indole is easily detected by the addition of Kovac's reagent and the appearance of a dark cherry-red layer (Figure 50).

Detecting the deamination of phenylalanine requires the use of phenylalanine agar. After incubation, ferric chloride is added directly to a growing culture. The development of a green or brown color indicates the presence of phenylpyruvic acid, a product of phenylalanine deamination (Figure 51).

Gelatin, a protein formed from collagen (an animal protein) can be hydrolyzed by extracellular enzymes (proteinases) produced by several microbial species. As microorganisms hydrolyze this protein, it changes from a solid to a liquid, thus its value as a solidifying agent is destroyed. Several tests for proteases are based on such gelatin liquefaction (Figure 52).

Many bacteria are capable of synthesizing acids from

## Carbohydrate, Nitrogen, and Other Metabolic Reactions

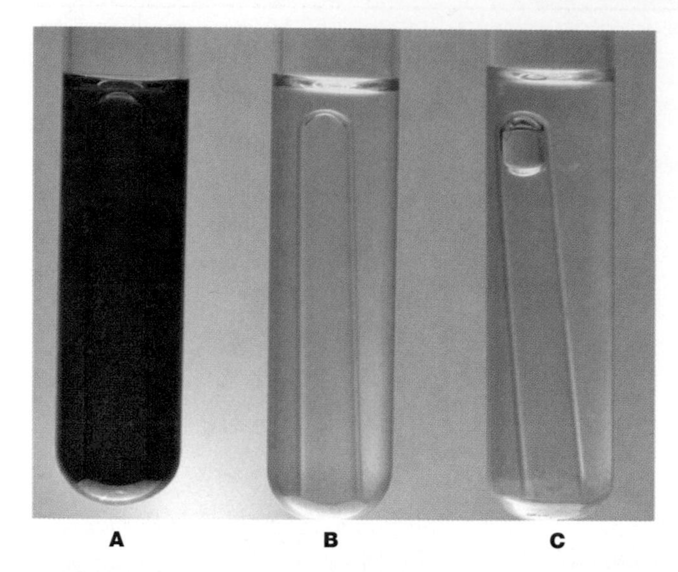


**Figure 46**
Carbohydrate fermentation employing Durham fermentation tubes. The indicator used is phenol red. (a) Uninoculated. (b) Acid production. (c) Acid and gas production. Note the collection of gas in the inverted vial.

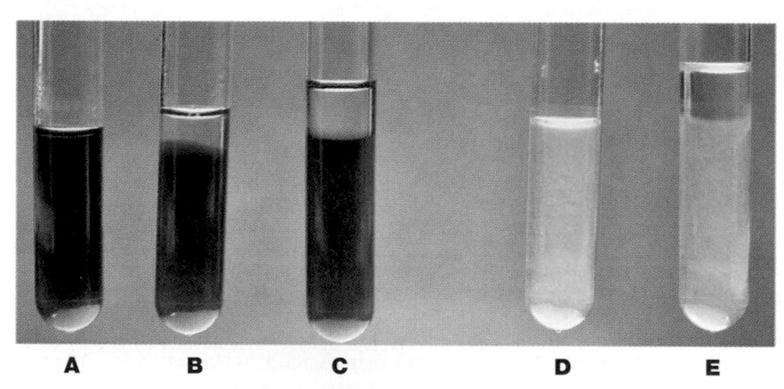


**Figure 47**
Carbohydrate oxidation or fermentation using tubes of glucose O/F agar. The indicator used is bromothymol blue. (a) Uninoculated. (b) Shows glucose oxidation. (c) Shows no glucose oxidation or fermentation,. (d,e) Show glucose oxidation and fermentation, respectively. Note that the fermentation reaction is detected specifically in the mineral oil–covered medium.

**Figure 48**
Methyl red test: Left tube, negative. Right tube, positive.

**Figure 49**
The Voges-Proskauer test. Left tube, negative. Right tube, positive.

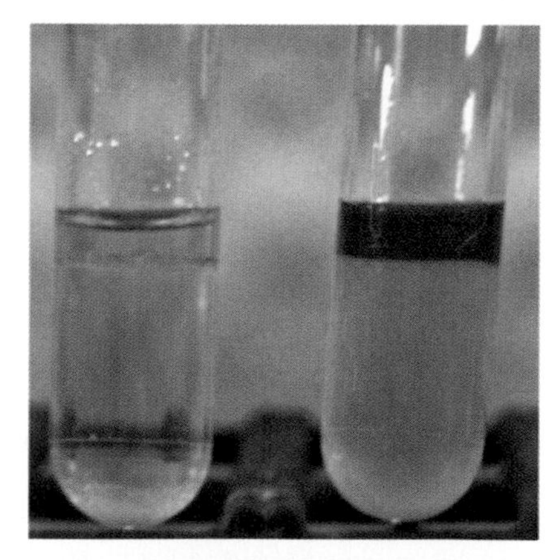

**Figure 50**
The Indole test. Left tube, negative. Right tube, positive.

by-products of carbohydrate and lipid metabolism when provided with ammonia as a nitrogen source. Some of these organisms are able to split the compound urea, a major organic waste product of animal metabolism, into ammonia and carbon dioxide (Figure 53). The medium becomes highly alkaline (basic) owing to ammonia production. The product, carbon dioxide, is incorporated into carbohydrate and nitrogen metabolism through a variety of important reactions.

A major aspect of metabolism is the generation of transport electrons, yielding the energy required for metabolism. One facet of this electron transport sys-

**Table 8** Summary of Selected Carbohydrate and Protein Intracellular Metabolic Reactions

| Test | Substrate(s) | Reagents and Incubation Times | Positive Reactions | Negative Reactions |
|---|---|---|---|---|
| Carbohydrate fermentation (Durham fermentation system) | Glucose, lactose, etc. | Phenol red indicator; 24–48 hours | Acid (yellow color); gas (gas bubble in inverted Durham tube) | Alkaline (red color) no gas (no bubble in Durham tube) |
| Methyl red | Glucose | Methyl red indicator; 48 hours | Acid production (red color) | No acid production (any color other than red) |
| Voges-Proskauer | Glucose | Alpha-naphthol and KOH plus creatine; 48 hours | Acetoin produced (red color) | No acetoin produced (any color other than red) |
| Indole | Tryptophan | Kovac's reagent; 24–48 hours | Indole production (dark red) color in layer on top of broth | Negative indole production (any color other than dark red) |
| Gelatin hydrolysis | Gelatin | None; refrigeration after incubation is used to test for liquefaction; 48 hours | Gelatin medium is liquid after 30 minutes of refrigeration | Gelatin medium is solid after 30 minutes of refrigeration |
| Phenylalanine deaminase | Phenylalanine | Ferric chloride; 24–48 hours | Formation of green or brown color upon ferric chloride addition | No color change |
| Urease | Urea | Phenol red; 24–48 hours | Red color formation | Color other than red |
| Nitrite reduction | Nitrate | Sulfanilic acid and dimethyl-alphanaphthyl-amine; 24–48 hours | Red color after addition of reagents | No red color |

tem is the ability to reduce nitrate to nitrite as shown in the following reaction. Some microorganisms may go beyond nitrite to nitrogen gas.

$$\underset{\text{Nitrate Ion}}{NO_3^-} + 2H^+ + 2e^- \longrightarrow \underset{\text{Nitrite Ion}}{NO_2^-} + H_2O$$

$$\xrightarrow{7H^+ + 7e^-} \underset{\text{Nitrous oxide}}{N_2O} \xrightarrow{2H^+ + 2e^-} \underset{\text{Nitrogen gas}}{N_2}$$

**Nitrate Reduction.** The enzyme responsible for the reaction is **nitrate reductase.** After incubation, a test for the presence of **nitrate** is performed by the addition of **sulfanilic acid** first and then **demethyl-alpha-naphthylamine.** The appearance of a red color is a positive test. If a red color does not occur, an additional test is necessary to determine whether nitrate was not reduced to nitrite (the absence of nitrate reductase) or whether nitrate was converted to gaseous nitrogen by means of **denitrification** (the presence of nitrite reductase). What actually occurred is determined by adding a small amount of powdered zinc to the culture. The zinc will reduce nitrate to nitrite, producing a positive (red) reaction. This reaction will show that the organism did not possess the necessary enzyme. A negative result at this point will confirm the presence of nitrate reduction (Figure 54). The presence of gaseous nitrogen can be demonstrated with the use of an inverted, medium-filled vial and appears as a bubble.

Table 7 summarizes the tests and reactions of nitrogen metabolism presented here.

## Oxygen Utilization: Oxidase and Catalase Activities

Cytochrome *c* and catalase are two enzymes involved in the use of oxygen by aerobically respiring bacteria. Cytochromes are enzymes that have nonprotein, iron-containing portions called **heme** and are tightly bound in prokaryotic plasma membranes.

The oxidase test is used to detect the presence of **cytochrome *c*.** The test uses the reagent tetramethyl-p-phenylenediamine, which can give its electrons (electron donor) to the oxidized form of cytochrome *c*. The resulting oxidized form of this reagent forms a dark blue, violet, or purple color (Figure 55), whereas the reduced form is colorless. Commercially available reagent-containing swabs such as **Oxyswab** and disposable slides such as the **DrySlide Oxidase** make testing for cytochrome *c* relatively simple (Figure 55b-c).

The oxidase test is of value in detecting the presence of cytochrome *c*–containing gram-negative cocci belonging to the genera *Moraxella (Branhamella)* and *Neisseria.* In addition, the test can distinguish between oxidase-negative, gram-negative enteric bacterial rods

## Carbohydrate, Nitrogen, and Other Metabolic Reactions

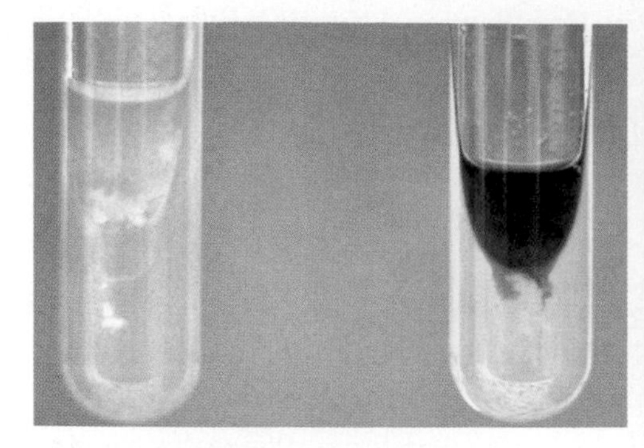

**Figure 51**
Phenylalanine agar (PA). This medium is used to detect the presence of phenylalanine deaminase. The formation of a green or dark brown color in the medium upon the addition of ferric chloride is a positive reaction.

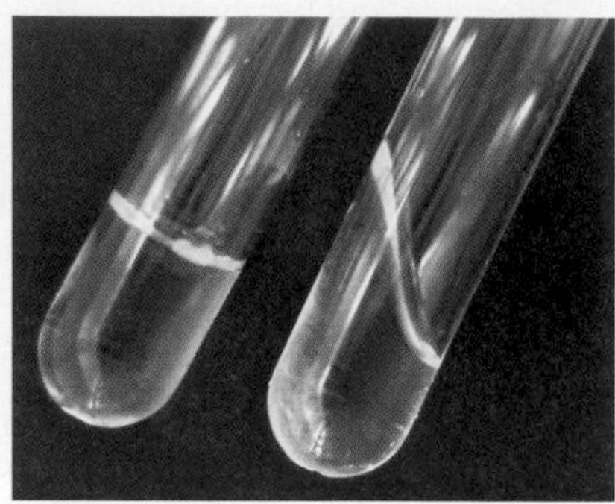

**Figure 52**
Gelatin hydrolysis. A positive result is indicated by a liquefaction of the gelatin agar (right tube). A negative reaction is indicated by the persistent solid form of the medium (left tube).

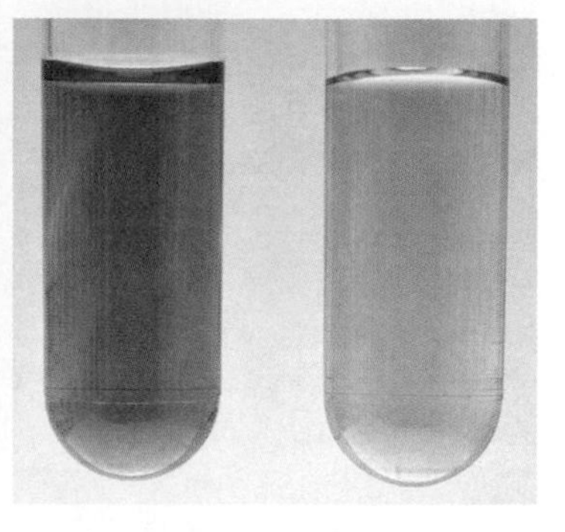

**Figure 53**
Urease activity. A positive reaction is indicated by the red color of the medium. A negative reaction is represented by an orange color. *Proteus* species generally produce positive results. The indicator in this medium is phenol red.

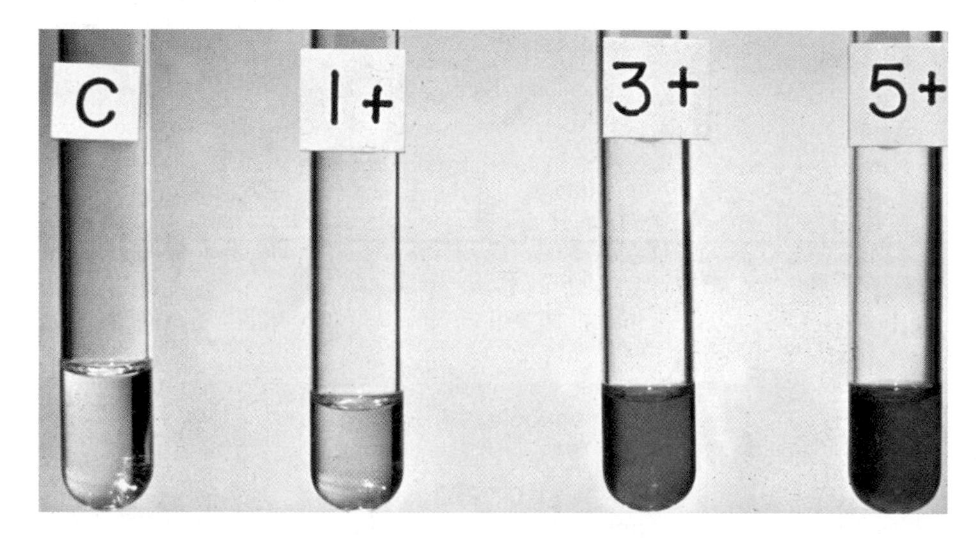


**Figure 54**
Nitrate reduction. A control tube is included with the positive nitrate reduction reactions to emphasize the range of color changes that can occur.

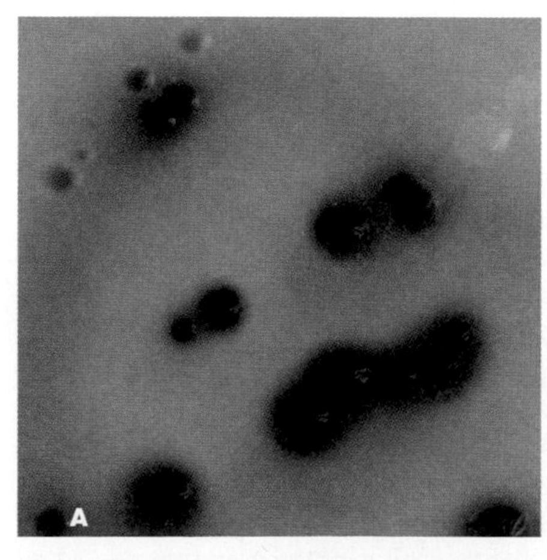


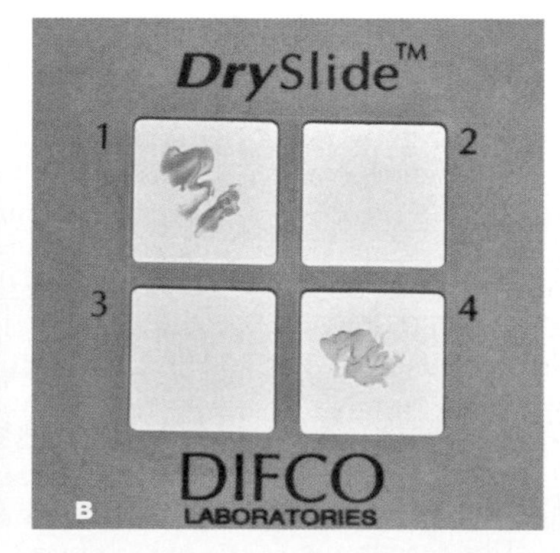


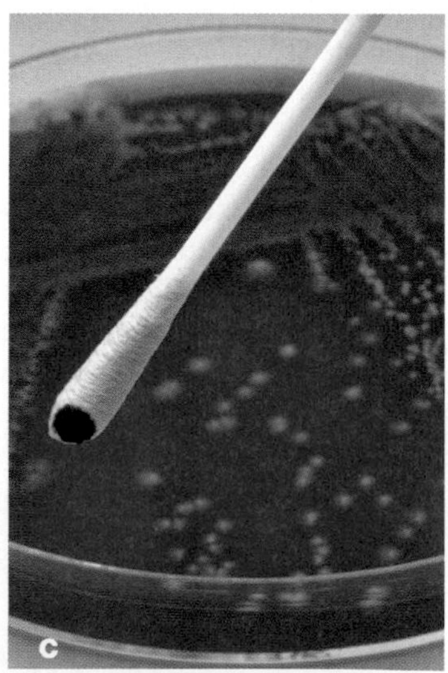


**Figure 55**
The oxidase test. (a) Colonies demonstrating a positive reaction are blue and turn darker later. Negative reactions are indicated by colorless colonies. (b) Results on a DrySlide. (Courtesy of Difco Laboratories, Detroit, MI). (c) A positive Oxyswab test. The reagent is contained in the swab material. (Courtesy of Remel, Lenexa, KS.)

and gram-negative rods belonging to the genera *Pseudomonas* and *Aeromonas*.

Toxic forms of oxygen such as the superoxide radical ($O_2$) are found in all liquid environments that contain dissolved oxygen. These chemicals are toxic to living cells, so microorganisms in aerobic environments must be able to detoxify them. Such microbes produce an enzyme, **superoxide dismutase** (SD), which adds protons to the superoxide radical to form hydrogen peroxide, as shown in the following equation:

$$\underset{\text{Superoxide}}{2O_2} \xrightarrow{2e^-} 2O_2 \xrightarrow{4H^+} \underset{\text{Hydrogen peroxide}}{2H_2O_2}$$

Because hydrogen peroxide also is toxic, microorganisms also must produce the enzyme **catalase** which is capable of decomposing the chemical into water and molecular oxygen:

$$2H_2O_2 \xrightarrow{O_2} 2H_2O$$

Catalase

Catalase is produced by all actively growing aerobic microbes. Because strict anaerobes lack the ability to use oxygen in their respiration, they also lack catalase.

The test for catalase is simple to perform. All it requires is the addition of a few drops of 3% hydrogen peroxide directly to a young broth culture (Figure 56a), or to colonies on an agar surface (Figure 56b), or to a clump of cells on a glass slide. The vigorous evolution of oxygen bubbles is a positive result. Cultures growing on blood agar are not tested because the agar medium itself may produce a positive reaction.

The catalase test is very useful in differentiating among bacteria that have similar morphological features but differ in their metabolic activities. Table 9 summarizes the features of the oxidase and catalase tests.

## Motility Media

Bacterial motility can be shown with the aid of several types of motility media. The composition of these preparations is such that they offer no more resistance to movement during incubation than would a broth culture. Motility Medium S, which contains 2, 3, 5-triphenyltetrazolium chloride (TTC), is used. Motility can be recognized by the presence of a diffuse red growth away from the line of inoculation, or stabline (Figure 57). Nonmotile organisms grow only along the stabline.

## Differential Test Patterns: The IMViC Test

The IMViC pattern of reactions reflects the enzymatic makeup or biochemical fingerprint of the microorganism being studied. For example, *Escherichia coli*, a normal inhabitant of the intestinal tracts of humans and lower animals, resembles *Enterobacter aerogenes*, a bacterial species that is widely distributed in nature, especially on plants and plant products. Both organisms are gram-negative, and they are similar in morphological and cultural characteristics. These bacteria can be differentiated by means of the IMViC set of tests (Figure 58). Each letter in the series (except *i*) refers to a separate procedure. *I* stands for the indole test; *M*, the methyl red test; *Vi*, the Voges-Proskauer reactions; and *C*, the citrate test. The small *i* after the *V* is used to make pronunciation easier. The indole, methyl red, and Voges-Proskauer tests were described earlier.

The remaining test in the IMViC series, the **citrate test**, determines the presence of enzymes that enable citrate to enter the cell, which can then use the citrate as the sole source of carbon for its metabolism and growth. Two media can be used for the citrate test: **Koser citrate** and **Simmons citrate**. The reaction in Koser citrate medium (a clear colorless liquid) is positive if the preparation becomes cloudy after incubation (Figure 59a). A light inoculum must be used with Koser citrate. Simmons citrate agar (Figure 59b) contains the pH indicator bromthymol blue, which is green under acidic conditions and dark blue when the medium becomes alkaline. Organisms utilizing citrate produce an alkaline reaction.

**Table 9** Summary of Oxidase and Catalase Tests

| Test | Substrate, Detects Presence of | Reagents and Incubation Time | Positive Reaction | Negative Reaction |
|---|---|---|---|---|
| Oxidase (cytochrome *c*) | No substrate; enzyme cytochrome *c* | Tetramethyl-phenylenediamine; depending on procedure, 24–48 hours may be necessary | Formation of a deep violet or purple color within 60 seconds after addition of reagent | No major color change after addition of reagent |
| Catalase | No substrate; enzyme catalase | Hydrogen peroxide; cultures 18–24 hours old | Formation of visible bubbles after addition of reagent | No visible bubbles formed after addition of reagent |

## Microbial Reactions in Multiple-Test Media: Litmus Milk; Triple Sugar Iron Agar; and Sulfide, Indole, Motility Medium

Investigators have long been interested in reducing the expense and drudgery associated with microbiological testing methods. The efforts of many individuals have resulted in the development of at least two general categories of improved biochemical testing procedures and materials: (1) several test substrates combined in one or two tubes that are inoculated and incubated in the conventional manner and (2) separate biochemical tests contained in miniaturized, multi-compartment devices that are inoculated by other than conventional methods but are incubated according to standard practices.

The first category includes several widely accepted media such as litmus milk (Figures 60 and 61), triple sugar iron agar (Figure 62), and sulfide, indole, motility medium (Figure 63).

Very few media, after inoculation and incubation, can yield as much information on microbial reactions as litmus milk. The medium includes materials that most bacteria require for growth and, specifically, the substrates lactose and casein. Litmus serves as an indicator of the organism's acid or alkali production and oxidation and reduction activities. Characteristic reactions observed with litmus milk are as follows (Figures 60 and 61).

1. Acid and acid curd formation
2. Alkaline conditions
3. Rennin curd production
4. Peptonization
5. Litmus reduction
6. Gas formation

Production of acid by an organism is a function of its ability to utilize lactose, one of the most abundant sugars in milk. Acid production is demonstrated when a litmus medium changes from blue to pink. If the organism is able to produce a considerable amount of acid, an insoluble complex of calcium and casein may be formed, resulting in a curd. An organism that cannot attack lactose might well use the milk proteins as a source of nitrogen and carbon, and an alkaline reaction might result. This reaction is indicated by an intensification of the blue color in the litmus medium. If the amount of acid or alkali produced is low, no apparent change in the medium may be observed. However, a rennin curd may develop even with low acid or alkali production if the organism has enzymes that can produce the insoluble calcium-casein complex. This type of curd usually retracts and yields a grayish liquid known as **whey.** Acid or rennin curds are quite palatable dairy products known as cottage cheese. Either type of curd can be digested by proteinases, a process known as **peptonization** (or **solubilization**). Peptonization is usually characterized by a reduction in curd size and the formation of a brownish liquid. This reaction can be observed easily when it occurs along with the curd. Reduction of the litmus by oxidation-reduction activities of the bacteria can be shown by loss of the litmus color. The reaction is quite apparent when a pink acid curd begins to turn white. The change develops at the bottom of the tube and moves upward. Another readily apparent characteristic is gas production, which can be observed as holes or tears in curds or as separated curd strands.

Triple sugar iron agar is a characteristic preparation used to differentiate gram-negative enteric organisms by their ability to ferment dextrose (glucose), lactose, or saccharose (sucrose) and to reduce sulfites to sulfides. The medium is dispensed as agar slants. It is inoculated by the insertion of an inoculating needle into the bottom, or butt region, of the tube and, as the needle is withdrawn from this area, the slant surface is streaked. Several types of reactions can be recognized (Figure 62 and Table 10).

When a culture ferments only dextrose the concentration of which is one-tenth of lactose or sucrose, the bottom portion of the medium will turn yellow from the acid produced. Because its concentration is low, glucose is quickly utilized. In the event that the organism in the culture cannot ferment either of the other two carbohydrates in the medium, growth in the butt stops. However, the organisms on the slant continue to utilize, as a source of energy, the peptone in the medium or the intermediate products of dextrose fermentation, or both. Use of these intermediate products results in their reduction, and the use of peptone results in the eventual secretion of ammonia into the medium. The ammonia will neutralize the intermediate products, causing the slant to turn red.

The entire medium will turn yellow and remain so if the culture can utilize lactose, sucrose, or both, in addition to dextrose. Apparently, the organisms in such cases will not exhaust the available fermentable carbohydrates. The peptone, or the intermediate end-products, are not limited. Gas production can be detected when holes in the agar are formed or the medium is broken into several fragments.

Hydrogen sulfide ($H_2S$) production by a culture results in the blackening of the medium. This color is caused by the production of $H_2S$ from an ingredient of the medium, sodium thiosulfate, which, when combined with another component of the medium, ferrous ammonium sulfate, results in the formation of the black insoluble compound ferrous sulfide.

Sulfide, indole, motility (SIM) medium can be used to determine hydrogen sulfide production, indole production, and motility. The presence of $H_2S$ is indicat-

## Oxygen Utlization, Motility, and IMViC Reactions

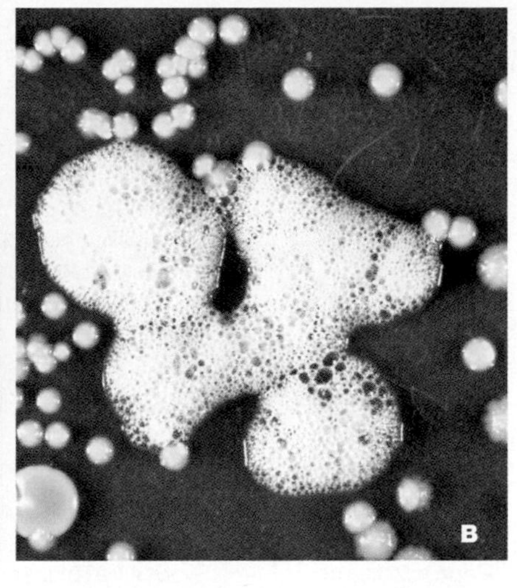


**Figure 56**
The catalase reaction. Catalase is an enzyme that catalyzes the breakdown of hydrogen peroxide ($H_2O_2$), thereby releasing oxygen gas. Formation of a white froth if a few drops of 3% $H_2O_2$ are added to a microbial colony or to a broth culture is a positive reaction. (a) A positive reaction in a broth culture. (b) The agar plate reaction.

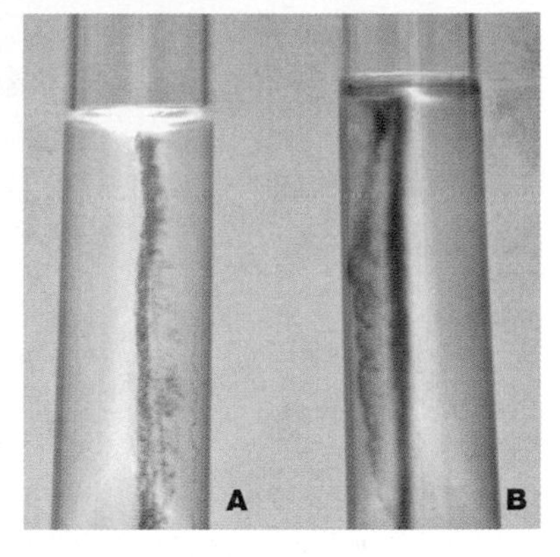


**Figure 57**
Motility medium. This preparation, to which a dye called 2,3,5-triphenyltetrazolium chloride is added, can be used to detect motile organisms. (a) In a nonmotile culture, growth appears only along the line of inoculation. (b) In a motile culture, growth spreads from the line of inoculation, making most of the medium turbid.

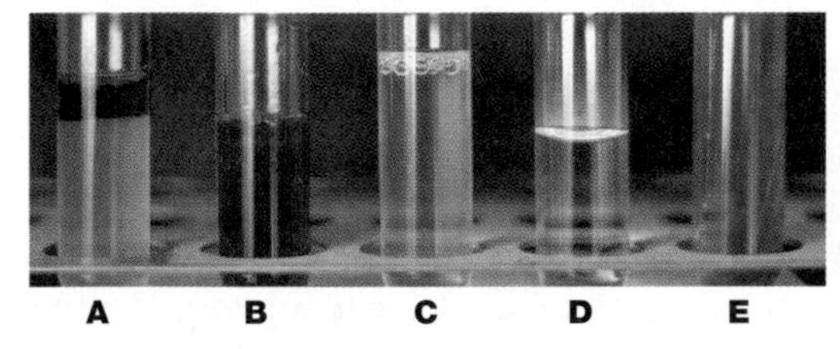


**Figure 58**
IMViC reaction results. The tests are from left to right: indole, positive; methyl red, positive; Voges-Proskauer, negative; citrate (Koser's medium), negative; (e) citrate (Simmons medium), negative.

**Figure 59**
The citrate test. (a) Koser's citrate. Left tube, negative; right tube, positive. (b) Simmons citrate. Left tube, negative; right tube, positive.

## Multiple-Test Media Reactions and Miniaturized and Rapid Microbiological Systems

**Figure 60**
Selected litmus milk reactions. (a) Uninoculated blue medium. (b) An alkaline reaction (dark blue to purple), indicating the inoculated organism's use of the milk proteins as a source of carbon and nitrogen. (c) An acid result (pink), demonstrating the production of a considerable amount of acid and lactose fermentation. (d) Acid production at the surface (pink) and the beginning of litmus reduction (white portion of the medium). The reaction is caused by the oxidation-reduction activities of the inoculated bacteria. This reaction begins at the bottom of the tube and spreads upward. (e) Litmus reduction. (f) Acid clot formation with extreme gas production causing stormy fermentation. (g) Casein curd formation and peptonization (proteolysis). With peptonization, a reduction in curd size occurs and a brownish, purple, or clear supernatant, called whey, is formed.

**Table 10** Summary of Selected Multiple Test System Biochemical Reactions

| Test | Substrate(s) and Incubation Times | Reagents(s) | Positive Reaction | Negative Reaction |
|---|---|---|---|---|
| Litmus milk | Lactose and/or casein; 24–48 hours | Litmus indicator | Acid (pink)<br>Alkaline (light to dark blue)<br>Litmus reduction (white)<br>Curd: milk changes to solid state with fluid (whey)<br>Proteolysis (decrease in milk turbidity and eventual formation of clear brownish or purple fluid) | No change (light blue) |
| Triple Sugar Iron Agar | Dextrose, lactose, sucrose; 24–48 hours | Phenol red | Acid from dextrose (yellow bottom)<br>Acid from lactose and sucrose (yellow slant) | Alkaline, no change (red bottom)<br>Alkaline (red slant) |
| | Sodium thiosulfate; 24–48 hours | Ferrous ammonium sulfate | $H_2S$ positve (blackening of medium) | $H_2S$ negative (no blackening of medium) |
| Sulfide, indole, motility medium | Sodium thiosulfate; 24–48 hours<br>Tryptone; 24–48 hours<br>None; 18–24 hours | Peptonized iron<br>Kovac's reagent<br>None | $H_2S$ positive (blackening of medium)<br>Indole positive (red layer)<br>Motility (turbidity throughout medium) | $H_2S$ negative (no blackening of medium)<br>Indole negative (no red layer)<br>No motility (growth only on line of inoculation) |

ed by a blackening along the line of inoculation in the medium. Quite often, the dark color diffuses when a motile organism is involved. The presence of indole is detected by the use of Kovac's reagent. Motility is recognized by the appearance of a diffuse growth or turbidity spreading from the inoculation site (Figure 63).

## Miniaturized and Rapid Microbiological Systems

Improved biochemical procedures include several commercially available miniaturized multiple test systems and devices designed to facilitate the identification of bacterial cultures. Complete instructions together with accurate identification keys based on established biochemical reactions are provided by the respective manufacturers. Two systems—the **Enterotube II** (Figure 64) and the API 20E (Figure 65)—consist of manufacturer-selected combinations of substrates that not only are ready for use but require minimal storage space and are stable at either room or refrigerator temperatures for significant time periods. The carefully selected tests in each case also form the bases for computer-developed identification systems. These two systems emphasize the importance placed on the need for rapid analysis of test results for the identification of clinically isolated bacteria.

The Enterotube II, which is used for the identification of lactose-fermenters, incorporates conventional media into a single, ready-to-use, multicompartment tube with an enclosed inoculating rod (Figure 64). The unique inoculating arrangement permits simultaneous inoculation of all compartments and the performance of 15 biochemical tests. The Enterotube II is used by first removing the end caps of the system to expose a long inoculating rod. The rod is used to pick an isolated colony from a selective and differential agar, or related plate medium (Figures 38–40). The rod is then drawn through a series of substrate-containing agar compartments in the plastic tube of the system. The end caps are then replaced, and the system is incubated. After incubation, appropriate reagents are introduced into the chambers requiring them. The color changes resulting from the biochemical activity of unknown bacterial cultures used as inocula are interpreted and used in their identification.

The API system employs a plastic strip holding 20 miniature compartments, or cupules. Each cupule contains a dehydrated substrate for a different test (Figure 65). At least 22 standard biochemical tests can be performed, including *o*-nitrophenyl-β-D-galactosidase (ONPG), arginine dihydrolase, lysine and ornithine decarboxylases, citrate utilization, hydrogen sulfide ($H_2S$) production, urease, tryptophan deaminase, indole production, acetoin production, gelatinase, and the fermentation of glucose, mannitol, inositol, rhamnose, sucrose, melibiose, amygdaline, and arabinose.

The dehydrated substrates are inoculated with a bacterial suspension and subsequently incubated according to a procedure described by the manufacturer.

## Multiple-Test Media Reactions and Miniaturized and Rapid Microbiological Systems *(Continued)*

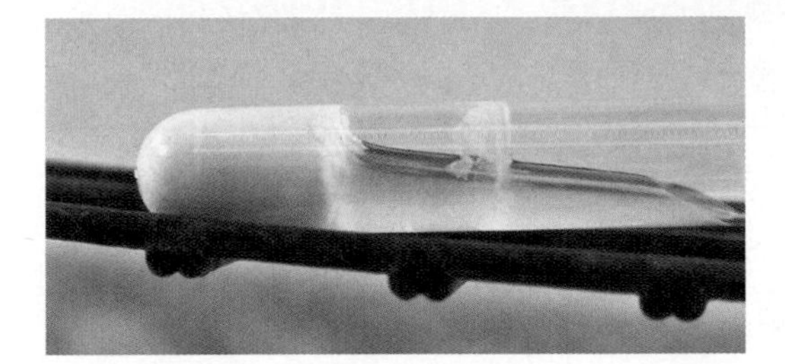

**Figure 61**
Coagulation and peptonization. Note the solidification of the milk protein and the clear liquid, whey.

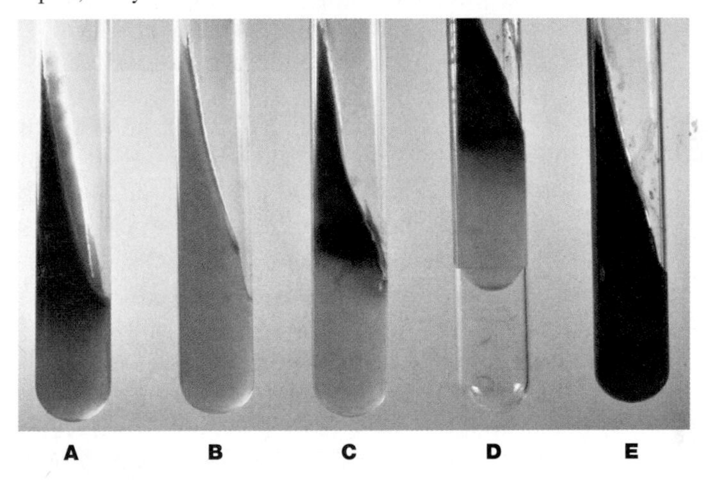


**Figure 62**
Selected triple sugar iron agar medium reactions. Alkaline reactions (Alk) are indicated by a red coloration; acid (A) production by a yellowing of the medium; gas formation by the presence of air pockets in the preparation; and the presence of $H_2S$ by a blackening of the medium. The reactions shown are as follows:

| | *Tube A* | *Tube B* | *Tube C* | *Tube D* | *Tube E* |
|---|---|---|---|---|---|
| Slant | Alk | Acid | Alk | Alk | Alk |
| Butt | A | A | A | A | Alk |
| Gas | Absent | Absent | Absent | Present | Absent |
| $H_2S$ | Absent | Absent | Present | Present | Absent |

*Note:* The dark color in tube e is caused by the purple pigment of the organism. Acid in glucose (bottom) must be produced before $H_2S$ can occur.

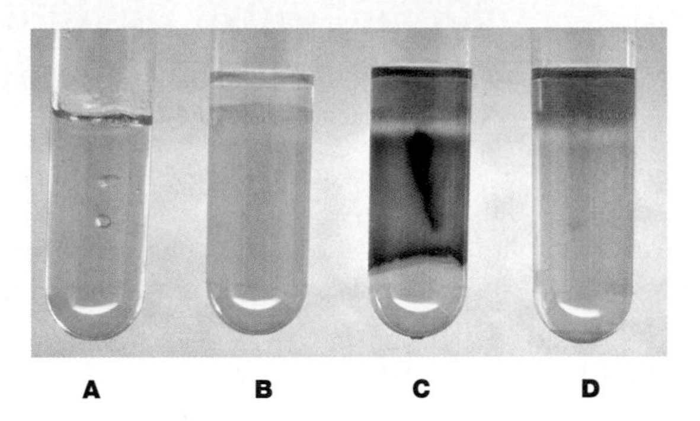


**Figure 63**
Bacto-SIM medium reactions. (a) Uninoculated. (b) Positive for motility (general turbidity in the medium). (c) Positive for hydrogen sulfide production (blackening of the medium), for indole (red layer), and positive for motility (cloudy). (d) Positive for indole, and for motility.

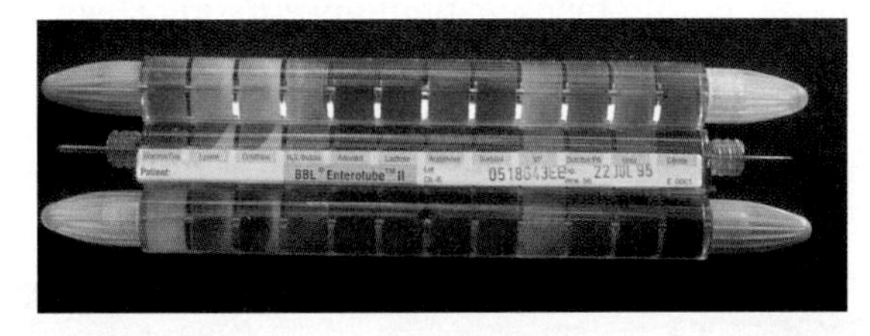

**Figure 64**
Enterotube II system results obtained after inoculation and incubation (bottom tube). The top and center systems are uninoculated. Note the color changes in the bottom system.

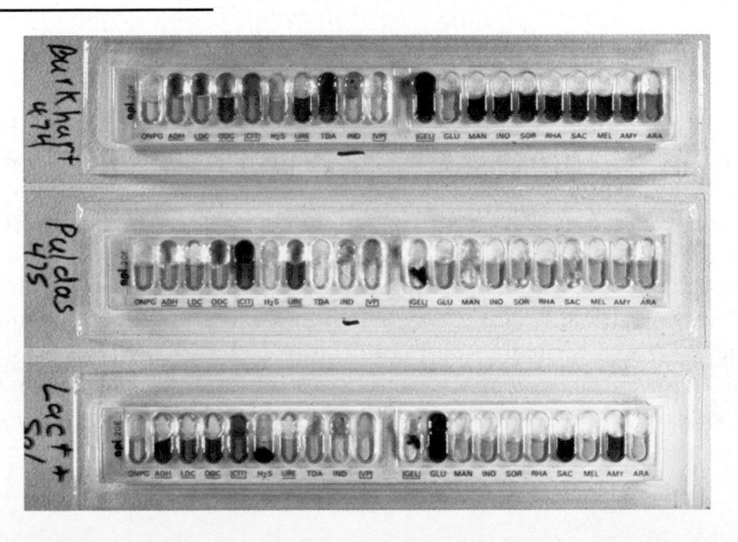


**Figure 65**
The rapid and ready-to-use API 20 Enteric (E) System uses a combination of substrates contained in individual compartments known as *cupules.* With the API 20E, 22 standard biochemical tests can be performed. Note the plastic strip with the cupules and the plastic incubation tray and lid. The system can be labled on the end of the incubation tray.

# Representative Bacteria, Their Distinctive Properties, and Selected Disease States

The classification of bacteria changes as the understanding of them improves. The presentation of bacteria here follows the classification scheme currently accepted by most microbiologists, namely *Bergey's Manual of Systematic Bacteriology*.

## The Spirochetes

The spirochetes are helical (spring-shaped) motile organisms. Their morphological features and a unique type of movement distinguish them from other bacteria. The spirochetes are unicellular and gram-negative. The size range of spirochetes is 0.1–3.0 µm × 5–250 µm. Because some spirochetes are very thin they can be seen only by dark-field microscopy (Figure 77), electron microscopy (Figure 67), or special staining techniques that increase their dimensions to bring them within the resolving power of the bright-field microscope. Dark-field microscopy depends on the reflection of light from the surfaces of particles or cells.

Spirochetes may be anaerobic, facultively anaerobic, or aerobic. Some spirochetes are free-living, whereas others are members of the normal microbiota of humans and lower animals. Three general—*Borrelia, Leptospira,* and *Treponema*—include causative agents of human and zoonotic diseases. A **zoonosis** is an infectious disease transmissible to humans from a lower animal host by natural means.

### Borrelia

#### *Borrelia burgdorferi*

Lyme disease (Figure 66) has emerged as the leading arthropod-borne disease in the United States. *Borrelia burgdorferi* (Figure 67) is the causative agent of the disease and related disorders.

**Transmission.** Various *Ixodes* ticks including *I. dammini* (Figure 68), *I. pacificus*, and *I. scapularies* transmit the disease agent. Small mammals, particularly mice, are important hosts for these ticks and are critical for maintaining *B. burgdorferi* in nature. Deer are important hosts for adult tick stages.

**Morphology and Cultural Properties.** *Borrelia burgdorferi* is a large helical cell, 0.2–0.5 µm wide and 30–20 µm long, with 3–10 loose spirals. It is gram-negative but is most easily shown by other staining procedures such as Giemsa and immunofluorescence. *Borrelia burgdorferi* is microaerophilic and can be grown on various types of blood agar media (Figure 69). The organism exhibits beta hemolysis, which can be made more pronounced with hot-cold incubation (incubating first at 34°C and then overnight at 4°C).

**Pathology and Clinical Features.** Lyme disease has been divided into three stages:

1. **Early Localized Stage.** Develops within a few days to weeks after infection. **Erythema (er'-i-THĒ-ma) migrans**, the characteristic lesion, appears as a large round or oval reddened area with definite borders at the site of the tick bite (Figure 66). Fever, headache, enlarged local lymph nodes, muscle and joint pain, and fatigue are typical symptoms.
2. **Disseminated or Spreading Stage.** Occurs within days to months. The appearance of various-sized reddened areas is typical. The symptoms and signs found with early Lyme disease also can be present. Facial paralysis may occur.
3. **Late Stage.** Appears within months to years. The effects of the disease include long-lasting arthritis and central nervous system involvement.

**Diagnosis.** Finding spirochetes in blood specimens taken from individuals in the early stage of Lyme disease is the principal basis for diagnosis. Cultural isolation from typical lesions and immunological tests such as the enzyme-linked immunoabsorbent assay (ELISA) also are of value (see Immunology section).

#### *Borrelia recurrentis*

The causative agent of **endemic relapsing fever** is *Borrelia recurrentis* (Figure 70).

**Transmission.** Endemic relapsing fever is transmitted to humans by soft-bodied ticks of the genus *Ornithodoros,* and **epidemic relapsing fever** transmitted by body lice of the genus *Pediculus* (Figure 71).

**Morphology and Cultural Properties.** *Borrelia recurrentis* is a large spirochete, 0.3 µm wide and 10–30 µm long. It is gram-negative, but as is the case with the other *Borrelia* the organisms can be seen more easily with other staining procedures. *Borrelia recurrentis* is microaerophilic and can be cultured on various types of artificial media containing long-chain fatty acids.

**Pathology and Clinical Features.** After an incubation period of about 7 days, a massive number of spirochetes appear in the blood, producing a high fever accompanied by a severe headache, muscle pain, chills, and general weakness. The fever period lasts approximately one week and ends suddenly with the development of an adequate immune (protective) response. Usually, less severe relapses (reoccurrences) occur 2–4 days later.

**Diagnosis.** Finding *B. recurrentis* among blood cells (Figure 70) in stained smears prepared from blood taken during the fever period is characteristic. Special immunological tests also may be helpful.

## Leptospira

### *Leptospira interrogans*

Leptospirosis (lep´-tō-spī-RŌ-sis), a worldwide disease of a wide range of animals, is caused by *Leptospira interrogans* (Figure 72).

**Transmission.** Humans acquire the disease mainly by ingesting water contaminated with infected animal urine.

**Morphology and Cultural Properties.** *Leptospira interrogans* is a thin spirochete, 0.15 μm wide and 5–15 μm long. It consists of fine, closely wound spirals with hook-like ends. Immunofluorescence staining and dark-field microscopy produce the best microscopic views. *Leptospira interrogans* is aerobic and can be cultured in certain enriched semisolid media.

**Pathology and Clinical Features.** Leptospirosis usually occurs in two phases:

1. **Phase 1**. Develops after an incubation period of seven days with the appearance of an influenza-like illness. Typical symptoms include fever, chills, headache, and muscle pain.
2. **Phase 2**. Usually lasts three or more weeks. Typical symptoms include muscle aches and headaches, accompanied by signs such as rash, yellowing of body tissues (jaundice), and laboratory evidence of liver and kidney damage. In the most severe form of the leptospirosis, known as Weil's disease, extensive damage to the liver, kidneys, and blood vessels occurs.

**Diagnosis**. *Leptospira* can be isolated from the blood or from other body fluids such as urine during the first weeks of the disease. Specimens must be used immediately for culturing.

## Treponema

### *Treponema pallidum*

The causative agent of syphilis, a sexually transmitted disease (STD), is *Treponema pallidum*. It was first recognized in the late fifteenth and early sixteenth centuries, when it spread through Europe, often with fatal effects.

**Transmission.** Syphilis is typically a sexually transmitted disease, but it can also be spread nonsexually through blood transfusions or nonsexual bodily contact, and a fetus can acquire it *in utero* or during the passage through the birth canal of an infected mother **(congenital syphilis).**

**Morphology and Cultural Properties.** *Treponema pallidum* is a slim spirochete, 0.15 μm wide and 5–15 μm long, with a characteristic corkscrew appearance (Figures 77 and 78). Dark-field microscopy and immunofluorescence techniques are routinely used to show the organisms in specimens. The usual bacteriological stains are not effective. However, the depositing of silver on spirochete surfaces can be used to demonstrate organisms in infected tissues (Figure 78). Special tissue culture methods can be used to grow *T. pallidum*.

**Pathology and Clinical Features.** Syphilis can be divided into four stages (Figures 73–76):

1. **Primary syphilis**. Appears within 21 days after infection as a hardened, swollen sore called a **chancre** (SHANG-ker) at the site of infection; if untreated, this lesion heals within 3 to 8 weeks.
2. **Secondary (disseminated) syphilis**. Occurs 2 to 10 weeks after primary chancre has healed. Signs of infection include blisterlike eruptions on the body (Figure 74), involving the face, palms, and soles of the feet, and painless warty growths on the genitalia and surrounding areas.
3. **Latent syphilis**. Occurs after the secondary stage. No evidence of infection is apparent other than positive blood tests.
4. **Tertiary stage**. Appears in the untreated as early as 5 years after infection, but generally after 15 to 20 years. Extensive tissue destruction occurs in the cardiovascular and nervous systems (Figure 75), but may involve any part of the body. The **gumma** (Figure 76) a firm, white lesion measuring up to 40 μm may be found in the body tissues and dramatically emphasizes the effects of *T. pallidum* infection.

**Diagnosis.** Dark-field and immunofluorescent microscopy (Figure 77) can detect *T. pallidum* in primary and secondary lesions. Most cases of syphilis, however, are diagnosed by tests that demonstrate the presence of antibodies in a patient's blood. These include the fluorescent treponemal antibody absorption test (FTA-ABS), and the micro-hemagglutination test for *T. pallidum* (MHAP-TP).

## The Spirochetes
## *Borrelia, Leptospira, and Treponema*

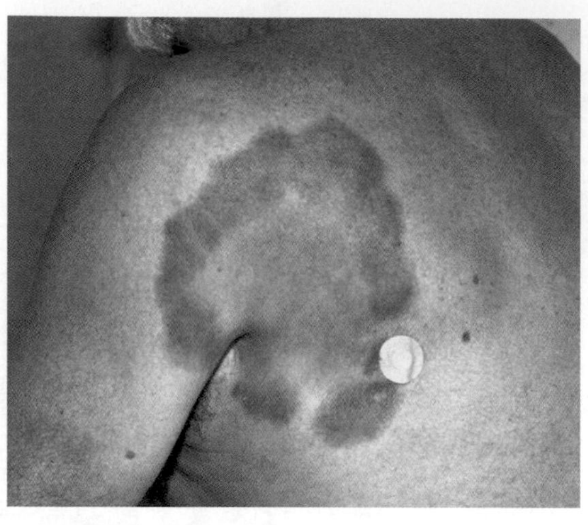

**Figure 66**
Erythema migrans, the typical lesion of early localized Lyme disease. The lesion develops at the site of the tick bite and occurs in 60–80% of cases. (From J. Cote, *Internat. J. Dermatol.* 30(1991): 500–501.)

**Figure 67**
A scanning electron micrograph of *Borrelia burgdorferi,* the causative agent of Lyme disease and related disorders. (Courtesy of Dr. F.-R. Matsucha, Free University of Berlin.)

**Figure 68**
A female deer tick, *Ixodes dammini,* which transmits *Borrelia burgdorferi,* the cause of Lyme disease. (Courtesy of Dr. F.-R. Matsucha, Free University of Berlin.)

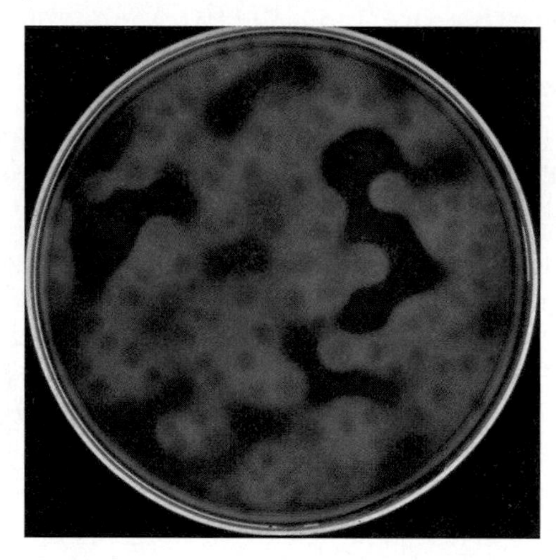

**Figure 69**
Hemolytic (blood agar plate) activity of *Borrelia burgdorferi* after incubation at 34°C and transfer to 4°C for overnight incubation. (From L.R. Williams and F.E. Austin. *Inf.* Immunol. 60(1992): 3324–3330.)

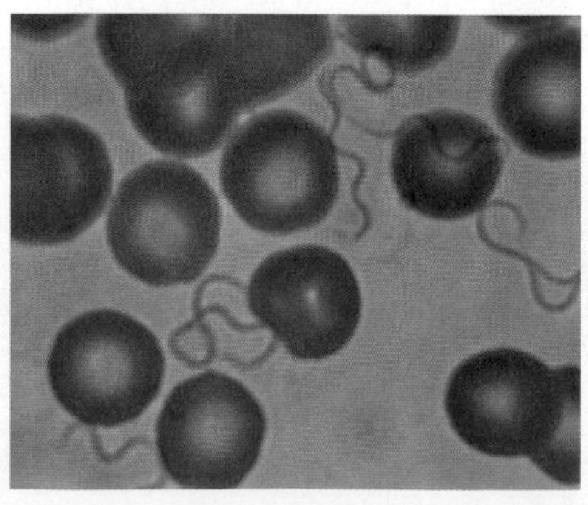

**Figure 70**
A microscopic view of *Borrelia recurrentis,* the causative agent of epidemic relapsing fever.

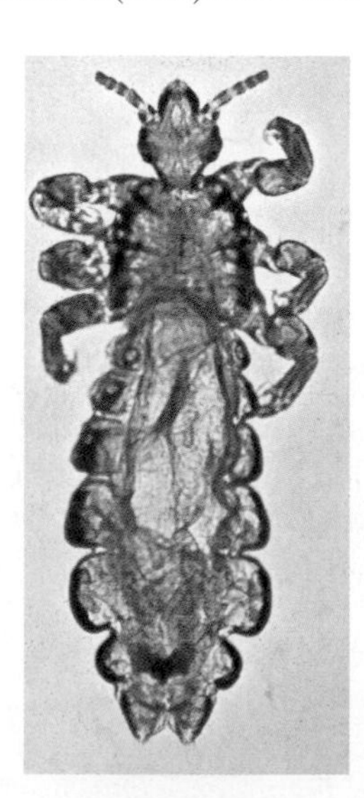

**Figure 71**
The human louse *Pediculus humanus,* known to transmit relapsing fever and other diseases.

## The Spirochetes *(continued)*

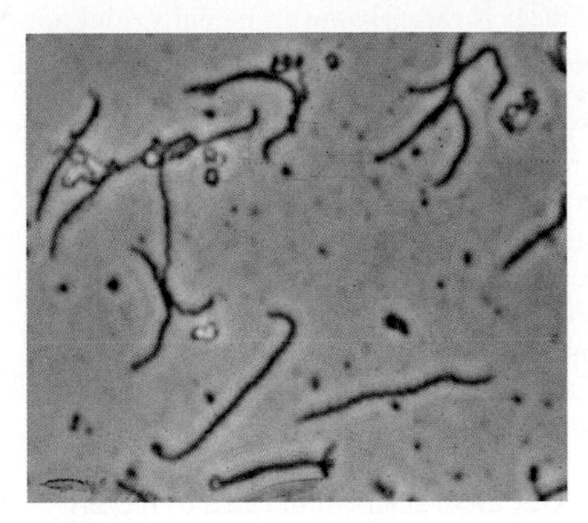

**Figure 72**
A microscopic view of *Leptospira interrogans,* the cause of leptospirosis.

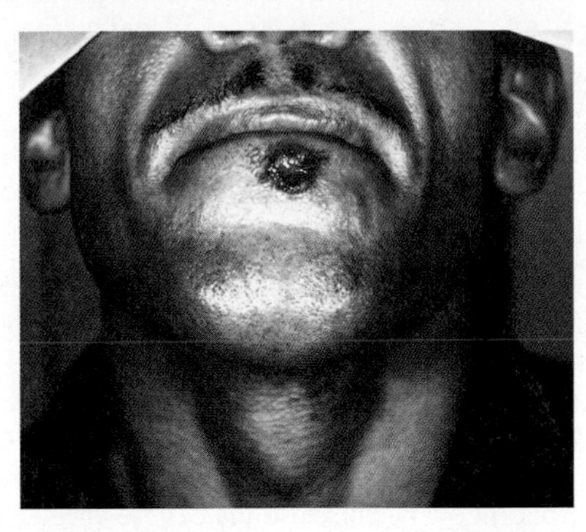

**Figure 73**
The primary sore, or chancre, of syphilis. This lesion contains hundreds of *Treponema.* (Courtesy of U.S. Public Health Service).

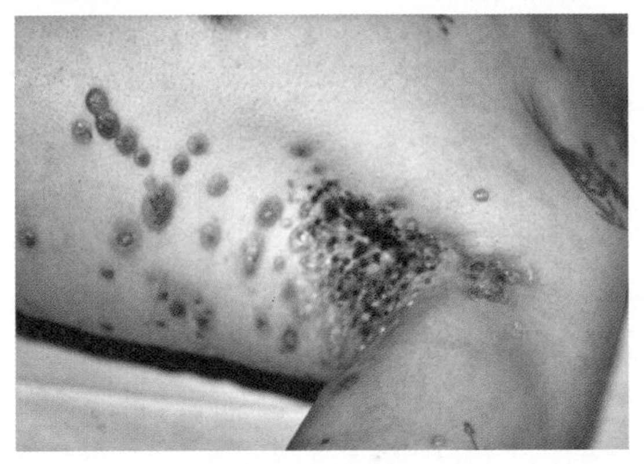

**Figure 74**
An example of the rash that may develop in secondary syphilis. (From P. Lawrence and N. Saxe. *Clin. Exper. Dermatol.* 17 (1992):44–48.)

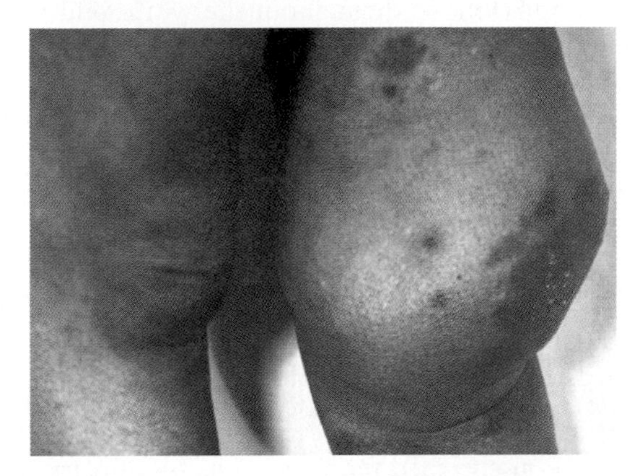

**Figure 75**
Charcot knee, a complication of tertiary syphilis involving the cardiovascular system.

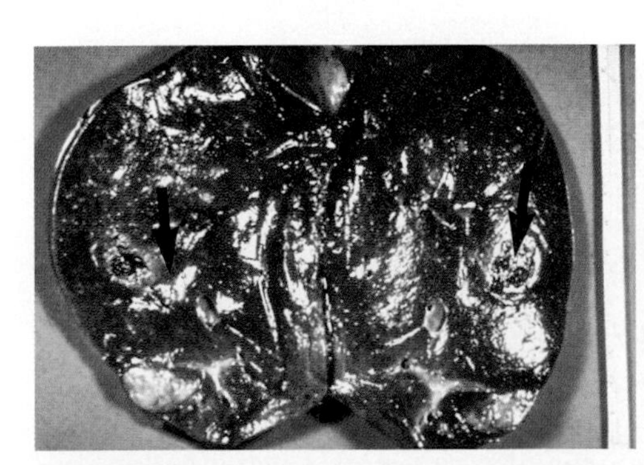

**Figure 76**
Gummas in the liver. Two gummas are evident (arrows). One shows extensive tissue injury.

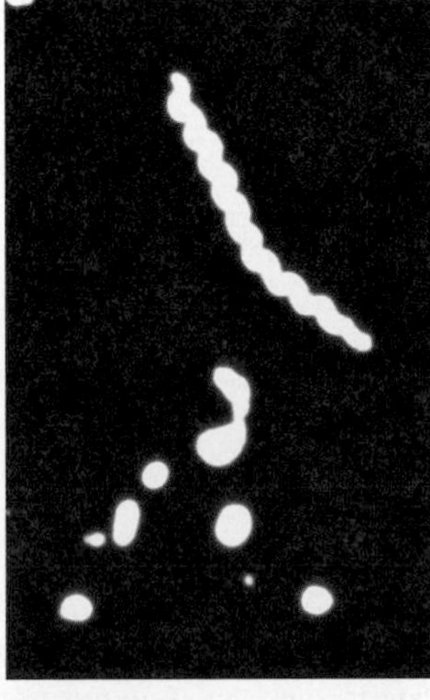

**Figure 77**
The spirochete *Treponema pallidum* revealed by dark-field microscopy. Organisms appear bright in a dark background.

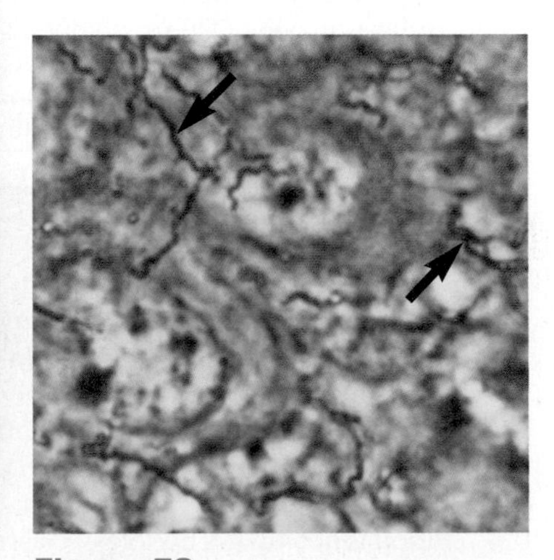

**Figure 78**
*Treponema pallidum* (arrows) in tissue.

## Aerobic/Microaerophilic, Motile, Helical/Vibroid, Gram-Negative Bacteria

### *Helicobacter*

***Helicobacter pylori***

A common pathogen of humans, *Helicobacter pylori* is the principal cause of chronic inflammation of the stomach (gastritis) and peptic ulcer disease (Figure 79). It is also a risk factor for gastric cancer, even though most infections are without symptoms. Once established, most infections last for years and rarely cure spontaneously. They usually can be cured by antimicrobial therapy. Infection also increases the risk of other diseases.

**Transmission.** *Helicobacter pylori* infections occur in human populations throughout the world and in all populations studied. Occurrence of infections increases with age. Person-to-person transmission appears to account for infections within families. Although the exact means by which infection is acquired is not well understood, fecal-oral or oral-oral transmission is likely.

**Morphology and Cultural Properties.** *Helicobacter pylori* is a gram-negative, spiral, curved or straight rod, 0.5–1.0 μm wide and 2.5–5.0 μm long (Figures 80 and 81). The organism is motile and microaerophilic and grows best in an atmosphere of 5% $O_2$ with 5–10% $CO_2$ and on enriched media such as blood agar. The organism also can grow under aerobic conditions with high humidity (Figure 82). Optimal temperature is 37°C. Colonies are nonpigmented, translucent, and 1–2 mm in diameter. *Helicobacter pylori* rapidly hydrolyzes urea in laboratory tests by means of an unusual urease enzyme. There is speculation that *H. pylori*'s urease production serves as a significant virulence factor by neutralizing the acidity in the stomach to allow the pathogen to reproduce more easily in the tissues of the stomach. Isolated *H. pylori* are oxidase and catalase positive.

**Pathology and Clinical Features.** *Helicobacter pylori* is a highly successful microbe, infecting most of the world's population. Once established in the stomach, the organism persists for decades, seemingly unaffected by the acid and gastric movement (**peristalsis**) in its environment. In most cases, *H. pylori* silently coexists with its host, causing no symptoms or signs. But this relation is not always harmless. Gastric cancer develops in approximately 1% of infected persons, and 20% have duodenal ulcers.

Gastric cancer is a leading cause of death from cancer in the developing world, where infection with *H. pylori* in early childhood is the rule. In infected children, duodenal ulcer is rare. However, by early adulthood, the infection acquired in childhood has frequently progressed to a pathologic condition in which there is a patchy loss of gastric glands that secrete protein and acid (**multifocal atrophic gastritis**). The decrease in acid secretion opens the door to both stomach (gastric) ulcers and cancer. Childhood infection predisposes patients to gastric ulcers and stomach cancer, but it inhibits the development of duodenal ulcers.

In industrialized countries, deaths due to gastric cancer, gastric ulcers, and duodenal ulcers are decreasing. This decrease is believed to be due to improvements in hygiene and economic conditions that have interfered with the transmission cycle of the organism to the point where *H. pylori* is now hard to find in children.

**Diagnosis.** The urea-breath test, a clinical procedure, can be used to identify infected individuals. However, tissue specimens obtained by biopsy are necessary for tissue and microbiological examination and confirmation of *H. pylori* infection. Several serological tests, including the ELISA procedure, have been introduced to aid in diagnosis. Newer diagnostic tests are under development.

### *Spirillum*

***Spirillum volutans***

Frequently found in stagnant freshwater environments, *Spirillum volutans* is a nonpathogen.

**Morphology and Cultural Properties.** *Spirillum volutans* is a gram-negative rigid helical cell, 1.4–1.7 μm wide and 14–60 μm long. The organism moves by means of large tufts of flagella located at both ends of a cell (Figure 83). *Spirillum volutans* is microaerophilic, but it can grow in special liquid media containing supplements. The organism is oxidase positive and catalase negative. It does not break down carbohydrates.

## Aerobic/Microaerophilic Gram-Negative Rods and Cocci

### *Alcaligenes*

***Alcaligenes faecalis***

A common inhabitant of the intestinal tract, *Alcaligenes faecalis* can also be found in water and in soil. Strains of this organism have been recovered from pussy ear discharges, blood, urine, and sputum specimens, and it has been associated with certain hospital-acquired (**nosocomial**) infections.

## Aerobic/Microaerophilic, Motile, Helical/Vibroid, Gram-Negative Bacteria

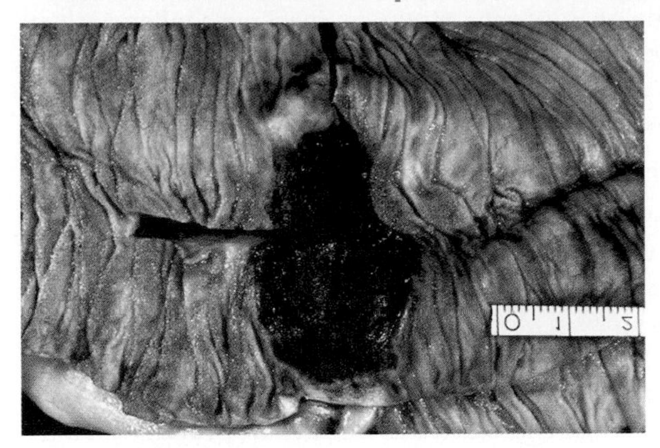

**Figure 79**
A large, well defined ulcer. (From M. Koona and associates. *Acta Pathol. Jpn.* 42(1992): 267–71.)

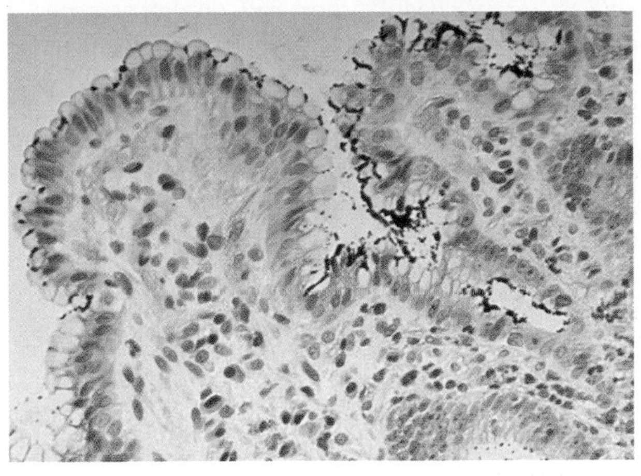

**Figure 80**
A cluster of *Helicobacter pylori* (red cells) along the lining of the stomach (1,000×).(Unpublished photo, courtesy of Dr. L.P. Andersen.)

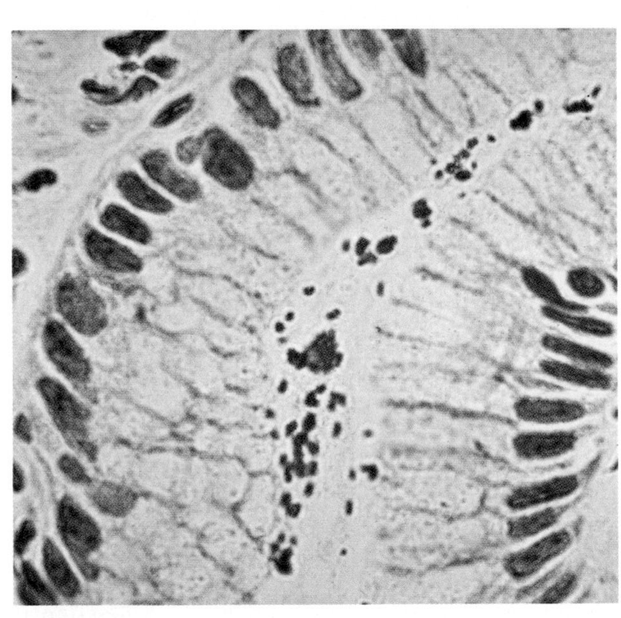

**Figure 81**
*Helicobacter pylori* (red cells) within the human intestine (1,000×.) (Courtesy of Dr. L.P. Andersen).

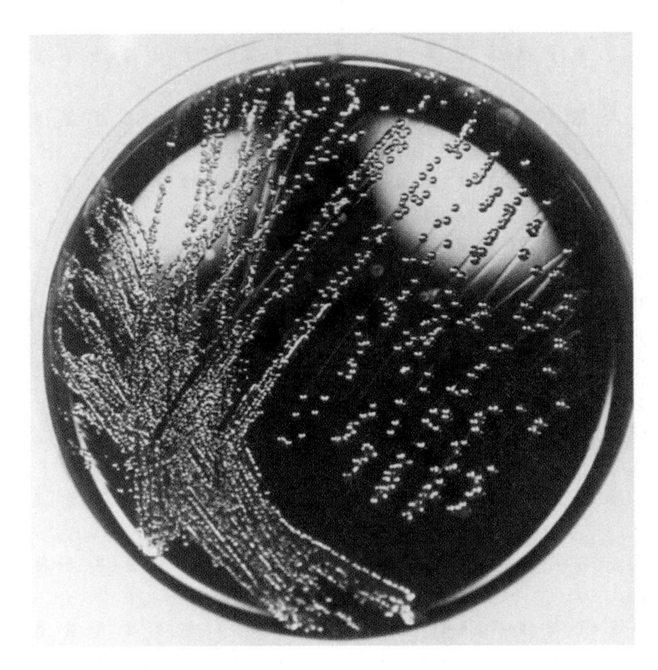

**Figure 82**
Colonies of *Helicobacter pylori* growing aerobically on an agar plate. (From H.X. Xia, C.T. Keane, and C.A. O'Morain. *European J. Clin. Microbiol. & Inf. Dis.* 13(1994):406.)

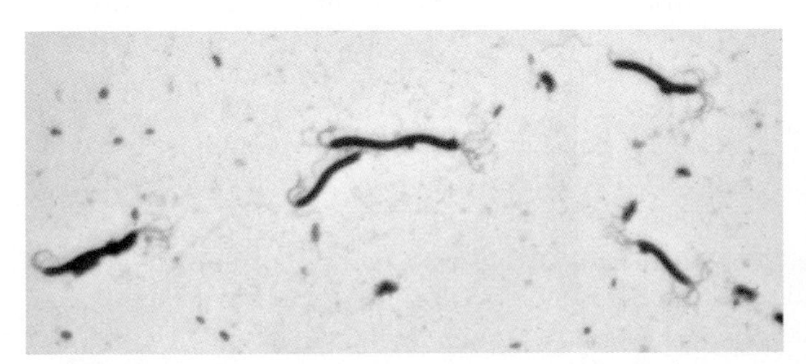

**Figure 83**
A photomicrograph of *Spirillum volutans.* In addition to the spiral morphology of this organism, the arrangement of two or more flagella at one or both ends of a cell is evident.

**Morphology and Cultural Properties.** *Alcaligenes faecalis* is a gram-negative short rod, or almost a coccobacillus, 0.5–1.0 μm wide and 0.5–2.6 μm long (Figure 84). The organism is motile and obligately aerobic. The growth temperatures range from 20° to 37°C. Colonies on nutrient agar are nonpigmented, glistening, and convex (Figure 85a). *Alcaligenes faecalis* is nonhemolytic (Figure 85b) and grows well on MacConkey agar (Figure 85c). The organism is oxidase and catalase positive.

**Pathology.** *Alcaligenes faecalis* occasionally causes opportunistic infections.

## *Bordetella*

### *Bordetella pertussis*

The causative agent of the respiratory disease **pertussis** (whooping cough) is *Bordetella pertussis*. The organism attaches to and grows in and on the ciliated cells of the mucous membranes of the human respiratory tract.

**Transmission.** *Bordetella pertussis* is spread by airborne droplets to individuals in close contact with infected persons in the early stages of illness. The human-to-human transmission is well established.

**Morphology and Cultural Properties.** *Bordetella pertussis* is a gram-negative cocco-bacillus, 0.2–0.5 μm wide and 0.5–2.0 μm long. Cells are generally arranged singly or in pairs.

The organism is nonmotile and strictly aerobic. The optimum growth temperature ranges from 35° to 37°C. Colonies are glistening, smooth, convex, and pearl-like. They are grown on specially prepared media (Figure 86), such as charcoal blood agar medium, which may contain antibiotics such as cephalosporin to inhibit organisms from the normal microbial inhabitants of the respiratory tract. Colonies appear after 3 to 6 days.

**Pathology and Clinical Features.** Pertussis follows a prolonged course consisting of three overlapping stages:

1. **Catarrhal stage**. Large amounts of watery discharge from the nose.
2. **Paroxysmal coughing stage**. Sudden and recurring violent coughing episodes up to 50 times a day for 2 to 4 weeks. The characteristic whoop, vomiting, and large amounts of mucous are typical of this stage.
3. **Convalescent stage**. A lessening of the signs and symptoms takes place over a 3–4-week period.

**Diagnosis.** Clinical diagnosis is confirmed by isolation of *Bordetella pertussis*. Specimens taken early in the course of the disease provide the best results. Fluorescent-antibody testing also is effective.

## *Brucella*

Brucellosis, also known as undulent fever and Malta fever, is an acute or chronic recurring disease spread from lower animals to humans. The disease brucellosis exists worldwide, especially in countries of the Mediterranean basin, the Arabian gulf, the Indian subcontinent, and parts of Mexico and Central and South America. There are four species of *Brucella* that are pathogenic for humans, and these have a limited number of preferred natural host animals: *B. abortus* (cattle), *B. melitensis* (goats and sheep), *B. suis* (swine), and *B. canis* (dogs). The disease is an important cause of abortion and sterility, and it decreases milk production in farm animals.

**Transmission.** Common routes of infection include inoculation through cuts and abrasions in the skin or through the conjuctival sac of the eyes, inhalation of infectious aerosols, and ingestion into the gastrointestinal tract.

**Morphology and Cultural Properties.** *Brucella* species are gram-negative cocco-bacilli, 0.5 μm wide and 0.3–0.9 μm long, occurring singly and in pairs (Figure 87). The organisms are nonmotile and may be encapsulated.

*Brucella* species require an aerobic environment and enriched media such as blood agar. Smooth pinpoint colonies appear on primary isolation after 2 to 3 days of incubation. Differentiation among *Brucella* species can be on the basis of responses to dye-impregnated disks and biochemical tests. These organisms are usually catalase and oxidase positive.

**Pathology and Clinical Features.** In humans, the reticuloendothelial system is the main target of infection. If not treated adequately with antibiotics, the infection results in the formation of small tumors, which serve as sites for bacterial reproduction. Eventually, bacteria are released periodically into the circulatory system. Such recurrent episodes are mainly responsible for the recurring chills and fever of the clinical illness.

Brucellosis begins with feelings of general discomfort, chills, and fever for 7 to 21 days after infection. Drenching sweats in the late afternoon or evening are common and may continue for several weeks or even 1 to 2 years. Other findings include weight loss, body aches, loss of appetite and enlargement of lymph nodes, spleen, and liver.

**Diagnosis.** Definitive diagnosis requires isolation of *Brucella* from blood or biopsy specimens. Serological tests also are used.

## *Francisella*

*Francisella tularensis* is the causative agent of tularemia, or rabbit fever. The disease agents are harbored in the blood and other tissues of wild and domestic animals.

**Transmission.** *Francisella tularensis* can be acquired through the skin or the conjuctiva (lining of the eyelid) from handling of infected animals or animal products, from the bites of infected blood-sucking deer flies and wood ticks, by ingestion of improperly cooked contaminated meat or water, and by aerosol inhalation.

**Morphology and Cultural Properties.** *Francisella tularensis* is a gram-negative rod, 0.3–0.5 μm wide and 0.2 μm long.

The organism is nonmotile and aerobic, and it requires special media for isolation. Small, smooth, grayish colonies appear within 48 hours. Growth may also occur later and may require 3 weeks. Optimum temperature is 37°C. Biochemical testing is of little value in identification.

**Pathology and Clinical Features.** Organisms establish infections locally and in regional lymph nodes. Such infections often result in ulcer formation and regional enlargement of lymph nodes (**lymphadenopathy**). Two forms of lymphadenopathy, ulceroglandular (Figure 88) and **oculoglandular**, are generally accompanied by fever, nausea, vomiting, or abdominal pain.

**Diagnosis.** Immunofluorescent or agglutination tests are commonly used for disease diagnosis.

## *Legionella*

More than 30 species of *Legionella* have been identified. Of these, at least 12 species have been isolated from humans with respiratory disease. Attention here will be given to the most prevalent pathogen, *L. pneumophila*.

Two clinical forms of disease are caused by *L. pneumophila*: **Legionnaires' disease**, a pneumonia-like infection, and **Pontiac fever**, an influenza-like, self-limited illness. The organism was named after it was isolated from lung tissue (Figure 89) obtained from autopsies of individuals who had died during the epidemic at the 1976 American Legion Convention in Philadelphia.

**Transmission.** Airborne transmission by way of aerosols is the most common route of infection. Waterborne disease may occur, particularly in hospital and related environments. There is no known case of human-to-human transmission at this time.

**Morphology and Cultural Properties.** *Legionella pneumophila* is a gram-negative rod, 0.3–0.9 μm wide and 2–20 μm long. The organism is aerobic and motile. It does not ferment carbohydrates and requires the amino acid cysteine for growth. (Figure 90). Among the preferred media used are Mueller-Hinton agar with hemoglobin and IsoVitaleX (Figure 90) and buffered charcoal-yeast extract agar (Figure 91). Optimum incubation temperature is 35°C. Colonies generally appear within 3 to 5 days and are convex, gray, and glistening (Figure 91).

**Pathology and Clinical Features.** Fever, chills, general discomfort, and muscle pain usually are the first symptoms that appear after an incubation period of 2 to 10 days following *L. pneumophila* infection. Legionnaires' disease involves many body systems and causes other notable clinical features including coughing, breathing difficulties, chest and abdominal pain, vomiting, diarrhea, and mental confusion. Hospitalization is frequently required, and the illness may be fatal without antibiotic therapy.

Pontiac fever presents early symptoms similar to those found with Legionnaires' disease within 24 to 36 hours after exposure. This form of *L. pneumophila* infection is remarkable for the absence of pneumonia or fatalities.

**Diagnosis.** Diagnosis depends on isolation of the organism or its detection in respiratory secretions by direct immunofluorescence testing (Figure 92). Modern molecular methods involving nucleic acid probes also are of value.

## *Moraxella*

### *Moraxella (Branhamella) catarrhalis*

A common inhabitant of the human upper respiratory tract is the organism *Moraxella catarrhalis*. It is an important cause of otitis media (middle ear infection) in infants and children, lower respiratory tract infections in adults, and nosocomial respiratory tract infections. *Moraxella catarrhalis* has been isolated exclusively from humans.

**Transmission.** Aerosols are common routes of infection.

**Morphology and Cultural Properties.** *Moraxella catarrhalis* is a gram-negative coccus, occurring as single cells or in pairs, with the adjacent sides flattened, giving a coffee-bean appearance (Figure 93). The cocci usually measure 0.6–1.0 μm in diameter.

*Moraxella catarrhalis* is aerobic, but it may grow poorly under anaerobic conditions. The organism will grow on media such as nutrient and blood agars. Blood agar colonies are small, circular, convex, and usually grayish white (Figure 94). Optimal temperature is 33° to 35°C.

*Moraxella catarrhalis* is catalase and oxidase positive and does not produce acid from carbohydrates.

## Gram-Negative Aerobic/Microaerophilic Rods and Cocci

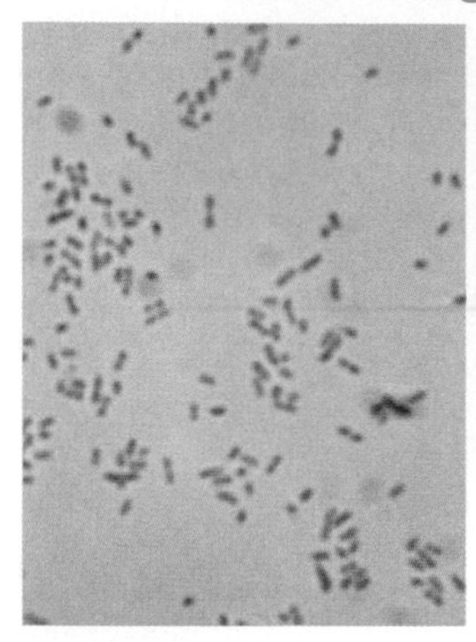

**Figure 84**
The gram-negative rods of *Alcaligenes faecalis* (1,000×).

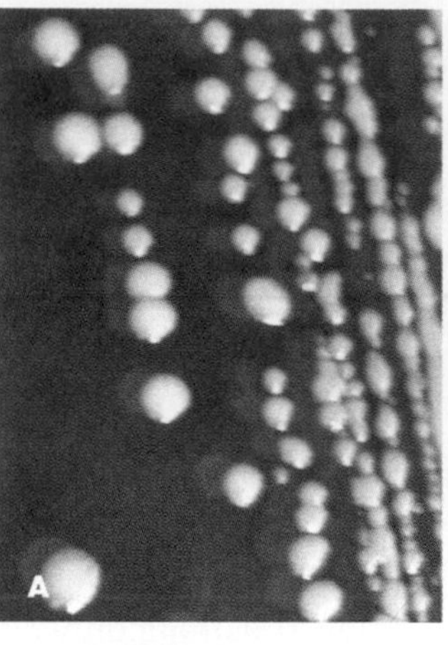


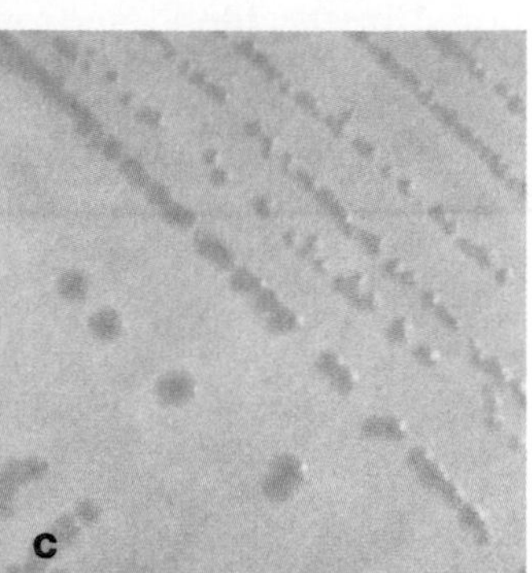


**Figure 85**
(a) *Alcaligenes faecalis* on nutrient agar. (b) *Alcaligenes faecalis* on blood agar. (c) Nonfermenting colonies of *Alcaligenes faecalis* on MacConkey agar.

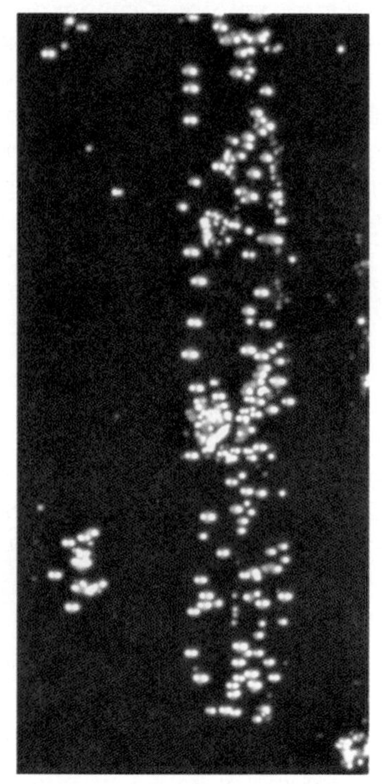

**Figure 86**
Colonies of *Bordetella pertussis* on charcoal blood agar.

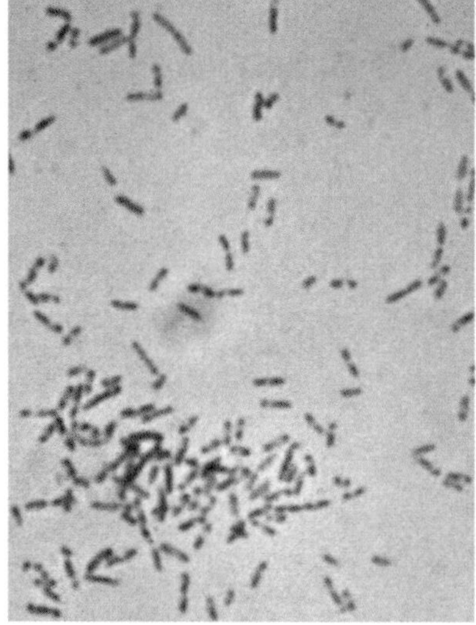

**Figure 87**
The gram-negative cocco-bacilli of *Brucella melitensis.*

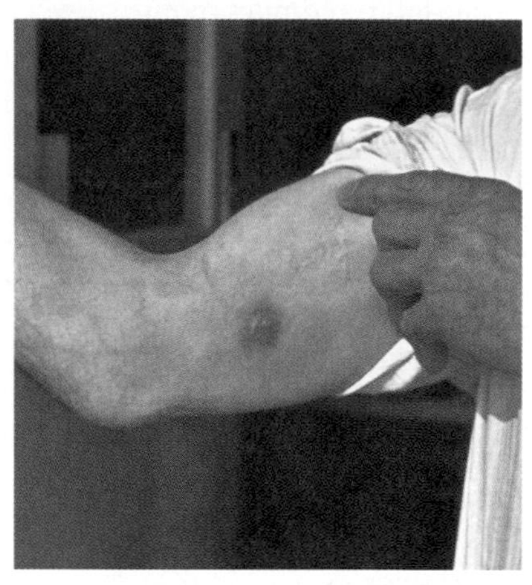

**Figure 88**
An enlarged lymph node in a patient with tularemia. (From Z. Cerny. *Internat. J. Dermatol.* 33(1994):468.)

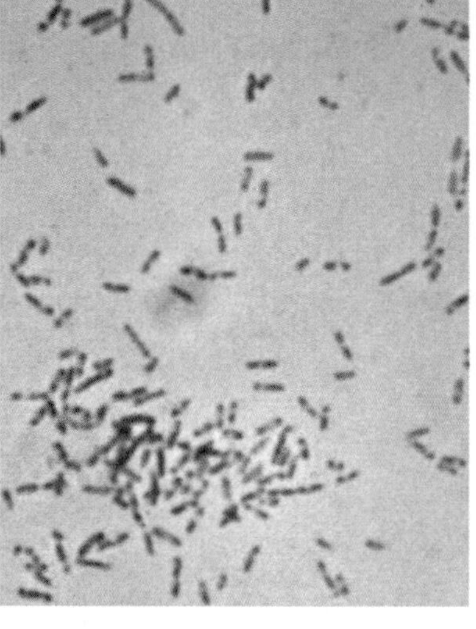

**Figure 89**
Gram stain results with *Legionella pneumophila.*

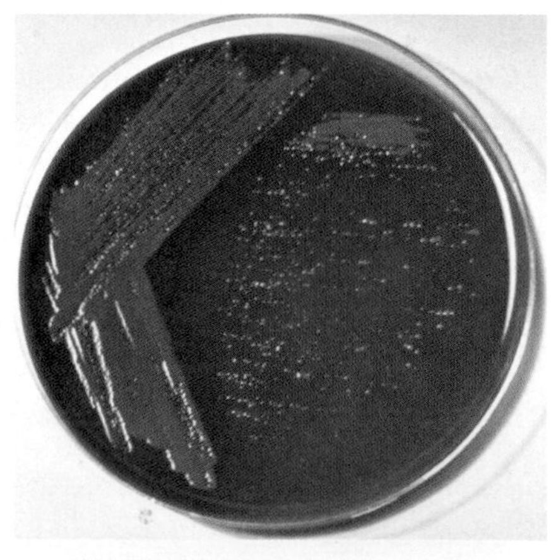

**Figure 90**
*Legionella pneumophila* colonies after 6 days of incubation on Mueller-Hinton agar supplemented with 1% hemoglobin and other growth factors.

**Diagnosis.** Isolation and biochemical tests are necessary for diagnosis.

## *Neisseria*

Most members of the genus *Neisseria* are gram-negative diplococci, measuring 0.6–1.0 µm in diameter (Figure 95). The adjacent sides of cell pairs are flattened, giving the characteristic kidney bean or coffee bean appearance in microscopic preparations (Figures 95 and 96).

*Neisseria* species have an optimal growth temperature of 35° to 37°C, and they are aerobic and oxidase positive (Figure 97b).

## *Neisseria gonorrhoeae*

The causative agent of the sexually transmitted disease gonorrhea (Figure 98) is *Neisseria gonorrhoea.* It is one of the most commonly reported bacterial infections in the United States. This organism is one of the most common causes of infertility in women worldwide.

**Transmission.** Infection is acquired by sexual contact, or an infected pregnant woman can transmit it to her newborn as it passes through the infected birth canal. Newborns not treated adequately may develop blindness of the newborn, or **ophthalmia neonatorum**.

**Morphology and Cultural Properties.** *Neisseria gonorrhoeae*, also commonly referred to as gonococcus, exhibits the typical gram-negative diplococcus appearance (Figures 95 and 96). The organism grows best on media enriched with blood or on special substances such as chocolate agar (Figure 97), and in the presence of 3 to 10% $CO_2$. Thayer-Martin chocolate agar containing specific antibiotics is widely used for isolation from clinical specimens, which include urethral, cervical, rectal, pharyngeal, and conjunctival discharges.

**Pathology and Clinical Features.** Gonorrhea has an approximate incubation period of 2 to 7 days. Infection without signs and symptoms occurs in about 10% of males and about 20 to 80% of females.

In males, the most common finding is an acute urethritis, which is accompanied by an abrupt and noticeable painful urination (**dysuria**) and a pussy urethral discharge (Figure 98a). Complications such as prostatitis, inflammation of the epididymis, and closure of the urethra (the canal for the elimination of urine extending from the bladder to the outside) are possible in untreated cases.

In females, the main site of urogenital disease is the lining of the cervix. Infection may be accompanied by a pussy discharge, painful urination, and a greater frequency of urination. Complications may lead to an inflammation of the uterine tubes and **pelvic inflammatory disease** (PID). Gonococci from cervical secretions may also contaminate the external region between the vaginal area and the anus **(perineum),** resulting in anorectal and neighboring gland infections.

In a small percentage of cases, *N. gonorrhoeae* may spread via the bloodstream to cause disseminated gonococcal infection (DGI). A sparse rash on the arms and legs and arthritis in one or more joints are typical of this condition.

Infants born to infected mothers may develop gonococcal ophthalmia (Figure 98c), also known as ophthalmia neonaturum, if preventive eyedrops or ointment is not administered at birth. This infection must be treated immediately or blindness can result. Non-sexually transmitted infections may occur in very young children, usually through accidental contamination with infectious discharges from infected individuals.

**Diagnosis.** The correct choice of specimens is critical to the successful isolation and identification of *N. gonorrhoeae.* The direct demonstration of the organism in white blood cells is diagnostic only when found in urethral discharges from males. Most other specimens require culture and biochemical testing.

Enzyme immunoassay, nucleic acid probes, and immunofluorescence tests (Figure 99) are available for the direct detection of gonococci.

### *Neisseria meningitidis*

Bacterial meningitis (an inflammation of the coverings of the brain and spinal cord) is a relatively common and devastating disease. *Neisseria meningitidis* is one of the three most common causative agents of the disease.

**Transmission.** *Neisseria meningitidis* colonizes the nasopharynx and spreads from person to person by means of respiratory droplets released during breathing, coughing, or sneezing. Humans are the only reservoirs for this organism.

**Morphology and Cultural Properties.** *Neisseria menigitidis* is, as are other members of the genus, a gram-negative, nonmotile diplococcus with the appearance of a kidney or coffee bean. The organisms are capsulated and have pili.

*Neisseria meningitidis* is a strict aerobe and grows best on chocolate agar in the presence of 3 to 10% $CO_2$. The organism produces acid from glucose and maltose.

**Pathology and Clinical Features.** *Neisseria menigitidis*, also known as the meningococcus, can gain entrance into the bloodstream, from where it can progress to the cerebrospinal fluid to cause meningitis. Organisms spread from the nasopharynx through the blood-

## Gram-Negative Aerobic/Microaerophilic Rods and Cocci *(Continued)*

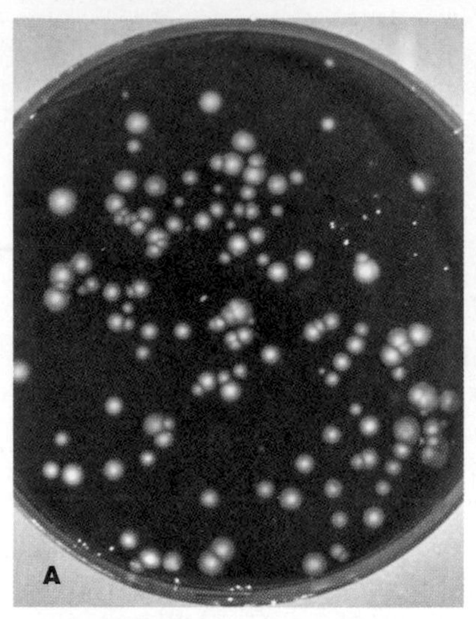


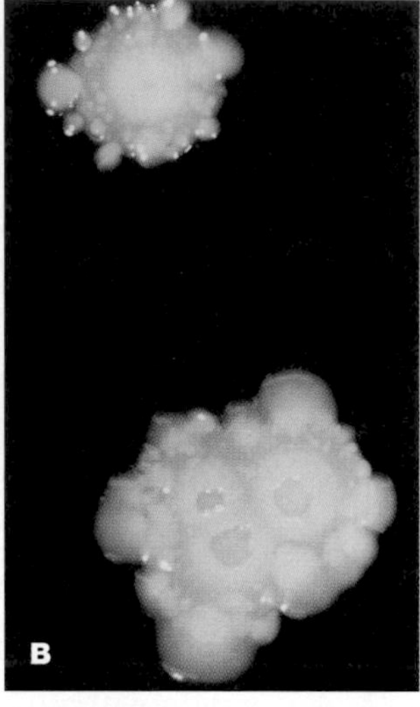


**Figure 91**
(a) *Legionella pneumophila* growing on buffered charcoal-yeast extract (BCYE) agar after 4 days of incubation. (b) A close-up of *Legionella pneumophila* isolated from a case of pneumonia after 6 days of incubation. (From M. Koide and A. Saito. *CID* 21(1995):199–201.)

**Figure 92**
Brightly fluorescing *Legionella* in a lung tissue suspension stained by immunofluorescence.

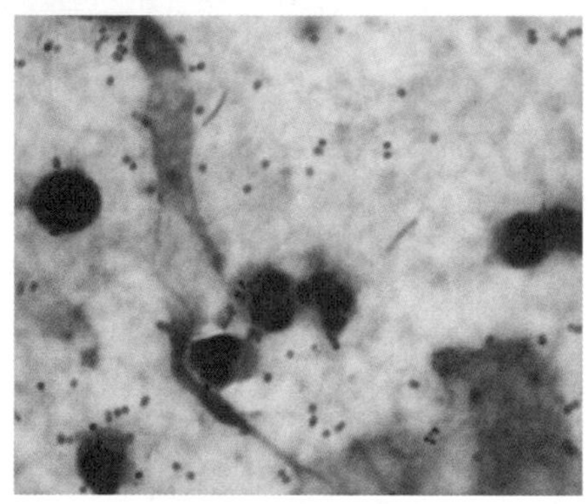

**Figure 93**
The results of a Gram-stained sputum specimen from a patient with chronic bronchitis caused by *Moraxella (=Branhamella) catarrhalis.* (From T.F. Murphy. *Microbiol. Revs.* 60(1996):267–79.)

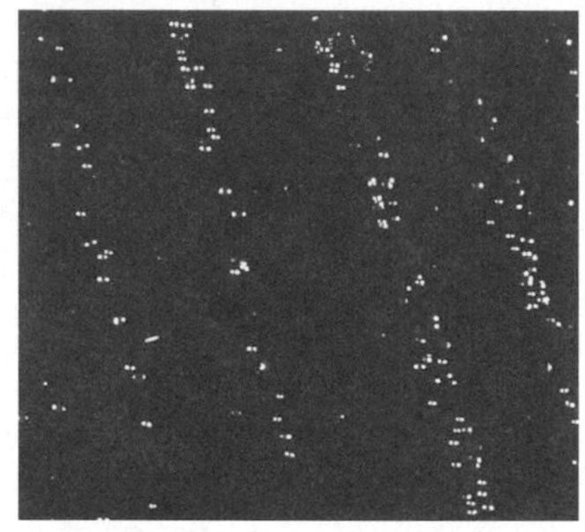

**Figure 94**
Colonies of *Moraxella catarrhalis* on blood agar.

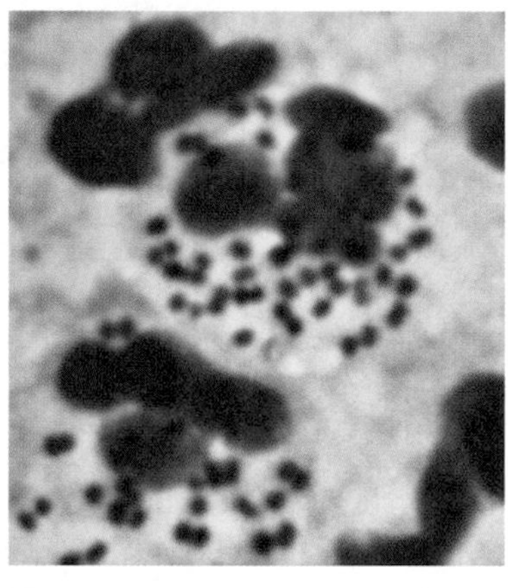

**Figure 95**
The "coffee bean" appearance of *Neisseria gonorrhoeae* from a urethral discharge. Note the presence of gram-negative diplococci in leukocytes.

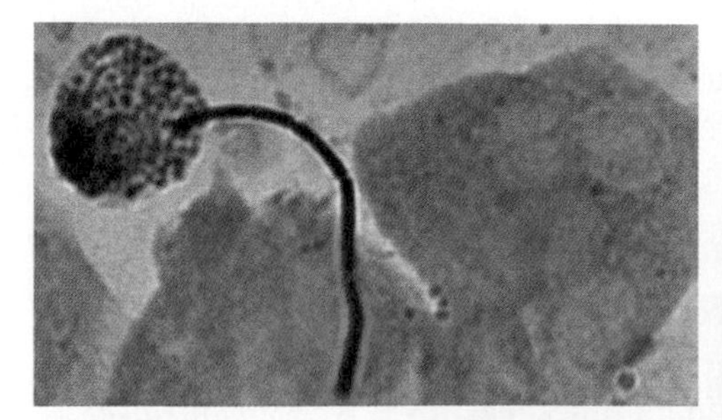

**Figure 96**
*Neisseria gonorrhoeae* in a cervical smear. Large cervical epithelial cells and a long gram-positive rod are also evident.

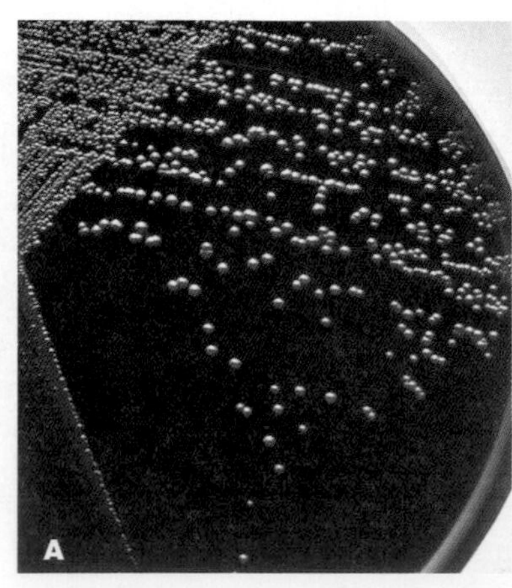

**Figure 97**
(a) Growth of *Neisseria gonorrhoeae* on a chocolate agar medium. (b) A positive (dark colonies) oxidase test.

stream to produce meningococcemia, meningitis, or both. The disease may be mild, or it may progress rapidly, resulting in death within a few hours. Meningococcemia causes severe damage to blood vessels, the most visible effect being small hemorrhages (**petichiae**) in the skin (Figure 100).

Other signs and symptoms include fever, stiff neck, vomiting, severe headache, convulsion, and progression to a coma within a few hours.

**Diagnosis.** Laboratory diagnosis involves obtaining specimens for direct microscopic examination and culture. A serological test, such as latex agglutination, for the detection of capsular polysaccharides is essential.

## Pseudomonas

### *Pseudomonas aeruginosa*

Among the most virulent opportunistic pathogens of humans is *Pseudomonas aeruginosa*, which colonizes and invades injured epithelial surfaces (including the lungs of intubated patients, corneal abrasions, and injured skin of burn patients). Seventy-five percent of all intensive care unit patients are colonized by this organism, and, not surprisingly, *P. aeruginosa* is the leading cause of mortality among cystic fibrosis patients. The treatment of *P. aeruginosa* infections is difficult, in part because this organism's antibiotic resistance is common and is becoming more widespread.

*Pseudomonas aeruginosa* is widely distributed in soil, water, sewage, intestinal tracts, and plants. The organism has been isolated from various materials including disinfectants, cosmetics, and foods.

**Transmission.** *Pseudomonas aeruginosa* is spread in a number of ways, including by contaminated fingers and instruments such as urinary catheters, endoscopes, and respiratory therapy equipment. Bathing or soaking in contaminated water also can serve to transmit the organism.

**Morphology and Cultural Properties.** *Pseudomonas aeruginosa* is a gram-negative rod, 0.5–1.0 μm wide and 1.5–5.0 μm long, usually occurring singly and in pairs (Figure 101).

The organism is motile and aerobic. It grows best at 37°C, but good growth can occur at 42°C. Large translucent, spreading colonies with irregular edges appear within 24 to 48 hours. *Pseudomonas aeruginosa* produces a blue-green pigment that diffuses into media (Figure 102b). Special media such as *Pseudomonas* P agar can be used to demonstrate pigment production (Figure 102b).

*Pseudomonas aeruginosa* grows at 42°C, oxidizes glucose, and does not produce acid from dissacharides such as lactose. Colonial growth appears within 48 hours and exhibits a variety of forms including smooth, rough, and mucoid and a small size. The variation exhibited by a single strain gives the false impression that several bacterial species are present.

**Pathology and Clinical Features.** *P. aeruginosa* has become a major cause of hospital-acquired infections. Infections with this organism rarely occur in persons with normal immune systems and defenses. Patients with extensive burns may become colonized with *P. aeruginosa* since burn injuries not only destroy the mechanical barrier of the skin but also seriously impair all other aspects of the immune system.

Skin lesions caused by *P. aeruginosa* appear as round, hardened, purple areas about 1 cm in diameter with an ulcerated center surrounded by a zone of redness. Inflammation of hair follicles, a skin condition and ear infections resulting from bathing in contaminated water and associated with hot tubs, and swimming pools and ear infections are among the few *Pseudomonas* infections occurring in healthy individuals.

**Diagnosis.** Most strains of *P. aeruginosa* are identified on the basis of its characteristic grapelike odor, colonial morphology, growth at 42°C, and the production of a water-soluble blue pigment, pyocyanin (Figures 102b,c).

### *P. fluorescens*

Another species of this genus rarely associated with opportunistic infections is *P. fluorescens*. Most isolates are from respiratory tract specimens. The organism has also been isolated from various sources in hospital environments, such as water sources, sinks, floor, and contaminated blood. *P. fluorescens* is commonly associated with the spoilage of foods such as meats and fish.

**Morphology and Cultural Properties.** *P. fluorescens* is similar to *P. aeruginosa* in microscopic properties. It grows on most media, forms small colonies, and characteristically produces the fluorescent pigment, fluorescein (Figures 102c and 103).

**Pathology.** While *P. fluorescens* is mainly an environmental contaminant, it can be an opportunistic pathogen for humans. The organism has been associated with urinary tract infections, pelvic inflammatory disease, and postoperative infections.

**Diagnosis.** Most *P. fluorescens* strains are identified on the basis of colonial morphology and the production of the fluorescent pigment.

## Gram-Negative Aerobic/Microaerophilic Rods and Cocci *(Continued)*

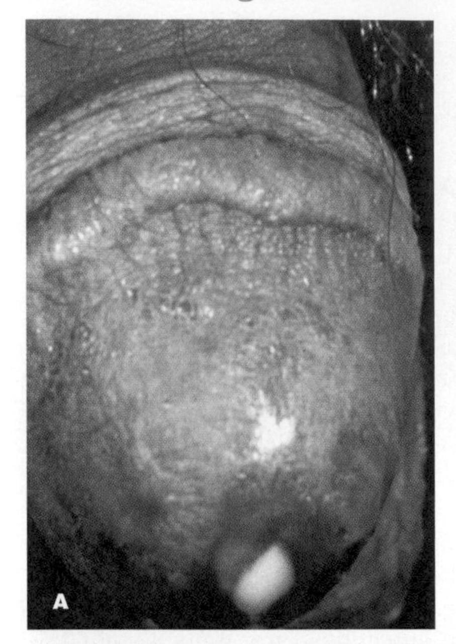


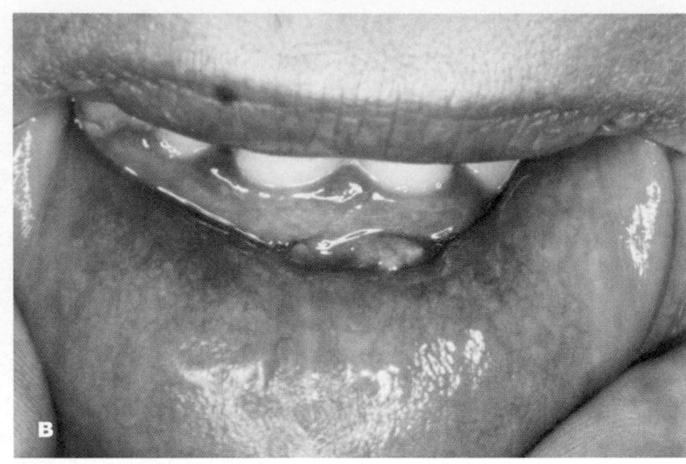


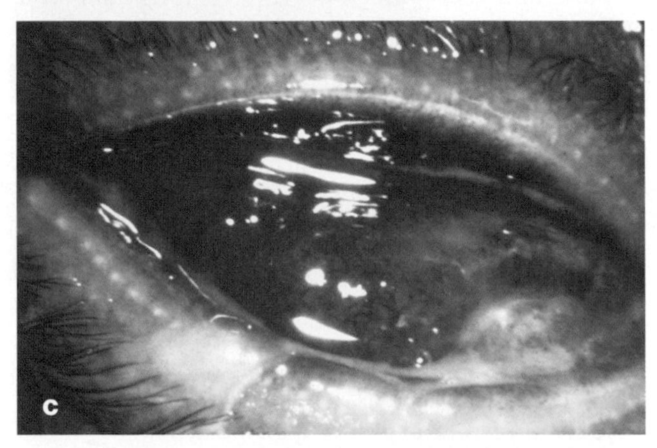


**Figure 98**
Clinical features of gonorrhea. (a) The typical discharge of this sexually transmitted disease in males. (b) An abscess (yellow growth) in the mouth of a patient. Such localized collections of pus in the mouth are not rare. (From D. Marini, S. Veraldi, and M. Innocenti. *Cutis.* 40(1987):363.) (c) A case of ophthalmia neonatorum caused by *Neisseria gonorrhoeae.*

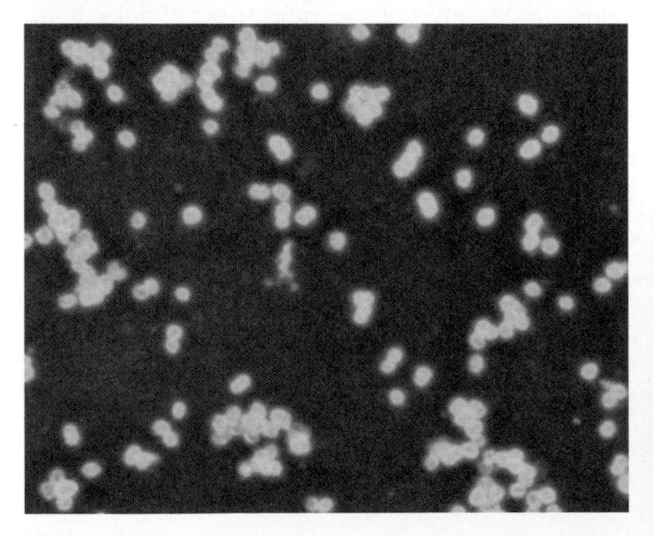

**Figure 99**
The results of the fluorescent antibody technique showing *Neisseria gonorrhoeae.* The diplococcus arrangement can be seen.

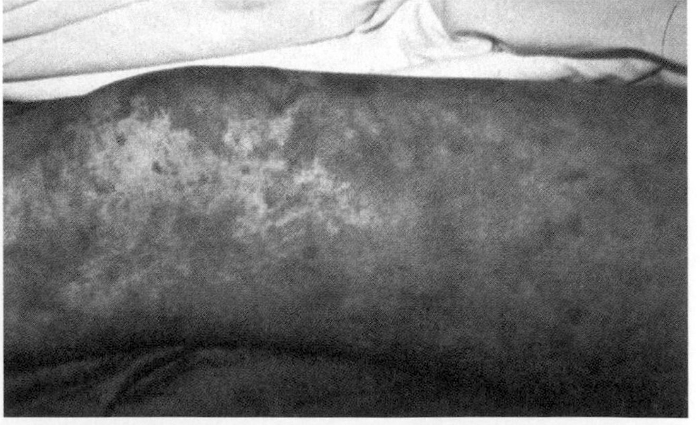

**Figure 100**
The petechial rash that commonly develops with the invasion (meningococcemia) of the bloodstream by *Neisseria meningitidis.* (From M. Barza. *NEJM* 328(1996):34.)

## Gram-Negative Aerobic/Microaerophilic Rods and Cocci *(Continued)*

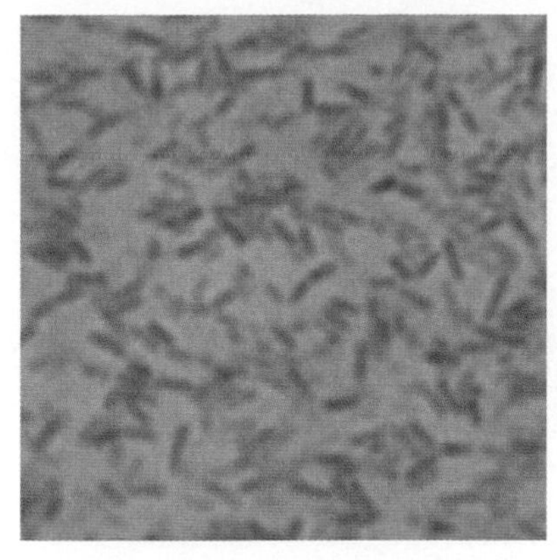

**Figure 101**
The gram-negative rods of *Pseudomonas aeruginosa.*

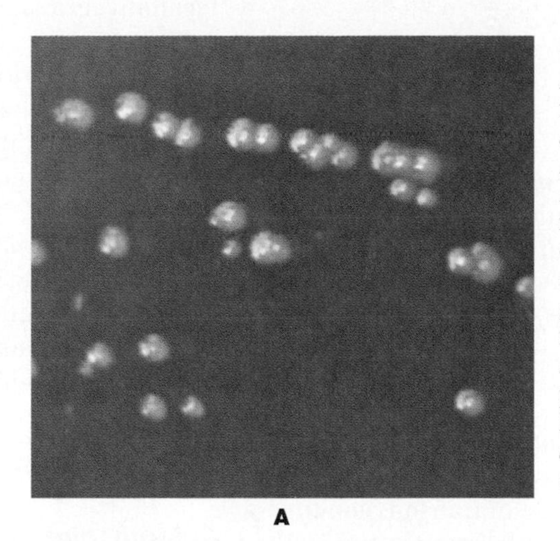

**A**

**Figure 102**
*Pseudomonas aeruginosa* cultural features. (a) A young culture of *P. aeruginosa* on nutrient agar. (b) *P. aeruginosa* growing on *Pseudomonas* P agar. This medium enhances the production of pyocyanin (a bluish pigment) and inhibits the production of the greenish yellow pigment fluorescein. The red-pigmented *Serratia marcescens* is shown for purposes of comparison. (c) A 72-hour broth culture of *P. aeruginosa* ( center) and the appearance of its typical pigment.

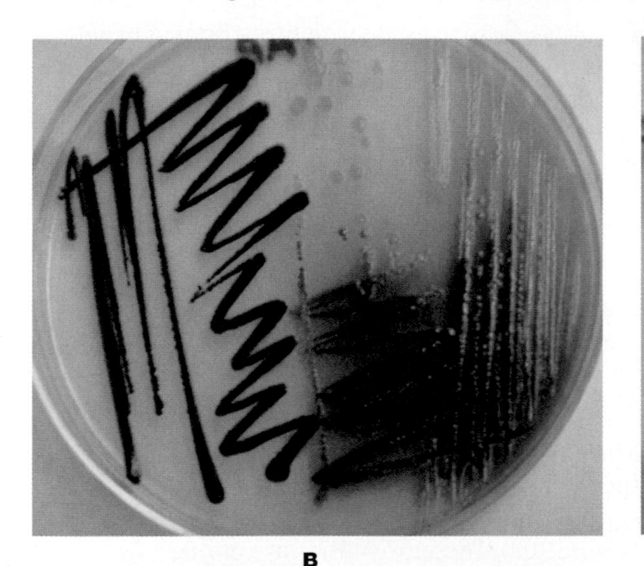

**B**

**C**

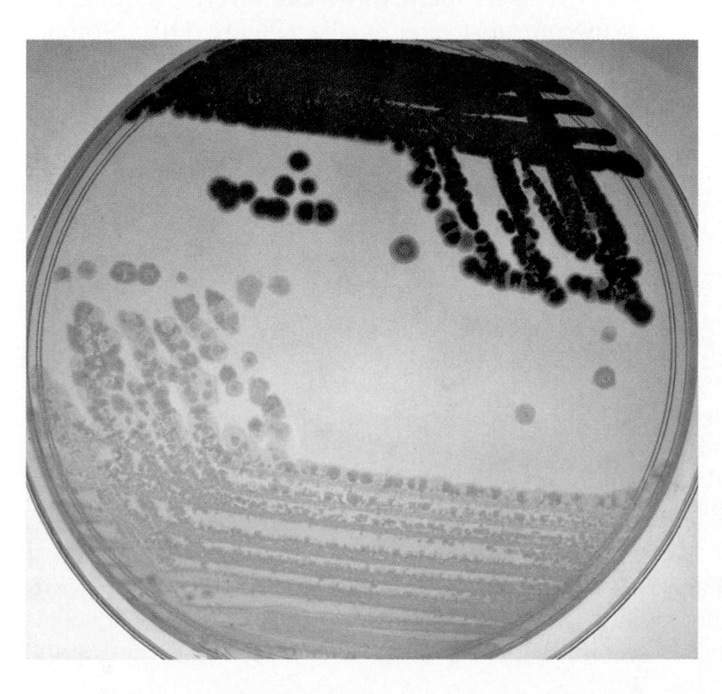

**Figure 103**
*P. fluorescens* growing on *Pseudomonas* F agar and producing its characteristic fluorescein (yellow fluorescent) pigment.

## Facultatively Anaerobic Gram-Negative Rods

### *Calymmatobacterium*

***Calymmatobacterium granulomatis***

The causative agent of granuloma inguinale, also known as donovanosis and granuloma venereum, is *Calymmatobacterium granulomatis*. The disease is chronic and involves the skin and the subcutaneous tissue of the genital, inguinal, and anal regions (Figure 104).

**Transmission.** *Calymmatobacterium granulomatis* is thought to be sexually transmitted. However, several findings do not support this mode of transmission totally. Granuloma inguinale is mainly found in tropical areas of the world.

**Morphology and Cultural Features.** *Calymmatobacterium granulomatis* is a gram-negative, pleomorphic rod, 0.5–1.5 μm wide by 1.0–2.0 μm in length (Figure 105). Material from specimens stained with Wright's, Giemsa, or other stains shows characteristic intracellular organisms within large mononuclear cells known as **Donovan bodies** (Figure 106).

Routine culture in embryonated eggs is neither practical nor highly successful. Specially prepared media containing growth factors have been used to grow *C. granulomatis.*

**Pathology and Clinical Features.** Granuloma inguinale begins as subcutaneous nodules that eventually erode and produce a clear, tumorlike, sharply defined lesion (Figure 104). Without treatment, the disease progresses by extension to neighboring skin areas and frequently spreads along the groin. Although such lesions generally are not painful, they can be mutilating.

**Diagnosis.** Diagnosis is usually based on finding the typical intracellular organisms (Donovan bodies) with large mononuclear cells in tissues known as histiocytes (Figure 106).

### *Citrobacter*

***Citrobacter freundii***

Occurring in the feces of humans and lower animals, *Citrobacter freundii* is probably a normal intestinal inhabitant. The organism also is found in soil, water, sewage, and food. *Citrobacter freundii* often is isolated from clinical specimens as an opportunistic pathogen.

**Morphology and Cultural Properties.** *Citrobacter freundii* is a gram-negative, straight rod, measuring about 1.0 μm in width and 2–6.0 μm in length and occurring singly or in pairs (Figure 107). The organism is motile, facultatively anaerobic, oxidase negative, and catalase positive.

It grows on various media including nutrient, blood, and eosin–methylene blue agars (Figure 108). Optimal growth temperature is 39°C. On blood agar, colonies are large, white, raised, smooth, and entire (Figure 108a).

*Citrobacter freundii* are positive for methyl red, nitrate, and citrate tests and negative for indole and Voges-Proskauer. A variety of sugars including glucose, mannitol, and maltose are fermented with the production of acid.

**Diagnosis.** Isolation and biochemical tests are necessary for diagnosis.

### *Enterobacter*

*Enterobacter* species are widely distributed in nature, occurring in soil, fresh water, sewage, and animal feces and on plants including vegetables. Several species are opportunistic pathogens. *Enterobacter aerogenes* and *E. cloacae* are the species most frequently isolated from clinical specimens and are important causes of hospital-acquired cases of bacteremia and contamination of wound infections in burn patients.

**Transmission.** *Enterobacter* species are found as contaminants of intravenous fluids and hospital equipment. Bacteremia can result from the use of these materials.

**Morphology and Cultural Properties.** *Enterobacter* species are small, straight, gram negative rods measuring 0.6–1.0 μm in diameter and 1.2–3.0 μm long (Figure 109). These organisms are motile.

*Enterobacter* species are facultatively anaerobic and grow on most agar media (Figure 110). Optimal growth temperatures range from 30° to 37°C. Colonies are generally cream to tan in color, entire, glistening, convex, and circular. Lactose positive (fermenting) colonies are produced on media such as eosin-methylene blue or endo agars (Figure 110b).

Glucose and other carbohydrates are catabolized with the production of acid and gas. The I (indole), M (methyl red), VP (Voges-Proskauer), C (citrate) pattern for *Enterobacter* species usually is –, –, +, + (see Figure 58), which is the opposite of the reaction pattern produced by *Escherichia coli*. Hydrogen sulfide, DNAse, and lipase are not produced.

### *Escherichia*

*Escherichia* species occur as normal members of the intestinal microbiota in the large intestine of most mammals. The most important of these is *E. coli*. This organism as well as several other related species commonly cause infections outside of the gastrointestinal

## Facultatively Anaerobic Gram-Negative Rods

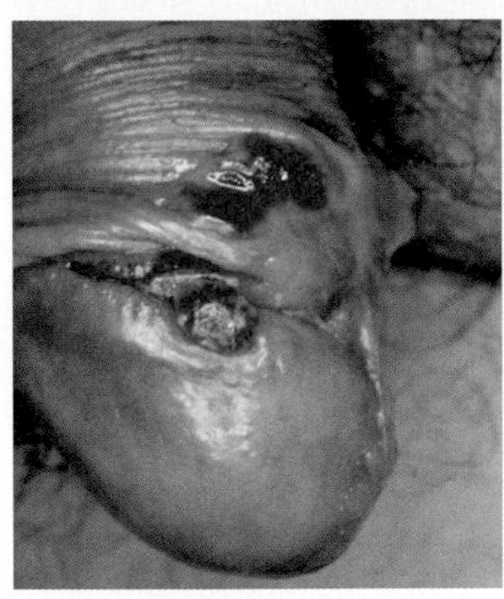

**Figure 104**
A case of granuloma inguinale, showing numerous ulcerated red nodules on the penis. (From P. Hacker and B.K. Fisher. *Internat. J. Dermatol.: 31(1992):696.)*

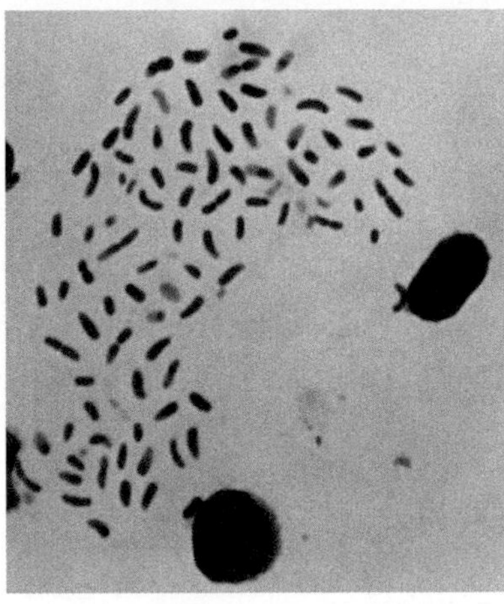

**Figure 105**
A rapid Giemsa (Rapidiff) stain of *Calymmatobacterium granulomatis* (from a 48-hour monocyte culture (1,000×). [Courtesy of A. Kharsany, B. Housen, M. Housen, H.A. Housen, P. Kiepiela, T. Caicker, and A.W. Sturm. Faculty of Medicine, University of Natal, Durban.

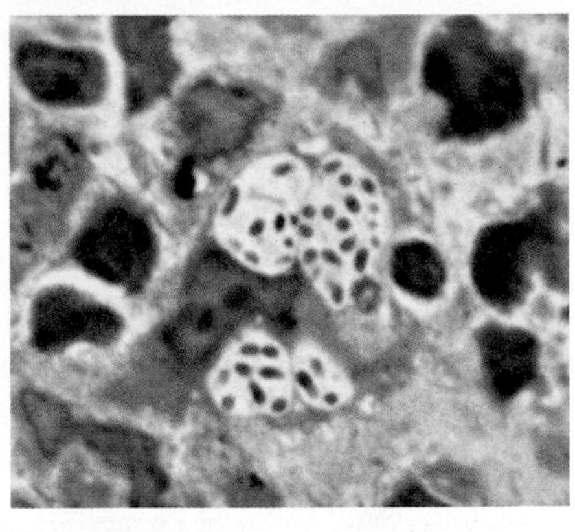

**Figure 106**
A microscopic view of a specimen from a case of granuloma inguinale showing Donovan bodies. (From P. Hacker and B.K. Fisher. *Internat. J. Dermatol.* 31(1996):696.)

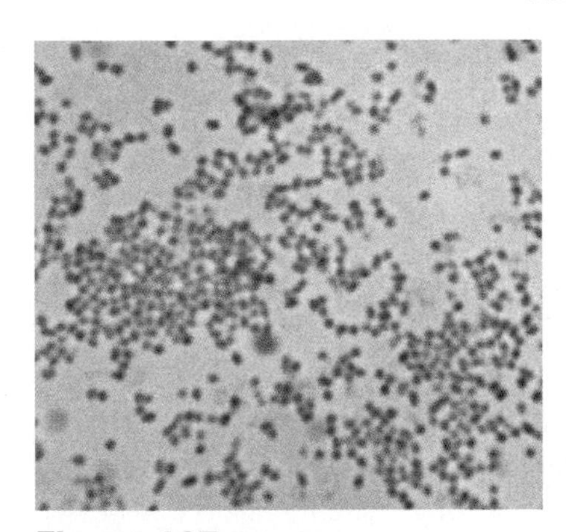

**Figure 107**
The small gram-negative rods of *Citrobacter freundii.*

**Figure 108**
Growth of *Citrobacter freundii* on various media. (a) Blood agar. (b) The lactose-positive reaction on eosin–methylene blue agar.

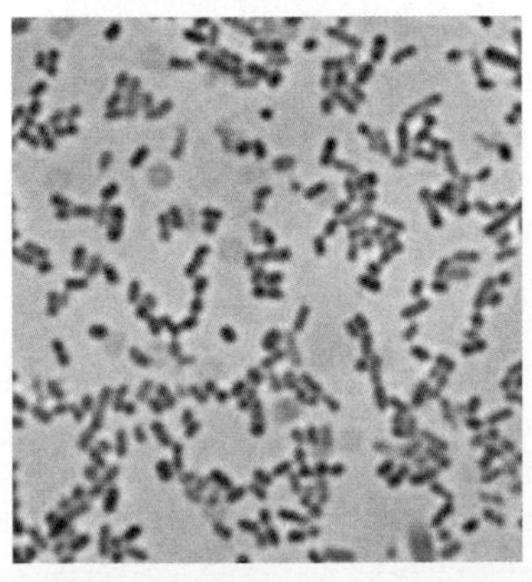

**Figure 109**
*Enterobacter aerogenes* Gram stain.

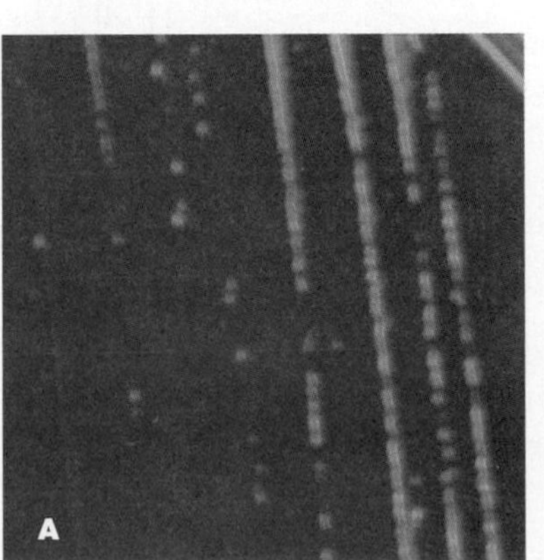


**Figure 110**
*Enterobacter aerogenes* on media. (a) Young culture on blood agar. (b) Colonies on eosin–methylene blue agar.

tract. Urinary tract infections, primarily of the bladder, are the most common, followed by respiratory, wound, bloodstream (sepsis), and central nervous system infections. Many of these infections, especially sepsis (Figure 114) and meningitis, are life-threatening and are often hospital-acquired.

### *Escherichia coli*

**Transmission.** *Escherichia coli* types, which include 0157:H7, associated with gastrointestinal and related infections are generally acquired by consuming contaminated water or food.

**Morphology and Cultural Properties.** *Escherichia coli* is a short, straight, gram-negative rod, measuring 1.10–1.5 μm in diameter and 2.0–6.0 μm in length (Figure 111). Cells occur singly or in pairs. *Escherichia coli* is motile and does not form capsules.

The organism is facultatively anaerobic and grows on a wide variety of agar media (Figure 112). Optimal temperature ranges between 30° to 37°C. Colonies on nutrient or blood agar are cream to tan in color, entire, convex, circular, and smooth. Dark purple colonies, typical of lactose fermentation results are formed on eosin–methylene blue agar (Figure 112c). The colonies of certain strains have a metallic sheen.

*Escherichia coli* is hydrogen sulfide, urease, and oxidase negative and catalase positive, and it catabolizes glucose and other carbohydrates with the formation of acid and gas. The I (indole), M (methyl red), VP (Voges-Proskauer), C (citrate) pattern of reactions is +, +, –, – (see Figure 58).

**Pathology and Clinical Features.** Several distinct types of *E. coli* are currently recognized to cause human diarrhea. These organisms can be grouped into several categories on the basis of pathogenic mechanisms: enteropathogenic *E. coli* (EPEC), enterohemorrhagic *E. coli* (EHEC), enteroinvasive *E. coli* (EIEC), enteroadherent *E. coli* (EAEC), and enterotoxigenic *E. coli* (ETEC).

ETEC strains are known to possess two pathogenic properties: an ability to adhere to intestinal tissue (Figure 113) and the production of two enterotoxins, one heat-labile and the other heat-stabile.

The signs and symptoms of mild infection include diarrhea, vomiting, chills, headache, and fever following an incubation period of 1 to 2 days. More serious infections produce severe abdominal cramps, toxemia, and watery stools consisting of blood and mucus. Severe dehydration, shock, and death also may occur.

**Diagnosis.** Isolation and biochemical testing are the main approaches used for identification. Specific selective and differential media and immunologic tests such as the particle agglutination procedure are used for the identification for *E. coli* 0157:H7 strains (Figure 112d).

## *Gardnerella*

### *Gardnerella vaginalis*

Found in the human genital and urinary tracts, *Gardnerella vaginalis* is associated with bacterial vaginosis. The condition also involves various obligate anaerobic bacteria.

**Transmission.** *Gardnerella vaginalis* is generally acquired through sexual activity.

**Morphology and Cultural Properties.** *Gardnerella vaginalis* is a gram-negative to gram-variable pleomorphic rod, about 0.5 μm in width and 1.5–2.5 μm in length. The organism is nonmotile and does not form capsules.

*Gardnerella vaginalis* is facultatively anaerobic. Special media and a $CO_2$ incubator are necessary for culture. Small, opaque, convex colonies typically appear after 48 hours of incubation. Optimal growth temperature is 35° to 37°C.

*Gardnerella vaginalis* is catalase and oxidase negative, produces acid but no gas from glucose and certain other carbohydrates, and does not reduce nitrates.

**Pathology and Clinical Features.** *Gardnerella vaginalis* infection produces a foul-smelling vaginal discharge with a pH ranging from 4.5 to 5.5. The organism is frequently isolated from women with infections of the mucous membrane lining the inner surface of the uterus.

**Diagnosis.** The direct microscopic examination of vaginal secretions is more relevant for the diagnosis of *G. vaginalis* infection than is the isolation of the organism from such specimens. A minimum diagnostic requirement from bacterial vaginosis usually includes the finding of: (1) excessive vaginal discharge, (2) vaginal pH of greater than 4.5, (3) vaginal epithelial cells covered by small gram-negative rods known as **Clue cells** (Figure 115), and (4) a fishy chemical odor in the potassium hydroxide test.

## *Haemophilus*

*Haemophilus* species are minute to medium-sized oval or rod-shaped gram-negative cells (Figure 116). Organisms are nonmotile and facultatively anaerobic. They produce acid from glucose and other carbohydrates and reduce nitrates to and beyond nitrites.

Almost all species require preformed growth factors found in red blood cells, such as hematin (X factor) and NAD or nicotinamide adenine dinucleotide (V factor). The V factor is found in a variety of biological

## Facultatively Anaerobic Gram-Negative Rods *(Continued)*

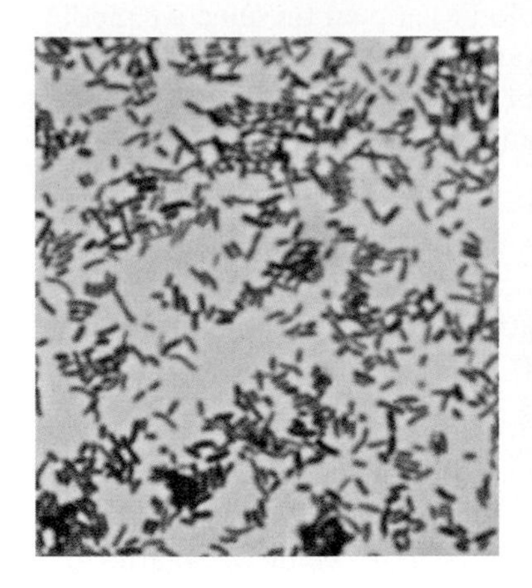

**Figure 111**
Gram-negative staining results with *Escherichia coli.*

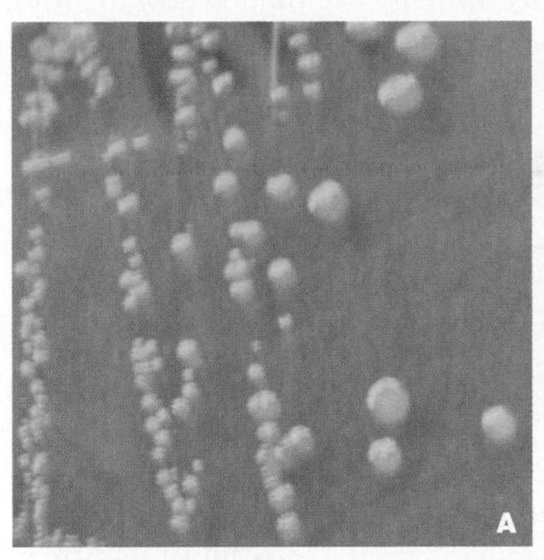


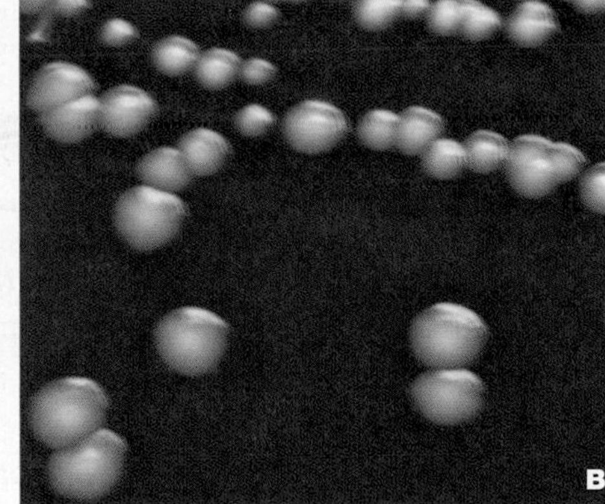


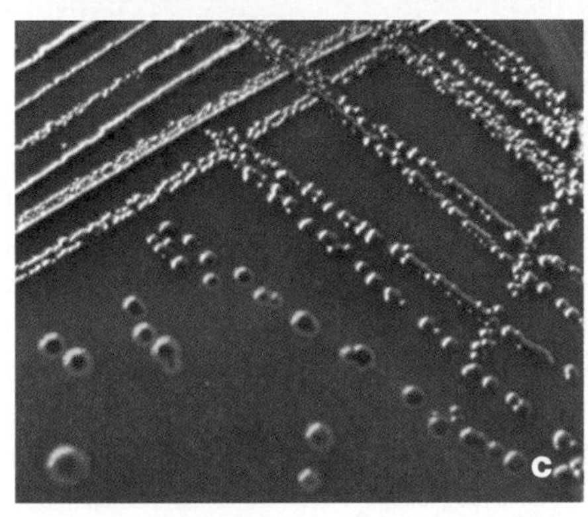

**Figure 112**
*Escherichia coli* on media after 48 hours of incubation. (a) Colonies on nutrient agar. (b) Colonies on blood agar. (c) The green metallic sheen produced by *E. coli* growing on eosin-methylene blue agar. Lactose fermentation also is shown. (d) Sorbitol-negative colonies (pale pink) typical of *Escherichia coli* 0157:H7. The medium here is MacConkey sorbitol agar.

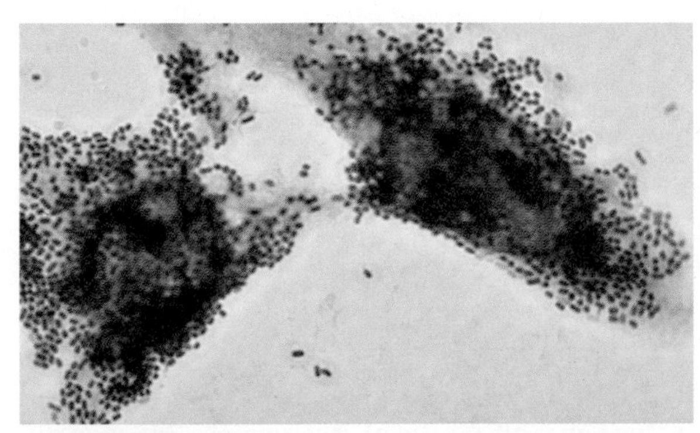

**Figure 113**
A light micrograph showing enteroadherent attached *Escherichia coli* (EAEC) to intestinal tissue cells (1,000×). (From A. Darfeuille-Michaud and Associates. *Inf. Immunol.* 58(1990):893–902.)

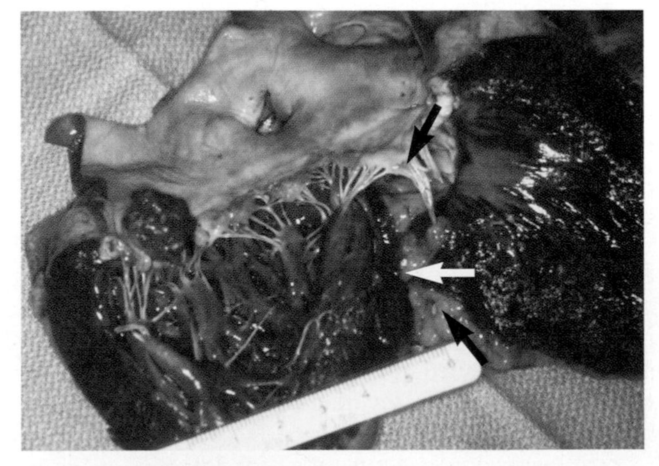

**Figure 114**
A large collection of fibrin and white and red blood cells known as a vegetation (arrow) on the mitral valve. This vegetation was found in a case of infective endocarditis caused by *Escherichia coli* (From C. Watanakuna Korn and J. Kim. *CID* 14(1992):501.)

materials and is produced by bacteria such as *Staphylococcus aureus* and by yeast. *Haemophilus influenzae* will grow on blood agar in the vicinity of *S. aureus* colonies producing this growth factor. This phenomenon is called **satellitism** (Figure 117).

***Haemophilus influenzae***
*Haemophilus influenzae* is known to cause a variety of infections, including meningitis, especially in children, otis media, pneumonia, and epiglottitis. The organism is part of the normal upper respiratory tract microbiota of humans.

**Transmission.** *Haemophilus influenzae* is usually transmitted by contact with secretions produced by sneezing or coughing, or found on the hands.

**Morphology and Cultural Properties.** *Haemophilus influenzae* is a nonmotile, frequently capsulated, gram-negative small rod, measuring less than 1.0 μm in width and variable in length. Growth is best on complex media such as chocolate agar (Figure 118).

**Pathology and Clinical Features.** *Haemophilus influenzae* is one of the common causes of an inflammation of the epiglottis (**epiglottitis**). The infection often develops quickly and if not diagnosed promptly, it may result in respiratory blockage and death. Typical signs and symptoms include sore throat, fever, difficulty in swallowing and breathing, and the appearance of a characteristic swollen epiglottis which looks like a bright-red cherry blocking the throat at the base of the tongue.

If *H. influenzae* colonizes the nasopharynx, infections from this site may extend to local tissues, the middle ear, invasion of the blood stream resulting in such conditions as meningitis, pneumonia, bronchitis, sinusitis, and osteomyelitis.

**Diagnosis.** Various specimens are used for direct microscopic examination or culture. These include blood, spinal fluid (Figure 116), bronchoalveolar lavage, and urine. Rapid immunologic tests for the detection of polysaccharide antigens also are used.

***Haemophilus ducreyi***
*Haemophilus ducreyi* is the causative agent of the sexually transmitted disease known as **chancroid** or **soft chancre** (Figure 119). Humans are the only known hosts for the disease. Chancroid was at one time limited in occurrence to tropical, undeveloped areas. However, since 1985, several large outbreaks have appeared elsewhere.

**Morphology and Cultural Properties.** *Haemophilus ducreyi* is a gram-negative slender rod occurring in pairs or chains, and measuring 0.5 μm in width and 1.5–2.0 μm in length. They may appear in arrangements described as "schools of fish."

After 48 to 72 hours of incubation small, flat, smooth and yellow to grey colonies appear on selective media. Such colonies are translucent to opaque. Additional cultural properties of *H. ducreyi* include positive reactions for oxidase and nitrate reduction.

**Pathology and Clinical Features.** The incubation period of this disease is about 2 to 5 days. Typically a localized, red, elevated area or papule forms at the site of infection. The lesion develops into a painful ulcer with sharp margins (Figure 119).

**Diagnosis.** Specimens for diagnosis should be taken for microscopic examination and culture. *Haemophilus ducreyi* appears in characteristic parallel rows. Various media with growth factors are of value in isolating the organism (Figure 120).

## *Klebsiella*

*Klebsiella* species are straight, gram-negative rods measuring 0.3–1.0 μm in diameter and 0.6–6.0 μm in length. Cells occur singly, in pairs, or in short chains (Figure 121). Organisms are nonmotile and capsulated (Figure 122).

*Klebsiella* species are facultatively anaerobic, oxidase negative, and catalase positive. They reduce nitrates and ferment most commonly tested carbohydrates, producing acid and gas.

These organisms can be found in association with human feces, soil, water, fruits, vegetables, and grains. Certain species are opportunists and are known to cause urinary tract infections and a variety of nosocomial infections, bacteremia, and pneumonia.

***Klebsiella pneumoniae***
*Klebsiella pneumoniae,* the most common species in the genus, causes lobar pneumonia, a disease also associated with other encapsulated organisms.

*Klebsiella pneumoniae* forms large mucoid colonies, especially on carbohydrate media.

**Diagnosis.** Diagnosis involves isolation and biochemical testing.

## *Morganella*

An opportunistic organism, *Morganella morganii* is found in the feces of several mammals and reptiles. The organism is known to cause infection in older patients with serious underlying disease. *Morganella morganii* has been isolated from respiratory and urinary tract infections, wounds, and cases of bacteremia.

**Morphology and Cultural Properties.** *Morganella morganii* is a gram-negative rod 0.6–0.7 μm in diameter and 1.0–1.7 μm in length (Figure 123). The organism is nonmotile.

The organism is facultatively anaerobic and grows

## Facultatively Anaerobic Gram-Negative Rods *(Continued)*

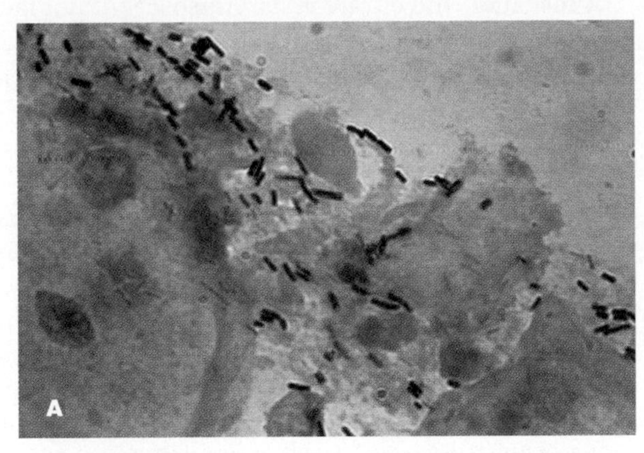


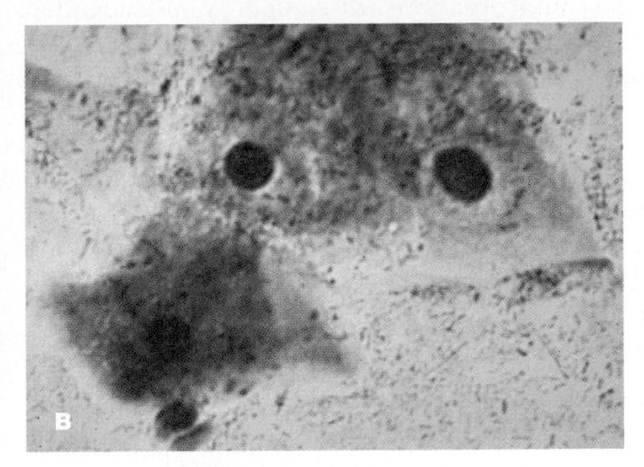


**Figure 115**
*Gardnerella vaginalis.* (a) A normal cervical smear showing epithelial cells and gram-positive rods. (b) Clue cells covered with the gram-negative rods of *G. vaginalis.*

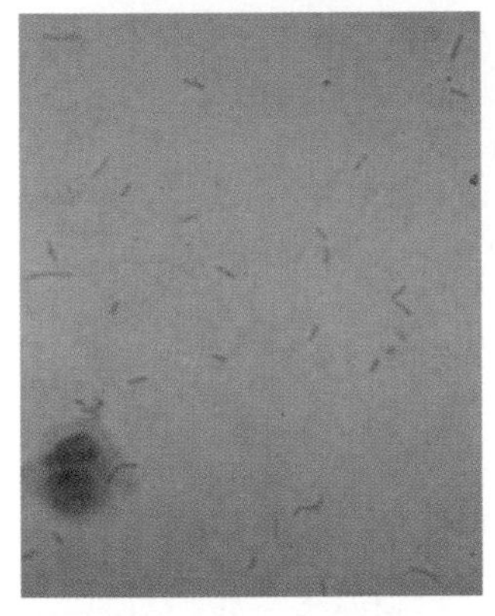

**Figure 116**
Gram-negative small rods or cocco-bacilli of *Haemophilus influenzae* in a spinal fluid specimen (1,000×).

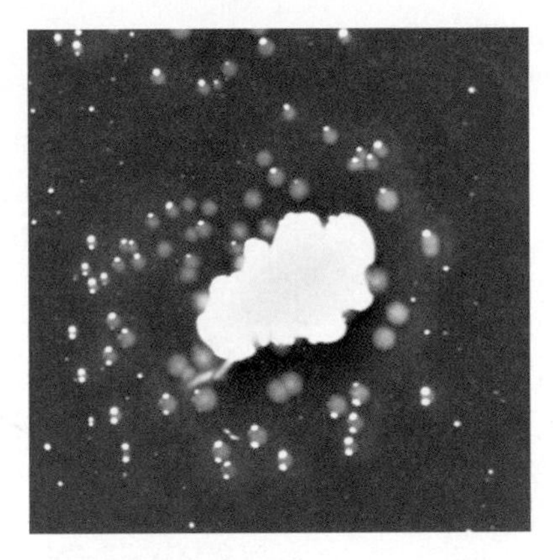

**Figure 117**
*Haemophilus influenzae* growing on a blood agar plate containing *Staphylococcus aureus* (central white growth) and exhibiting satellitism.

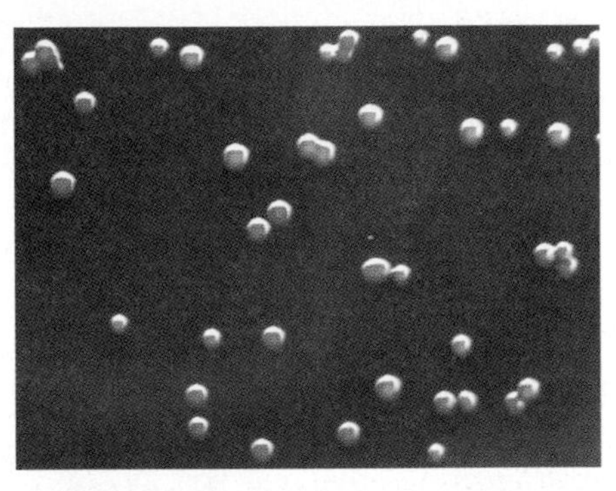

**Figure 118**
Colonies of *Haemophilus influenzae* on chocolate agar.

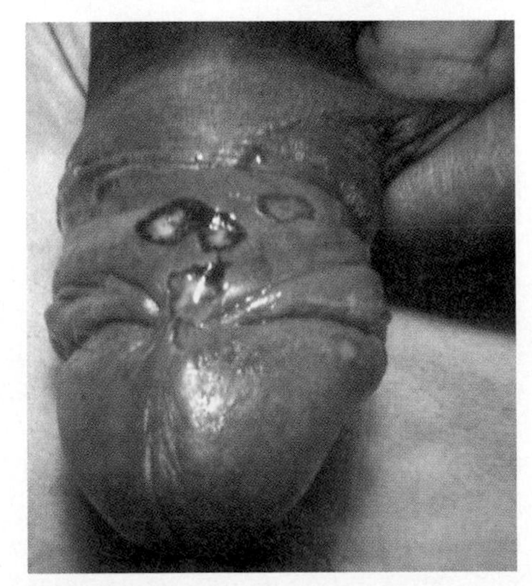

**Figure 119**
The clinical appearance of chancroid, a genital ulcer, that may be accompanied by painful, pus-filled enlarged lymph nodes.

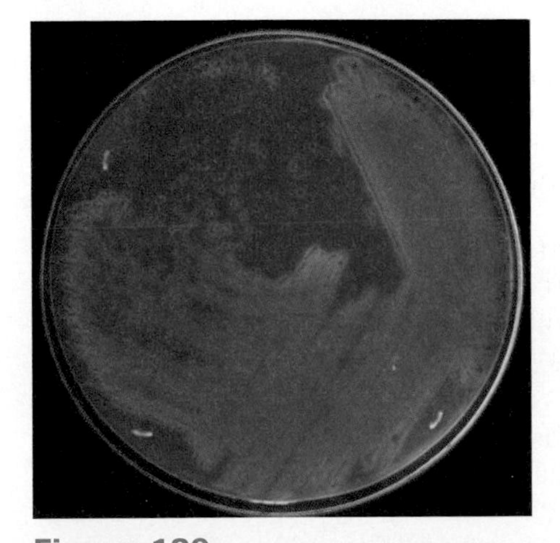

**Figure 120**
An entire plate and closer view of individual *Haemophilus ducreyi* colonies surrounded by zones of beta hemolysis. (From P.A. Totten, D. V. Norn, and W.E. Stamm. *Inf. Immunol.* 63(1995):4409–16.)

on a variety of media (Figure 124). Optimal temperature is 37°C. *Morganella morganii* is oxidase negative, catalase positive, and it hydrolyzes urea. Hydrogen sulfide is not produced. The organism's IMViC pattern is +, +, –, –. The only carbohydrates catabolized are glucose and mannose, with the production of acid and usually gas.

## Pasteurella

### *Pasteurella multocida*

One of several species found in the normal respiratory tract microbiota of some animals is *Pasteurella multocida*. This organism is by far the most common cause of an infected dog or cat bite (Figure 125).

**Transmission.** Humans are usually infected by the bite or scratch of a domestic cat or dog.

**Morphology and Cultural Properties.** *Pasteurella multocida* is a gram-negative, ovoid or rod-shaped cell measuring 0.3–1.0 μm in diameter and 1.0–2.0 μm in length (Figure 12.6). Cells occur singly or in pairs.

Organisms are nonmotile and facultatively anaerobic. *Pasteurella multocida* are oxidase and catalase positive, produce acid but no gas from most carbohydrates, and reduce nitrates. Methyl red and Voges-Proskauer reactions are negative.

**Pathology and Clinical Features.** Infection develops at the site of a bite or scratch, often within 24 hours. The typical infection involves the subcutaneous tissue (**cellulitis**) with the development of a well-defined reddened border.

**Diagnosis.** Culture of pus taken from a lesion provides the best approach to diagnosis.

## Proteus

*Proteus* species, which include **P. vulgaris** and **P. mirabilis**, occur in the intestines of a variety of mammals and also can be found in manure, soil, and polluted waters.

**Transmission.** *Proteus* species may be acquired by the ingestion of contaminated food or water.

**Morphology and Cultural Properties.** *Proteus* species are gram-negative, straight rods, 0.4–0.8 μm in diameter and 1.0–3.0 μm in length (Figures 127 and 128). These organisms occur singly and in pairs and are motile by means of peritrichous flagella. They generally produce concentric zones of growth (swarming) over most agar surfaces (Figure 129).

*Proteus* species are facultatively anaerobic and are positive for urea hydrolysis (see Figure 53) and phenylalanine deaminase (see Figure 52). These organisms are oxidase negative and catalase and methyl red positive, and they give variable results with indole, Voges-Proskauer, and citrate tests. Glucose, but not lactose, and a few other carbohydrates are catabolized, with the production of acid and usually gas. Hydrogen sulfide is produced by *P. vulgaris* and *P. mirabilis* (Figure 130).

**Pathology and Clinical Features.** *Proteus vulgaris* and *P. mirabilis* are associated with gastrointestinal infections, and they are often isolated from urinary tract and other extraintestinal infections such as septic lesions in burn patients. Urease production by *Proteus* species and *Morganella morganii* is thought to play a major role in the pathogenicity of these organisms, especially in the case of urinary tract infections.

**Diagnosis.** Culture and biochemical tests are used for identification.

## Salmonella

*Salmonella* species other than **S. typhi** (the causative agent of typhoid fever) can be found in the gastrointestinal tracts of both warm- and cold-blooded animals and in the environment. These microorganisms either cause disease or are carried without any apparent harmful effects on the host.

**Transmission.** Infections with **Salmonella** species for the most part are acquired through the ingestion of contaminated food or drink. Contaminated meat and dairy products are the most likely sources of disease agents, although uncooked eggs also may be responsible for foodborne infection. Infected food handlers and other persons can transmit *Salmonella* infection as well.

**Morphology and Cultural Properties.** *Salmonella* species are straight, gram-negative rods, 0.7–1.5 μm in diameter and 2.0–5.0 μm in length. The organisms are usually motile by peritrichous flagella (Figure 131).

*Salmonella* species are facultative anaerobes. Optimal growth temperature is 37°C. Organisms are oxidase negative and catalase positive, and they reduce nitrates. The IMViC pattern of reactions is –, +, –, +. Glucose and other carbohydrates, with the exception of lactose, usually are fermented. *Salmonella* forms distinctive colonies on various selective and differential media such as Hektoen enteric, *Salmonella-Shigella*, and XLT4 agars (Figure 132). Hydrogen sulfide is typically produced by various species (Figure 132).

**Pathology and Clinical Features.** From a clinical standpoint, most *Salmonella* infections can be divided into gastroenteritis (inflammation of the stomach and intestines), enteric fever, bacteremia, and the more serious typhoid fever.

## Facultatively Anaerobic Gram-Negative Rods *(Continued)*

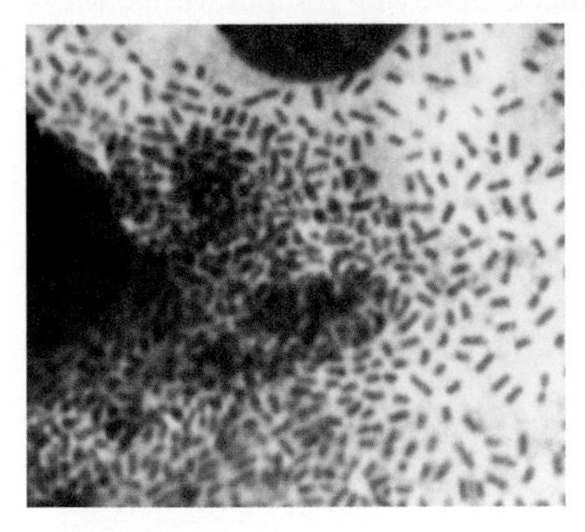

**Figure 121**
Rod-shaped *Klebsiella pneumoniae* (1,000×). (Courtesy of A. Faurse-Bonte, A. Darefeuille-Michaud, and C. Forester. Université d'Auvergne, France.)

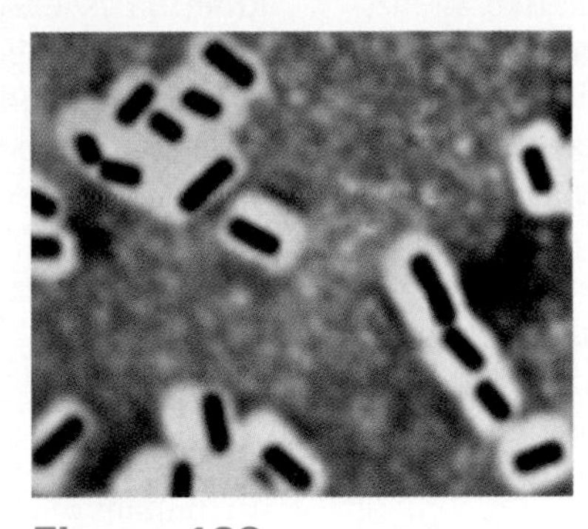

**Figure 122**
*Klebsiella pneumoniae* and its capsules.

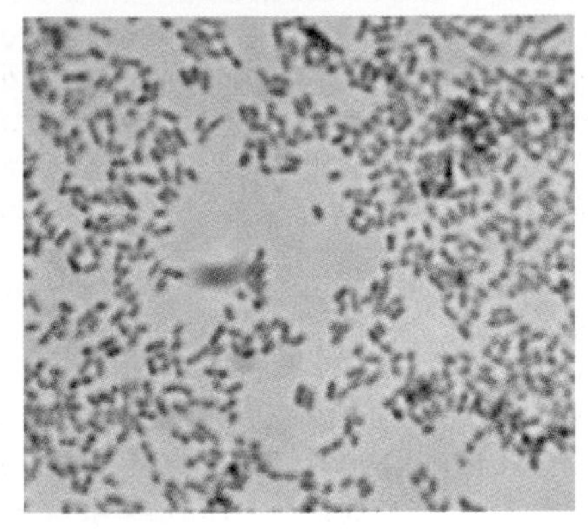

**Figure 123**
Gram stain of *Morganella morganii.* (1,000×).

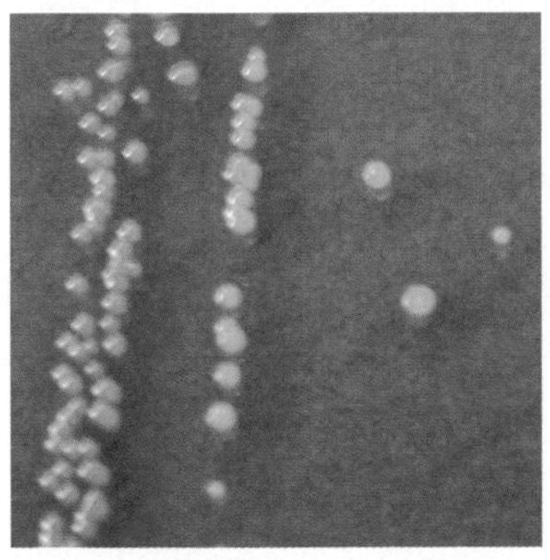

**Figure 124**
*Morganella morganii* colonies on nutrient agar.

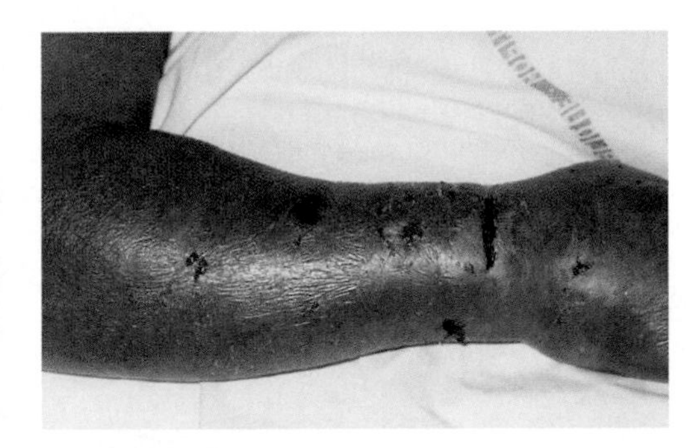

**Figure 125**
Wound infection and spreading inflammation involving subcutaneous tissue of the arm caused by *Pasteurella multocida.* (From O. Chosidow. *Internat. J. Dermatol.* 33(1994):471.)

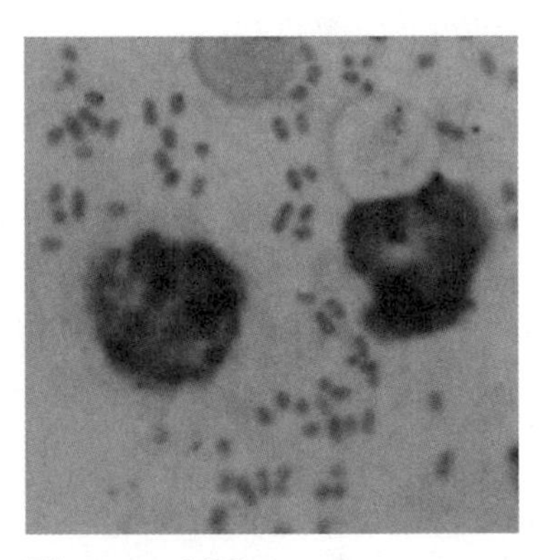

**Figure 126**
The bipolar gram-negative coccoid to rod-shaped *Pasteurella multocida.* (From O. Chosidow. *Internat. J. Dermatol.* 33(1994):471.)

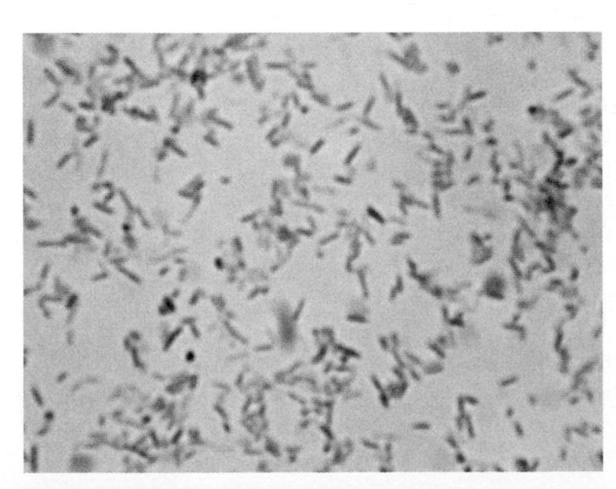

**Figure 127**
Gram stain of *Proteus vulgaris* (1,000×).

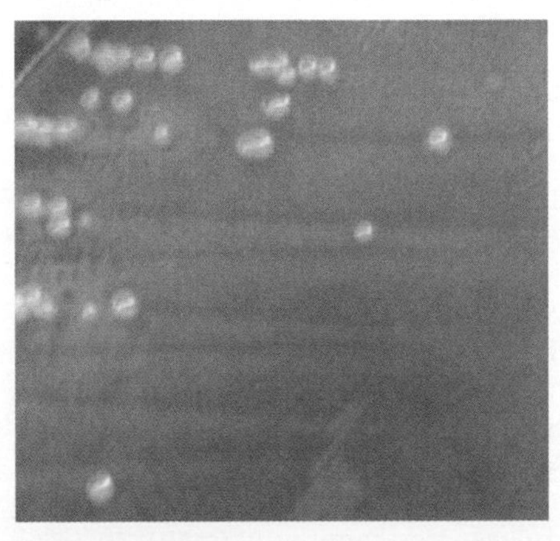

**Figure 128**
*Proteus vulgaris* colonies on nutrient agar.

Gastroenteritis has an incubation period of about 24 to 48 hours after ingestion of the causative agents. Episodes of nausea and vomiting are followed or accompanied by abdominal cramps and diarrhea. These signs and symptoms, especially diarrhea, may persist for 3 to 4 days. Fever does occur in about half of patients. *Salmonella typhimurium* and *S. enteritidis* are the major causes.

Enteric fever involves several organs, particularly the reticuloendothelial system, liver, spleen, and mesenteric lymph nodes. A prolonged fever is typical. *Salmonella typhi* and other *Salmonella* species are associated with enteric fever.

In typhoid fever, organisms penetrate the lining of the small intestine, enter the local lymph nodes, where they multiply, and eventually gain access to the bloodstream. By means of the bloodstream, the salmonellae are carried to most organs of the body, where they continue to multiply.

Typhoid fever has an incubation period ranging from 7 to 14 days. Common early signs and symptoms include headache, chills, fever, loss of energy, and generalized aches and pains. Constipation rather than diarrhea is typical. Organisms continue to be spread throughout the body, causing a bacteremia and a very high fever. Small purplish areas of bleeding (**petchiae**) called **rose spots** may develop in the skin around the trunk.

Systemic *Salmonella* infections are known to spread and may involve the heart and parts of the respiratory, skeletal, and central nervous systems.

**Diagnosis.** Various specimens are used for culture and diagnosis. These include stool for gastroenteritis, blood for enteric fever and bacteremia, stool and blood for typhoid fever, and sputum for pneumonia.

## *Serratia*

*Serratia* species are found in soil in water, on plant surfaces, in the digestive tracts of rodents and insects, and in human clinical specimens. The red-pigmented *S. marcescens* (Figure 133) is a major opportunistic pathogen for hospitalized persons, causing urinary tract infections and septicemia.

**Transmission.** The means of transmission include aerosols.

**Morphology and Cultural Properties.** *Serratia* species are gram-negative rods measuring 0.5–0.8 μm in diameter and 0.9–2.0 μm in length (Figure 134). These organisms move by means of peritrichous flagella.

*Serratia* species are facultative anaerobes. They grow well at temperatures ranging from 30° to 37°C, and they catabolize glucose and other carbohydrates, with the production of acid and often of gas. Pigment production is enhanced at lower temperatures. Most strains produce DNAse (see Figure 45), hydrolyze gelatin, and reduce nitrates. Voges-Proskauer and citrate tests are positive; indole is generally negative. The methyl red test is variable.

## *Shigella*

The four species of *Shigella*—*S. dystenteriae* (Shiga bacillus), *S. flexneri, S. boydii,* and *S. sonnei*—cause bacillary dysentery. These organisms are found only in humans and certain other primates. The Shiga bacillus causes the severest form of the disease.

**Transmission.** Most *Shigella* infections are acquired by the fecal-oral route, by the ingestion of contaminated food or water.

**Morphology and Cultural Properties.** *Shigella* are straight, short, gram-negative rods, 0.5–0.7 μm in diameter and 2.0–3.0 μm in length (Figure 135). Cells occur singly or in pairs. These organisms are nonmotile and do not form capsules.

*Shigella* species are facultative anaerobes. Optimal growth temperature is 37°C. Most species produce smooth, colorless, circular colonies. *Shigella* species are oxidase negative, catalase positive, and they reduce nitrates. Methyl red, Voges-Proskauer, and citrate reactions are negative. Indole production is variable. $H_2S$ and urease are not produced. Glucose and certain other carbohydrates, with the exception of lactose, are catabolized, with the production of acid.

**Pathology and Clinical Features.** Once organisms gain access to and invade the cells of the small intestine, they multiply and bring about the destruction of the intestinal mucous lining, causing ulcer formation, bleeding, and diarrhea. The combination of diarrhea with blood and mucus constitutes dysentery. Infected persons may experience abdominal pain and severe cramping. Shigellosis is usually self-limiting in about 2 to 5 days.

**Diagnosis.** Stools or rectal swabs obtained during bowel examination are used for the inoculation of selective and differential media. Biochemical testing and serological tests are used for identification.

## *Streptobacillus*

***Streptobacillus moniliformis***

The only member of this genus is *Streptobacillus moniliformis*. It is found in the throat and nasopharynx of mice and of wild and some laboratory rats. *Streptobacillus moniliformis* causes one form of rat bite fever in humans.

## Facultatively Anaerobic Gram-Negative Rods *(Continued)*

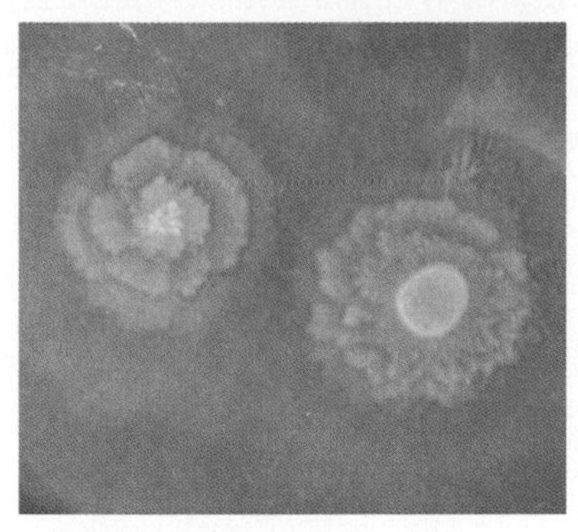

**Figure 129**
*Proteus vulgaris* exhibiting its characteristics swarming effect on trypticase soy agar.

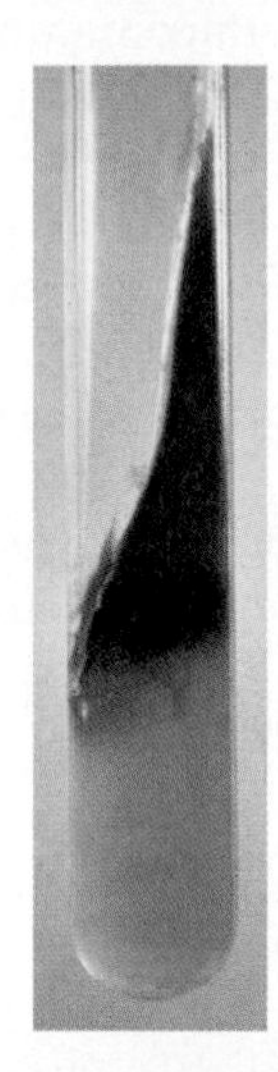

**Figure 130**
Hydrogen sulfide ($H_2S$) production by *Proteus vulgaris* in a triple sugar iron agar slant.

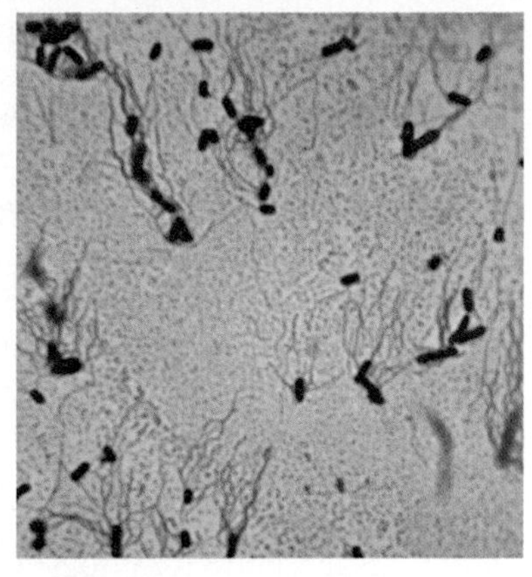

**Figure 131**
*Salmonella typhi* with peritrichous flagella.

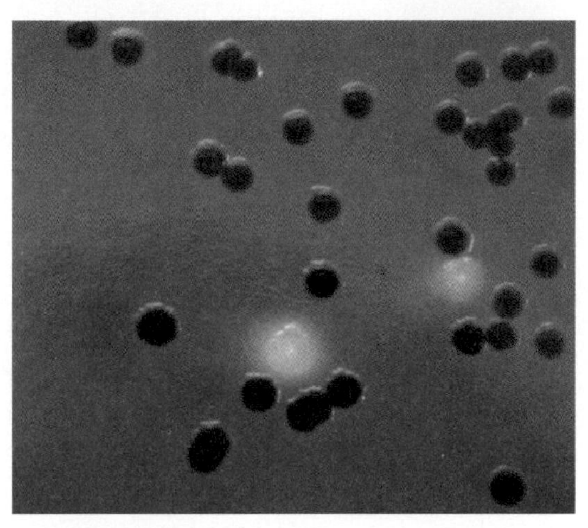

**Figure 132**
*Salmonella* species colonies (black) on XLT4 agar, showing hydrogen sulfide production. (Courtesy of Difco Laboratories Detroit, MI.)

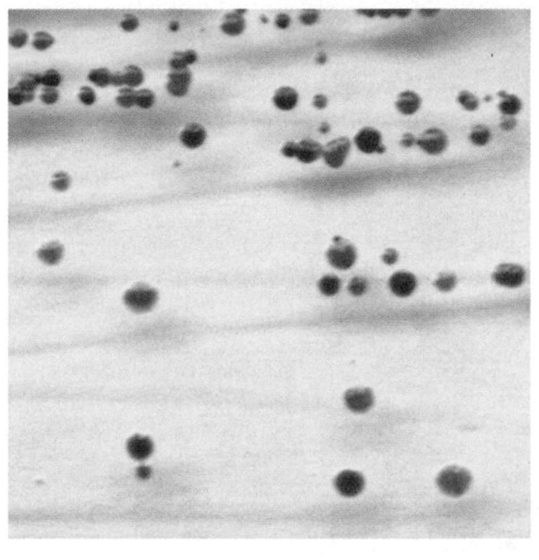

**Figure 133**
Red-pigmented colonies of *Serratia marcescens* on nutrient agar.

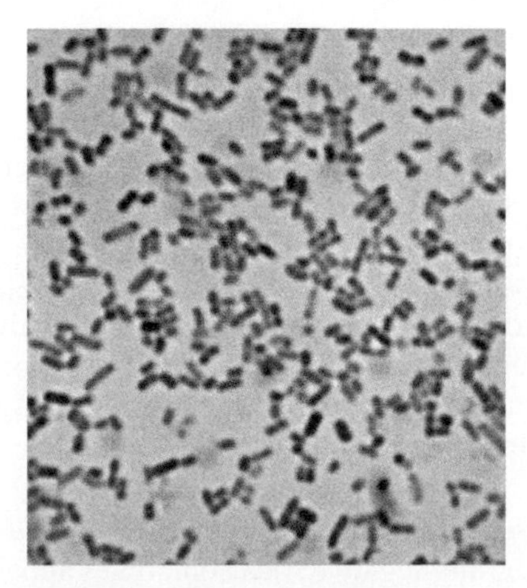

**Figure 134**
Gram-negative rods of *Serratia marcescens* (1,000×).

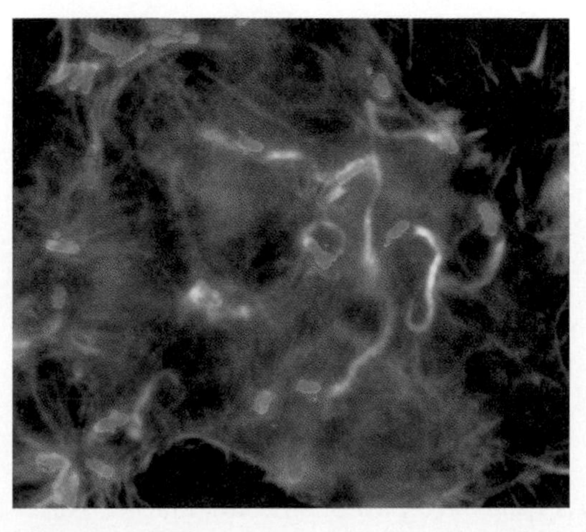

**Figure 135**
Tissue culture cells infected by *Shigella flexneri* shown by fluorescence labeling. (Courtesy of Dr. P.J. Sarsonetti, Unité de Pathogenie, Institut Pasteur.)

**Transmission.** Most human cases are acquired from bites of infected rats, mice, or cats. Infections also can result from the ingestion of milk contaminated with rat feces and by environmental exposure of persons working or living in rat-infested buildings.

**Morphology and Cultural Properties.** *Streptobacillus moniliformis* is a gram-negative to gram-variable rod, 0.1–0.7 µm in diameter and 1.0–5 µm in length. Cells occur singly or in long, wavy chains referred to as a **string of beads** 10–150 µm long (Figure 136). This organism is nonmotile.

*Streptobacillus moniliformis* is a facultative anaerobe and requires enriched media for best growth. The optimal temperature is 35° to 37°C. On blood agar, the organism forms small gray-white colonies. It is one of the few bacterial species that spontaneously converts to L forms (cells without cell walls). The organism is oxidase and catalase negative, and does not reduce nitrates. Glucose and other carbohydrates are fermented, with acid production only. All media for biochemical testing require blood, serum, or other enriching substances.

**Pathology and Clinical Features.** Illness occurs quickly after an incubation of less than 10 days. Usually, signs and symptoms include chills, fever, vomiting, and severe headache. A rash also may appear on the palms and soles. Arthritic involvement of the elbows, wrists, knees, and ankles is quite common.

**Diagnosis.** Blood, other body fluids, pus, and discharges from skin lesions are used for microscopic examination and culture. Biochemical tests and serological tests such as a tube agglutination are of value.

## *Vibrio*

**Morphology and Cultural Properties.** *Vibrio* species are straight or curved gram-negative rods, 0.5–0.8 µm in width and 1.4–2.6 µm in length (Figure 137). These organisms are motile by one or more polar flagella.

*Vibrio* species are facultative anaerobes. Optimal growth temperatures vary from 20° to 30°C . Glucose and most other carbohydrates are fermented, with the production of acid but not gas.

*Vibrio* species are commonly found in marine and estuarine environments and in the intestinal contents of marine animals. Some species are also found in freshwater environments.

Several species are pathogenic for marine vertebrates and invertebrates and for humans. The most notable of human pathogens are *V. cholerae,* the causative agent of cholera, *V. parahaemolyticus,* a major causative agent of food poisoning associated with contaminated fish or shellfish, and *V. vulnificus,* a cause of fatal septicemia.

### *Vibrio cholerae*

*Vibrio cholerae* is the cause of worldwide pandemics resulting in extremely dehydrating diarrheal disease.

**Transmission.** Cholera is acquired through the ingestion of contaminated food or water. Bathing, playing, and related activities in water contaminated by sewage are high-risk activities.

**Cultural Properties.** *Vibrio cholerae* grow on a variety of media. Optimal growth temperature is 35°C. Most species form 1- or 2-mm-wide colonies within 24 hours (Figure 138). *Vibrio cholerae* is oxidase positive.

**Pathology and Clinical Features.** *Vibrio cholerae* multiplies rapidly on the mucosal surface of the small intestine and produces its toxin, **choleragen.** The action of the toxin causes the release of fluids and electrolytes into the intestinal lumen. Within several hours to 3 days, infected persons experience a sudden onset of explosive, watery diarrhea with vomiting and abdominal pain. Up to 7 liters of liquid stool can be released by one person in a 24-hour period. Another effect can be the shredding of the intestinal lining, which appears as small white flecks resembling rice grains in stools. The term "rice-water stools" is used to describe the effect.

**Diagnosis.** Stool specimens are the only ones used for microscopic examination and culture. Serological tests are used for definitive identification.

## *Yersinia*

*Yersinia* species are primarily animal pathogens. The genus contains several important human pathogens including *Y. enterocolitica,* the cause of **enterocolitis** and other diseases, and *Y. pestis,* the cause of plague. Another species, *Y. pseudotuberculosis,* causes pseudotuberculosis, a disease resulting in inflammation and damage to lymph nodes, the spleen, and the liver. The clinical features of fever and abdominal pain often mimic acute appendicitis. In most cases, wild animals are possible sources of infection.

**Morphology and Cultural Features.** *Yersinia* species are small, gram-negative rods, sometimes appearing as cocco-bacilli. Cells measure 0.5–0.8 µm in width, and 1.0–3.0 µm in length. Retaining stain at the ends of cells produces a phenomenon called **bipolar staining,**

a common finding (see Figure 142a). Most species are nonmotile when grown at temperatures above 30ºC.

*Yersinia* species are facultative anaerobes and have an optimal growth temperature range from 28º to 30ºC. Organisms are oxidase negative and catalase and urease positive. They reduce nitrates and produce negative Voges-Proskauer and citrate tests when grown at 37ºC. Glucose and other carbohydrates are fermented, with the production of acid only. Hydrogen sulfide is not produced.

### *Yersinia enterocolitica*

*Yersinia enterocolitica* (Figure 139) along with *Y. pseudotuberculosis* is associated with a condition known as yersiniosis, a clinical condition that mimics acute appendicitis. This organism causes a wider variety of infections than other members of the genus.

**Transmission.** Infection is usually acquired by the ingestion of contaminated food or water.

**Pathology and Clinical Features.** The most common infection caused by *Y. entercolitica* is an enterocolitis, usually in children. Typical signs and symptoms include fever, abdominal pain, and diarrhea. This infection generally is self-limiting.

**Diagnosis.** Stool specimens can be used for diagnosis, but attempts to isolate the organisms have limited success.

### *Yersinia pestis*

*Yersinia pestis* (Figure 142) causes plague in both lower animals such as rodents (Figure 141a) and small mammals and humans.

**Transmission.** Plague has two major cycles, known as **urban** and **sylvatic**, and two major clinical forms, **bubonic** and **pneumonic**. The disease is transmitted to humans by the bite of infected fleas (Figure 141).

Pathology and Clinical Features. The incubation period for bubonic plague ranges from 2 to 7 days after the flea bite. The appearance of fever and the painful bubo (Figure 140), usually in the groin and less often in the armpit, signal the onset of the disease. If the person is not treated, the organisms gain access to the bloodstream and cause a fatal septic shock. Pneumonic plague develops in about 5% of victims. Fever, general loss of energy, and a tight feeling in the chest are early symptoms. The production of sputum, difficulty in breathing, cyanosis, and death occur later, as early as the second or third day of illness.

**Diagnosis.** Gram stains of material taken from a bubo or other specimens such as blood, sputum, and spinal fluid usually are used for identification (Figure 142). Similar specimens are used for culture.

## Anaerobic, Gram-Negative, Straight, Curved, and Helical Bacteria

### *Porphyromonas*

This genus *Porphyromonas* was newly created in 1988. It includes three former *Bacteroides* species: *Porphyromonas asaccharolytica, P. gingivalis,* and *P. endodontalis.*

**Morphology and Cultural Properties.** Members of this genus are short, gram-negative rods, 0.5–0.8 µm in diameter and 1.0–3.0 µm in length. Organisms are nonmotile and strict anaerobes. *Porphyromonas* species produce pigmented colonies ranging from buff to tan to black after 2 to 21 days incubation (Figure 143).

These organisms do not break down carbohydrates (asaccharolytic), a property that separates them from *Prevotella* species.

Certain *Porphyromonas* species are associated with gum disease and have been isolated from infected dental root canals.

### *Prevotella*

The genus *Prevotella* was created in 1990. Fifteen former *Bacteriodes* species were transferred to *Prevotella,* including several human pathogens, such as *P. denticola* (Figure 144), *P. intermedia* (Figure 145), and *P. melaninogenica* (Figure 146).

**Morphology and Cultural Properties.** *Prevotella* species are gram-negative, rods measuring 0.5–0.8 µm in diameter and 1.0–3.0 µm in length. These organisms are nonmotile and non-spore-forming. They tend to be pleomorphic.

*Prevotella denticola* is catalase and indole negative. It hydrolyzes esculin and ferments glucose, lactose, and maltose. *Prevotella intermedia* differs from *P. denticola* by being positive for indole and fermenting glucose and lactose only. *Prevotella melaninogenica* produces reactions identical to those of *P. denticola* with the exception of esculin hydrolysis. None of these species produces cellobiose.

**Pathology.** *Prevotella denticola* is commonly found in the human oral cavity, but its clinical significance is not known. *Prevotella intermedia* also has been isolated from the human oral cavity but also has been found in specimens from head and neck infections and from pleural infectious conditions.

## Facultatively Anaerobic Gram-Negative Rods *(Continued)*

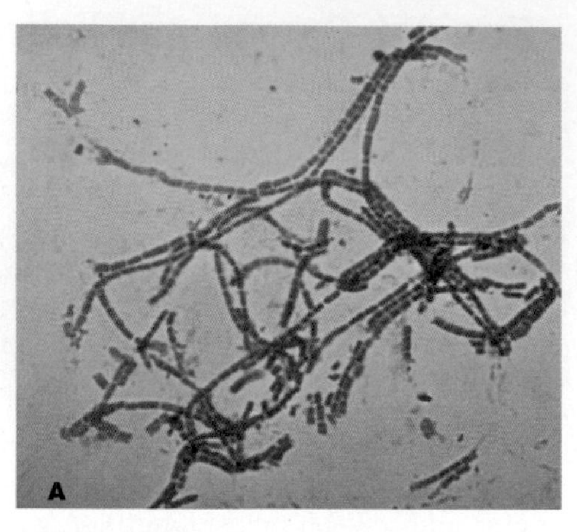


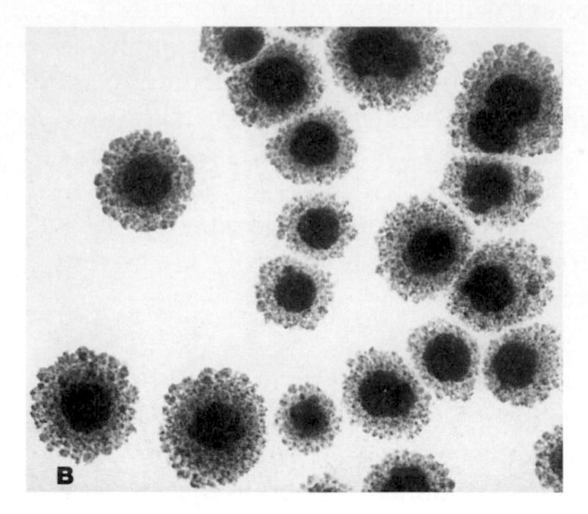


**Figure 136**
The long, wavy chains of the gram-negative *Streptobacillus moniliformis* (1,000×). (b) Colonies on Dienes agar (magnification 200×). (From M.R. Pins, J.M. Holden, J.M. Yang, S. Madoff, and M.J. Ferraro. *CID* 22(1996):471–76.)

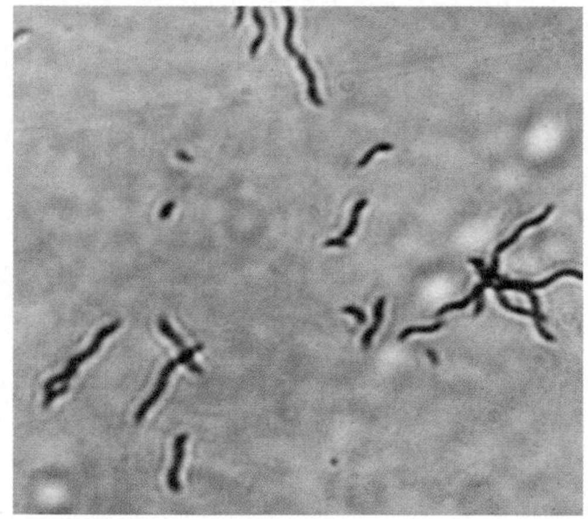

**Figure 137**
Gram-negative stain of *Vibrio cholerae* (1,000×).

**Figure 138**
*Vibrio cholerae* colonies grown on meat extract agar and incubated overnight at 37°C. (From R.A. Finkelstein, M. Boesman-Finkelstein, Y. Chang, and C.C. Hase. *In. Immunol.* 60(1992):472–78.)

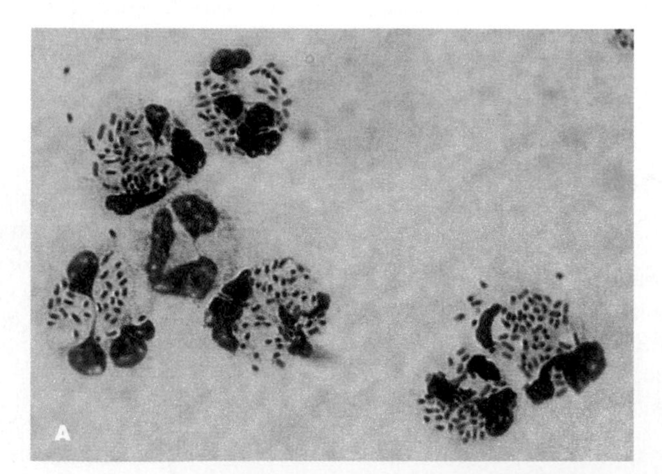


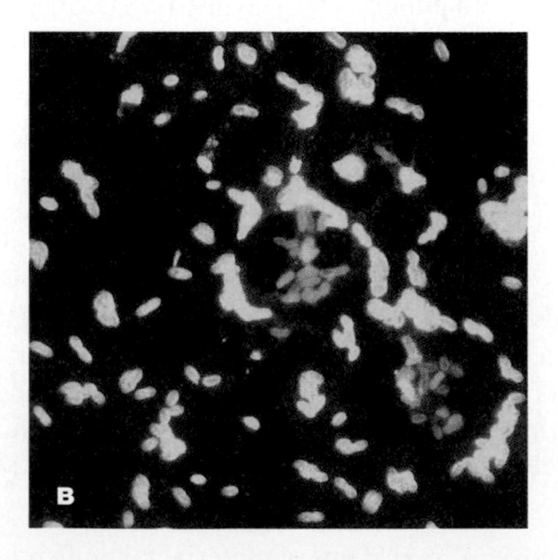


**Figure 139**
Two microscopic views of *Yersinia enterocolitica.* (a) Giemsa stain showing *Y. enterocolitica* ingested by polymorpholeukocytes. (b) Similar view of phagocytosis shown by double immunofluorescence. Intracellular pathogens are green, and extracellular forms are yellow. (From J.H. Ewald, J. Hessemann, H. Riidiger, and I.B. Autenrieth, *J. Inf. Dis.* 170(1994):140–50.)

## Facultatively Anaerobic Gram-Negative Rods *(Continued)*

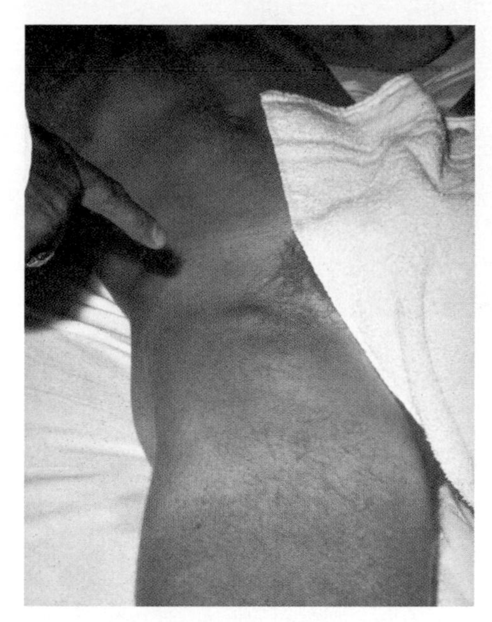

**Figure 140**
The appearance of a bubo, the first sign of the plague in humans.

**Figure 141**
Important players in the transmission of plague. (a) The rat, the source of *Yersinia pestis* among wild rodents. (b) The flea, the arthropod that transmits the disease agents among lower animals and to humans.

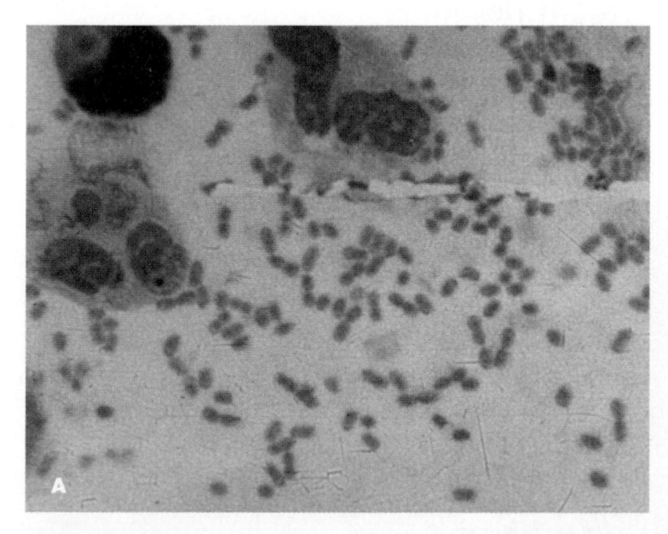

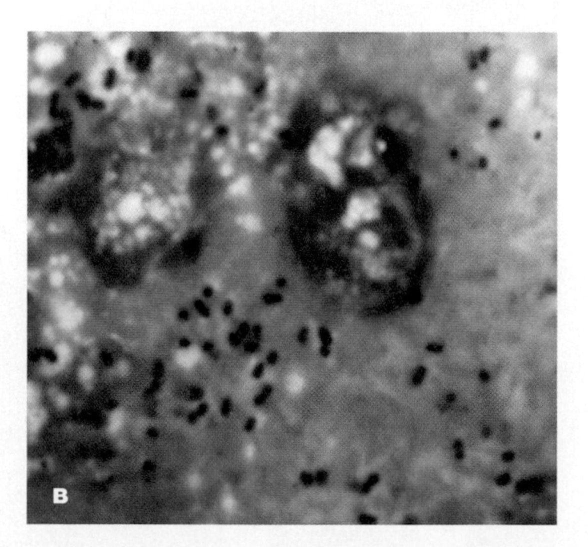

**Figure 142**
Two microscopic views of *Yersinia pestis.* (a) A Gram stain of spinal fluid from a patient with plague meningitis showing bipolar and pleomorphic gram-negative cells. (b) A Wayson stain of bubo material taken from a patient with bubonic plaque. (Courtesy of T. Butler, Division of Infectious Diseases, Texas Tech University Health Sciences Center, Houston, Texas.).

## Anaerobic, Gram-Negative, Straight, Curved, and Helical Bacteria

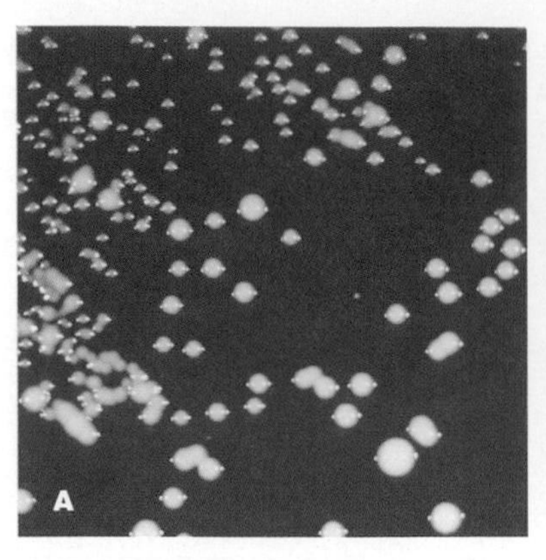


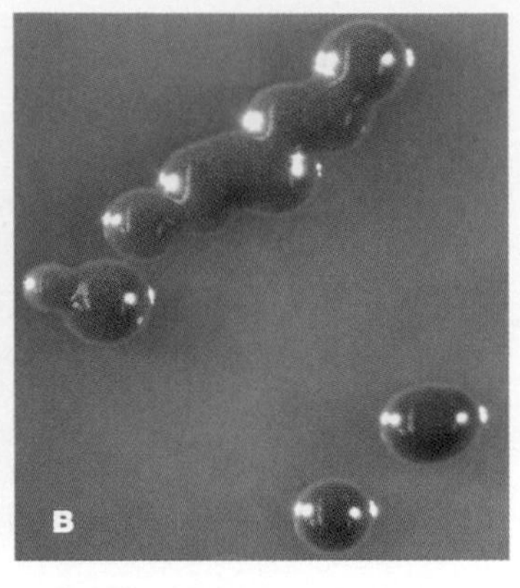


**Figure 143**
*Porphyromonas endodontalis.* (a) Colonies on *Brucella* agar. (b) Highly pigmented colonies on *Brucella* agar. (Courtesy of Dr. H.F. Somer, Anaerobe Laboratory, National Public Health Institute, Helsinki, Finland.)

**Figure 144**
*Prevotella denticola* colonies on a laked blood medium. (Courtesy of Dr. H.F. Somer, Anaerobe Laboratory, National Public Health Institute, Helsinki, Finland.)

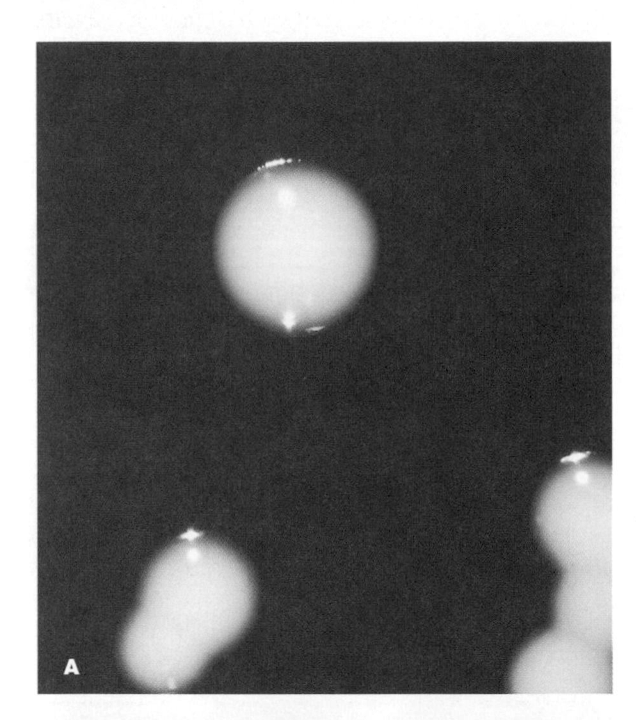


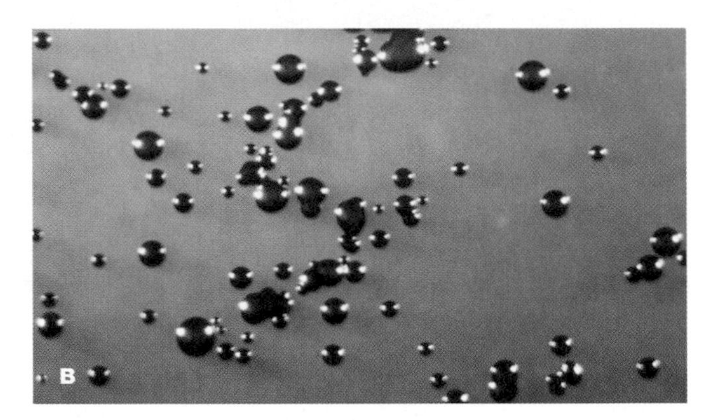


**Figure 145**
*Prevotella. intermedia* on two different agar media. (a) Colonies on *Brucella* agar. (b) Colonies on a laked blood medium. (Courtesy of Dr. H.F. Somer, Anaerobe Laboratory, National Public Health Institute, Helsinki, Finland.)

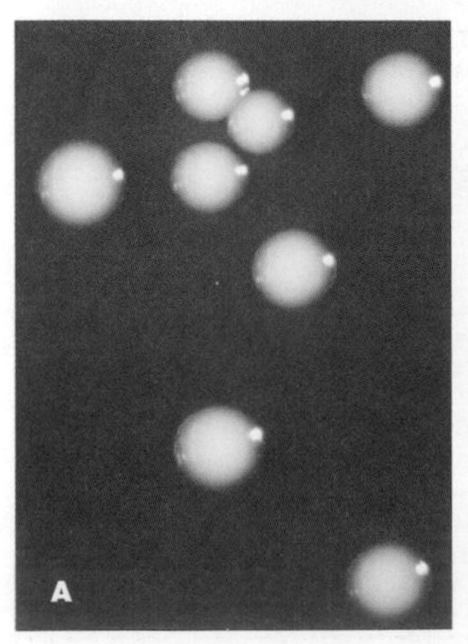


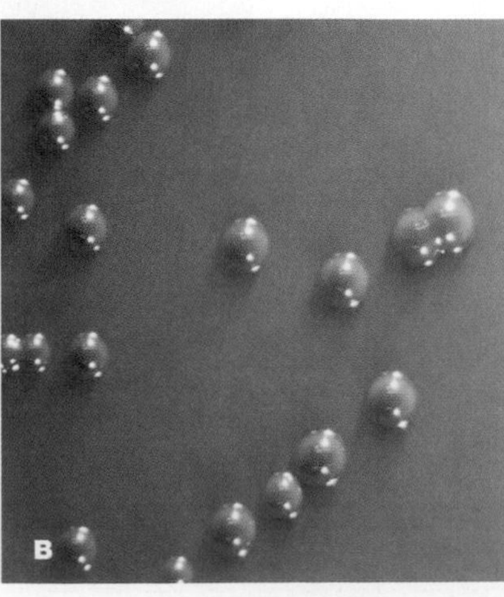


**Figure 146**
*Prevotella melaninogenica* on two different media. (a) Colonies on *Brucella* agar. (b) Colonies on a laked blood medium. (Courtesy of Dr. H.F. Somer, Anaerobe Laboratory, National Public Health Institute, Helsinki, Finland.)

## ▲ The Rickettsias and Chlamydias

### *Rickettsias*

Rickettsias are a diverse collection of obligate intracellular bacteria that grow and reproduce within eukaryotic host cells. They are the causative agents of a variety of zoonotic diseases called rickettsioses and include the genera *Coxiella, Ehrlichia, Orienta,* and *Rickettsia* (Figure 147).

Historically, members of the genus *Rickettsia* have been divided into the spotted fever group (SFG), the typhus group, and the scrub typhus group. Examples of species included in each of these groups are listed in Table 11. The associated diseases and the means of transmission also are indicated.

**Transmission.** *Rickettsia* and *Orienta* species are spread by the bite of infected ticks (Figure 148b) or mites or by the feces of an infected louse (Figure 149b) or flea.

**Morphology and Cultural Properties.** The rickettsiae are mainly gram-negative, often pleomorphic, small rods. Cells measure 0.25 μm in width and 0.5–1.25 μm in length (Figures 148a and 149b). Most rickettsiae can be isolated or cultured in embryonated eggs and cell cultures.

**Pathology and Clinical Features.** The severity of the different rickettsioses varies considerably. For example, rickettsialpox is a relatively mild disease that has never proved fatal (Figure 147), whereas Rocky Mountain spotted fever and epidemic typhus fever are serious, life-threatening illnesses with high fatality rates in untreated individuals.

**Diagnosis.** The rickettsioses are difficult to diagnose both clinically and in the laboratory. Laboratory diagnostic techniques include animal or cell culture inoculation with specimens and serological methods such as fluorescence-antibody tests (Figures 148a, 149a and 150), enzyme immunoassay, or the older Weil-Felix test. The last-named technique is used to show the presence of antibodies that agglutinate specific antigens of the bacterium *Proteus mirabilis*.

### *Coxiella*

*Coxiella burnetii* is the causative agent of Q fever. The disease is a zoonosis that is widespread in a variety of animals including cats, cattle, goats, and sheep. Organisms are found in their infected mammary glands, milk, urine, feces, and placentas. *Coxiella burnetii* is considered to be one of the most infectious of bacteria and among the most stable in the environment.

**Transmission.** Infection in humans occurs through inhaling *C. burnetii* in contaminated aerosols. Outbreaks of Q fever are also associated with domestic animals, especially infected newborns.

**Morphology and Cultural Properties.** *Coxiella burnettii* is an obligate intracellular, gram-variable coccobacillus. It is quite pleomorphic and measures 0.25 μm in width and 0.5–1.25 μm in length (Figure 150).

*Coxiella burnetii* can be cultured under laboratory conditions in embryonated chicken eggs or cell cultures.

**Pathology and Clinical Features.** After infection with *C. burnetii*, a variety of clinical syndromes may result. Some patients are entirely asymptomatic, but most exhibit a mild, flu-like illness. The major obvious clinical features of Q fever can be divided into two groups: acute and chronic. Acute Q fever is characterized by fever, general discomfort, headache, pneumonia, and hepatitis. The relative nonspecific nature of these signs and symptoms leads to difficulties in diagnosis.

**Table 11** Human Rickettsioses

| Group | Pathogen | Disease | Vector |
|---|---|---|---|
| Spotted fever | *Rickettsia rickettsii* (Figure 148a) | Rocky Mountain spotted fever | *Dermacentor* species (wood ticks; Figure 160b) |
| | *R. akari* | Rickettsialpox | *Allodermanyssus sanguineus* (mite) |
| Typhus | *R. prowazekii* (Figure 149a) | Epidemic typhus fever | *Pediculus humanus corporis* (the human body louse, Figure 161b) louse feces or bite |
| | *R. typhi* | Murine typhus fever | *Xenopsylla cheopis* (rat flea; Figure 141b) flea feces or bite |
| Scrub typhus | *Orienta (Rickettsia) tsutsugamuchi* | Scrub typhus fever | *Leptotrombidium akamushi* (mite) |
| Q fever | *Coxiella burnetii* | Q-fever | Inhalation of pathogen |

The development of chronic Q fever complicates the infectious process, which ultimately leads to valvular endocarditis (the most common result), hepatitis, and vertebral osteomyelitis.

**Diagnosis.** Although *C. burnetii* can be isolated by inoculation of embryonated eggs or by cell culture, diagnosis is generally made by serological methods demonstrating antibodies to specific proteins. Direct light or electron microscopic demonstrations of *C. burnetii* in cardiac valves by means of immunofluorescence or immunohistochemistry also are used.

## *Ehrlichia*

*Ehrlichia chaffenis* (Figure 151) and *E. sennetsu* are two species known to cause human ehrlichiosis. Other species cause several important veterinary diseases. Blood-forming cells are the major sites of invasion by ehrlichiae.

**Transmission.** Ehrlichiosis is transmitted by ticks.

**Morphology and Cultural Properties.** The genus *Ehrlichia* consists of small, obligate intracellular bacteria that average 0.5 μm–1.5 μm in length. The ultrastructural morphology is highly variable and ranges from cocci or cocco-bacilli to unusual shapes and forms. Ultrastructural studies have revealed a gram-ngeative-type cell wall. Specialized laboratories can be used for cell cultures, isolations, and cultivations.

*Ehrlichia* form **morulae**, large solid masses of cells. When stained, they appear as round or ovoid purple bodies measuring 2–5 μm in diameter and surrounded by a single membrane (Figure 151a). *Ehrlichia* also can cause cellular damage (cytopathic) that appears as clear areas (plaques) in cell cultures (Figure 151b).

**Pathology and Clinical Features.** Human ehrlichiosis ranges from asymptomatic illness or mild fever with spontaneous recovery to a severe disease with hepatitis and kidney failure. The disease has an incubation of about 2 weeks in adults. Its onset is usually abrupt, and individuals experience a fever, chills, muscle pain, and headache.

**Diagnosis.** Microscopic examination of Giemsa-stained blood smears may be of value in some cases. However, fluorescent-antibody tests using *E. chaffenis* as an antigen is most commonly the method of choice. An increase in antibody levels and the polymerase chain reaction also are of value.

## *Bartonella*

*Bartonella* (formerly *Rochalimaea*) species are now associated with several clinical syndromes including trench fever, bacillary angiomatosis (Figure 152), bacilliary peliosis hepatitis (Figure 153a), and cat scratch fever disease. These disease states are mainly found in but are not limited to adults infected with human immunodeficiency virus (HIV) and transplant patients.

**Morphology and Cultural Properties.** *Bartonella* species are nonmotile, gram-negative, slightly curved rods, 0.5–0.6 μm wide and 1.0–2.0 μm long (Figure 153b). Because these organisms can be cultivated in or on cell-free enriched media, they do not fit the obligate intracellular property of rickettsias (Figures 148a and 149b).

*Bartonella* species are catalase, oxidase, and urease negative. They also are nonreactive in carbohydrate test media.

**Diagnosis.** *Bartonella* can be detected with several techniques. These include staining tissue biopsy specimens, culturing on media such as chocolate agar (Figures 154 and 155) or trypticase soy, or brain-heart infusion agar supplemented with 5% sheep blood and inoculated with 5$CO_2$, and commercially available serological tests.

### *Bartonella henselae*

*B. henselae* is known to cause persistent bacteremia, bacillary angiomatosis, and cat scratch fever disease. Contact with cats and lowered resistance, as in the case of HIV infection, are known risk factors.

**Pathology and Clinical Features.** Bacillary angiomatosis (BA) can present in several ways. The commonest form is an enlarged, red elevated skin area that sometimes resembles a cranberry. This lesion is generally surrounded by numerous additional growths that can penetrate deeper into the skin. Bacillary angiomatosis has been reported to occur in every body organ system.

Cat scratch fever disease usually occurs as a self-limited infection with enlarged and inflamed lymph nodes in an area draining from a site of a cat scratch or bite. Some nodes may ulcerate.

### *Bartonella quintana*

*Bartonella quintana* was first identified as an important pathogen during World War I and as the cause of trench fever. This disease is transmitted by the body louse (see Figure 149b). *Bartonella quintana* also has been identified as a cause of bacillary angiomatosis, endocarditis, and bacteremia.

**Pathology and Clinical Features.** The incubation period of *B. quintana* infections varies from 5 to 20 days. Clinical features vary but generally include nonspecific signs and symptoms including fever, malaise, headache, bone pain, and a temporary macular rash.

## The Rickettsias and Chlamydias

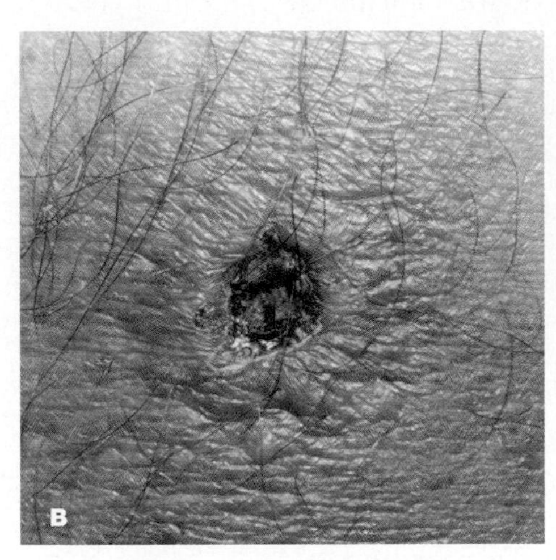

**Figure 147**
Clinical features of rickettsial infections. (a) A characteristic, painless lesion that can be found at the site of a tick bite. (b) The red, slightly elevated rash that commonly occurs. (From C.A. Kemper, A.P. Spivak, and S.C. Deresinki *CID* 15(1992):591–94.)

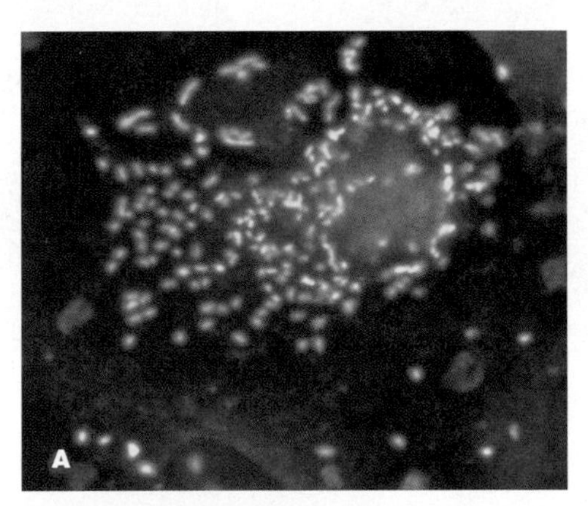

**Figure 148**
*Rickettsia rickettsii.* (a) The organism as shown by fluorescent-antibody staining. (Courtesy of Integrated Diagnostics, Inc.) (b) *Dermacentor andersoni*, the wood tick vector for Rocky Mountain spotted fever.

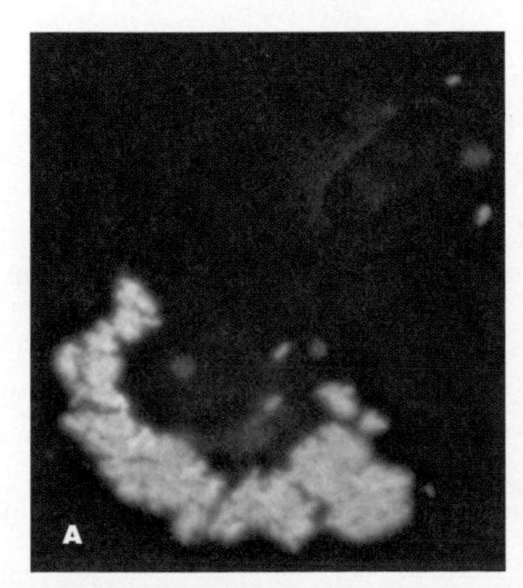


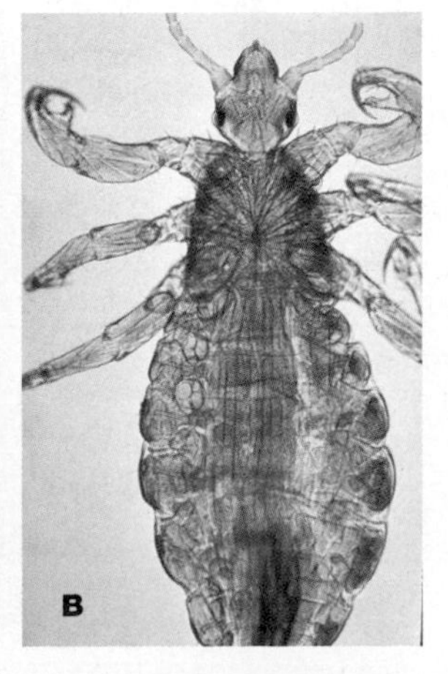


**Figure 149**
*Rickettsia prowazekii.* (a) The causative agent as shown by fluorescent-antibody staining. (Courtesy of Integrated Diagnostics, Inc.) (b) *Pediculus humanus corporis*, the body louse.

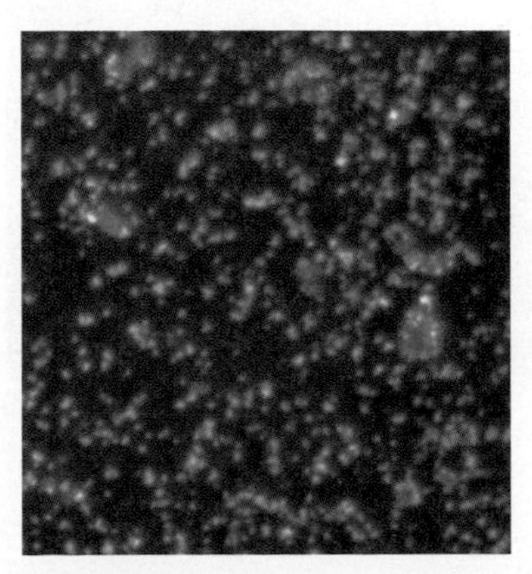

**Figure 150**
Fluorescent-antibody showing the presence of *Coxiella burnetii* (small, yellow-green cells). (Courtesy of Integrated Diagnostic, Inc.)

## The Rickettsias and Chlamydias *(Continued)*

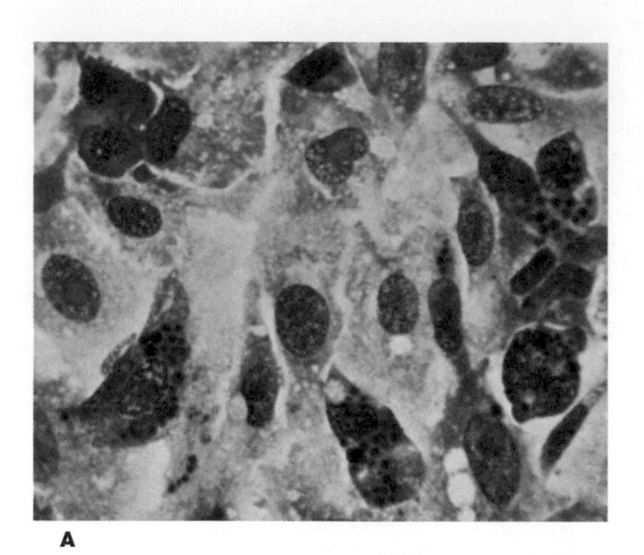

A

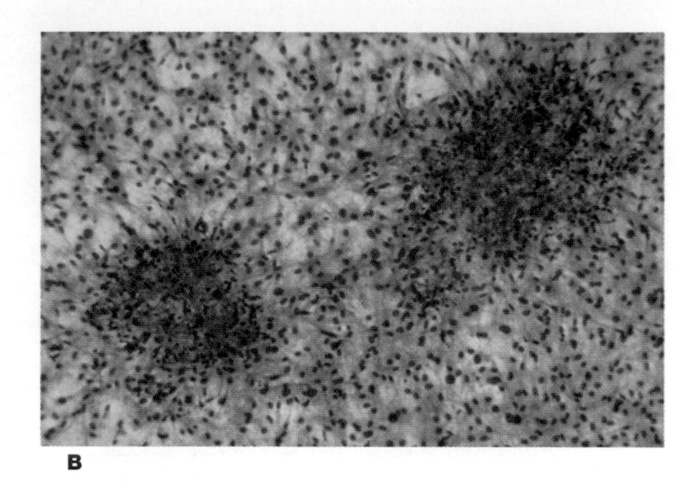

B

**Figure 151**
*Ehrlichia chaffenis.* (a) Infected cells with a large number of intracellular morulae (425×). (b) Overlapping infected tissue culture cells showing concentrated cytopathic effects (70×). (c) Neutral red staining of areas showing cellular destruction (plaques) in a tissue culture. The control tissue culture system shows no clear areas or evidence of infection. (2.2×). (From S.-M.Chen, V.L. Popov, H.-M. Feng, J. Wen, and D.H. Walker. *Inf. Immunol.* 63(1995):647-55)

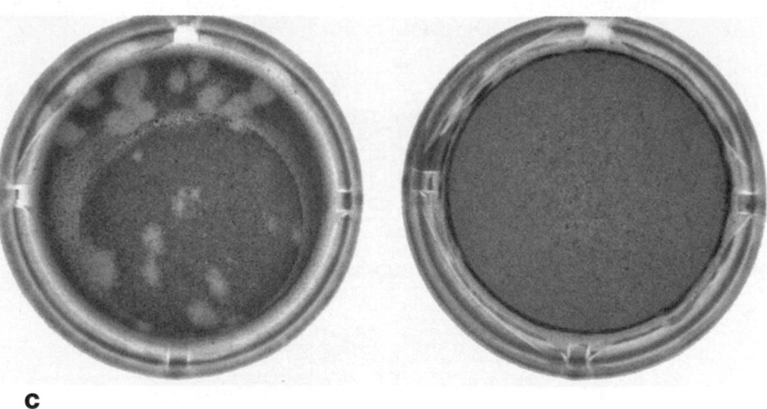

C

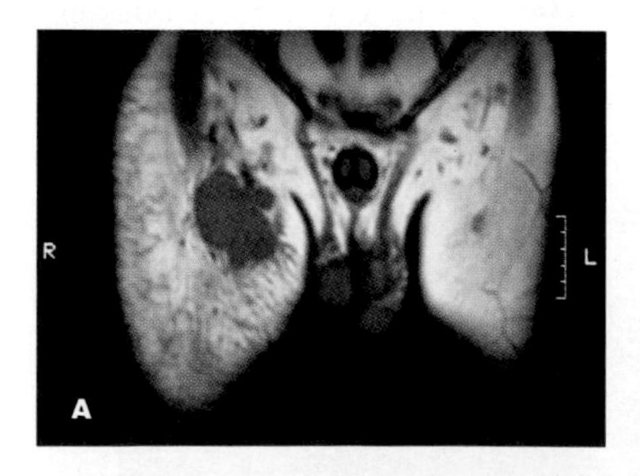


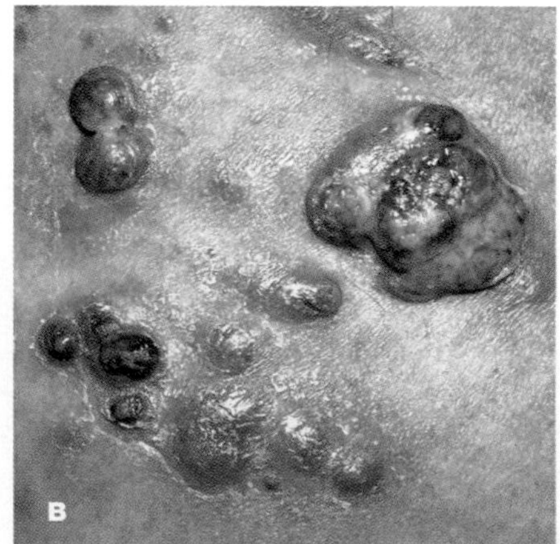


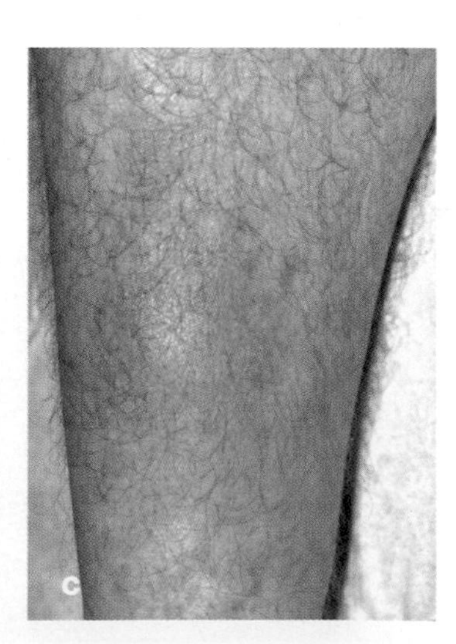

**Figure 152**
Clinical features of bacillary angiomatosis. (a) A magnetic resonance image showing a soft-tissue mass in front of femoral blood vessels and taken 8 months before diagnosis. (b) Multiple tender vascular, oozing skin lesions. (c) An erythematous, tender, raised lesion, called a plaque on the leg. (From J.E. Koeheler *et al.*, *NEJM* 327(1992):1625-31.)

## *Chlamydia*

The chlamydiae are a small group of coccoid cells that are obligate intracellular parasites of eukaryotic cells. Members of the genus cause several significant human diseases including trachoma, the leading cause of infectious blindness worldwide (Figure 156a), several common sexually transmitted diseases (Figure 156b), and acute respiratory infections. Arthopods neither transmit nor serve as hosts.

**Morphology and Cultural Properties.** Chlamydiae are gram-negative and are grown in cell cultures (Figure 158a, 158b). These organisms have a unique biphasic life cycle that alternates between an infectious, rigid, metabolically inactive form, the **elementary body,** and noninfectious, fragile, metabolically active form, the **reticulate body** (Figure 157).

**Diagnosis.** The clinical properties of chlamydial infections are often diagnostic. Laboratory confirmation may include immunofluorescence tests (Figure 158c and 161), nucleic acid probes (Figure 159), and certain serological tests.

### *Chlamydia trachomatis*

*Chlamydia trachomatis* causes trachoma (Figure 156a), inclusion conjunctivitis, and genital infections including inflammation of the urethra (nongonococcal urethritis) in men and acute inflammation of the uterine tubes (salpingitis) and inflammation of the cervix (cervicitis) in women. Specific strains cause lymphogranuloma venereum (LGV), a sexually transmitted disease (Figures 156b and 158b).

**Transmission.** *Chlamydia trachomatis* eye infections are spread by flies, contaminated fingers, and inanimate objects (fomites).

**Pathology and Clinical Features.** The clinical properties of trachoma include the development of follicles and an inflamed inner eyelid lining (conjunctiva) as shown in Figure 151a. The cornea may also become cloudy. Inclusion conjuctivitis is a much milder infection with a pussy discharge.

Lymphogranuloma venereum is a chronic, invasive, sexually transmitted disease that causes swelling and ulceration of regional lymph nodes (buboes).

### *Chlamydia pneumoniae*

*Chlamydia pneumoniae* is an important cause of human respiratory tract infections including sore throat, bronchitis, and pneumonia. Recently, it has been implicated in coronary artery disease.

**Transmission.** *Chlamydia pneumoniae* is spread from human to human by aerosols.

**Pathology and Clinical Properties.** *Chlamydia pneumoniae* typically causes hoarseness, prolonged coughing, and bronchial asthma, all of which damage the epithelial cells lining the respiratory system.

### *Chlamydia psittaci*

*Chlamydia psittaci* is widely distributed in nature, usually causing respiratory infections in many mammals and birds (Figure 155). The organism also has been associated with various forms of heart disease (Figure 161).

**Transmission.** Psittacine (tropical) birds are considered to be the major reservoirs, but human cases have been associated with infected pigeons, sparrows, ducks, cockatiels, and many other birds. Close or prolonged contact with infected birds is not necessary. Infections can be acquired from contact with a contaminated environment or materials such as a cage or bird droppings. Aerosols are the principal means of transmission.

**Pathology and Clinical Features.** An influenza-like illness known as psittacosis is the usual clinical condition caused by *C. psittaci.*

# Mycoplasmas

## *Mycoplasma*

Mycoplasmas are found worldwide. A significantly large number of distinct species have been described. They can be found in humans (Figure 162a), in other animals including arthropods, and in plants. Some species, such as those found in the genera *Mycoplasma* and *Ureaplasma,* are pathogenic for humans, and those in the genus *Spiroplasma* commonly colonize and infect arthropods (Figure 162b).

**Morphology and Cultural Properties.** Mycoplasmas are the smallest known free-living microorganisms, intermediate in size between bacteria and viruses. These organisms lack a cell wall (Figure 163), which is their single most distinguishing feature. Many of the biological properties of mycoplasmas are due to this feature.

The various mycoplasmal species are very different organisms and have unique metabolic properties and cultivation requirements. No single medium will adequately support the growth of all organisms. Many, such as *M. pneumoniae,* require cholesterol, and ureaplasmas require urea for growth.

Several mycoplasmas such as *M. fermentans* and *M. pneumoniae* exhibit unusual colonies that look like fried eggs (Figure 162).

### *Mycoplasma pneumoniae*

*Mycoplasma pneumoniae* causes disease of the upper and lower respiratory tract. The infection is called **primary atypical pneumonia** to distinguish it from other types of pneumonia.

## The Rickettsias and Chlamydias *(Continued)*

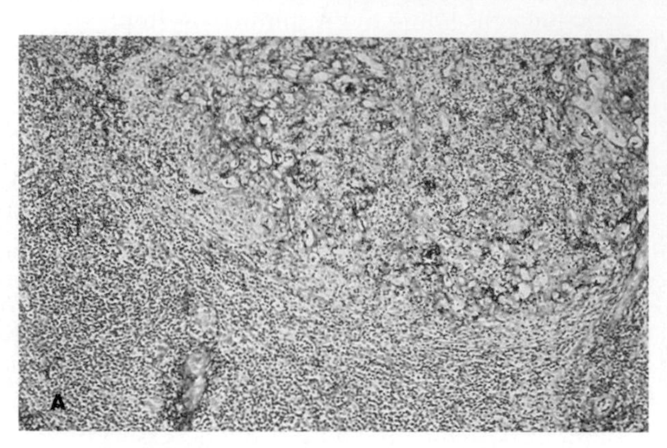


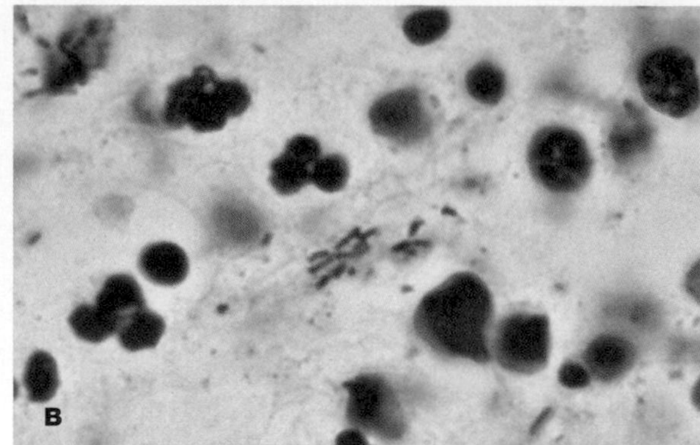


**Figure 153**
Microscopic features of *Bartonella henselae.* (a) Areas of tissue destruction in the liver of patients. (b) A higher magnification showing pleomorphic rods. (3,300×). (From D.A. Relman et al. *NEJM* 323(1990):1573–80.)

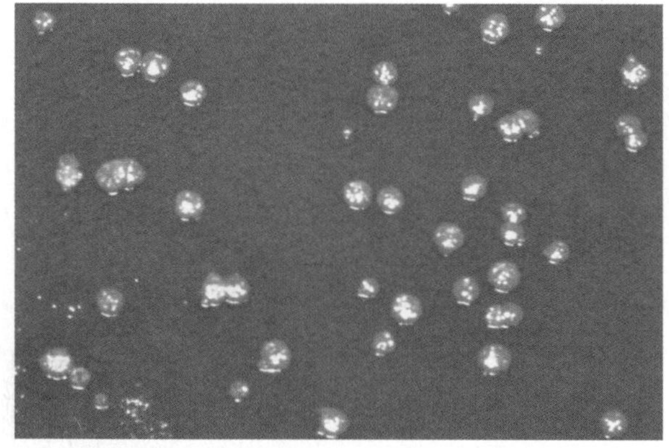

**Figure 154**
Primary culture of *Bartonella henselae* colonies on chocolate agar. (Courtesy of Drs. J.E. Koehler and J.W. Tappero. *NEJM* 337(1997):1876–1883.)

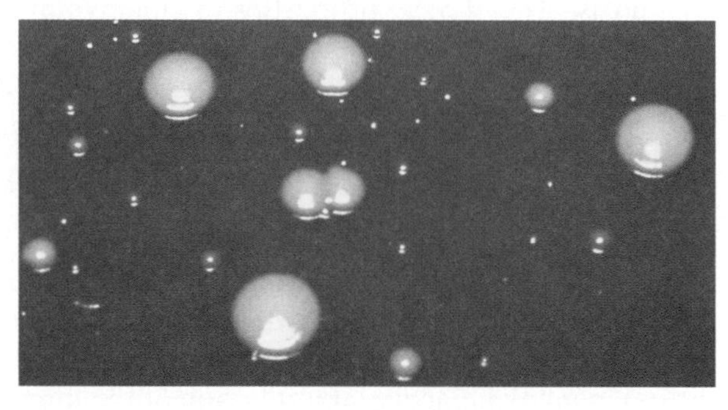

**Figure 155**
Primary culture of *Bartonella quintana* colonies on chocolate agar. (Courtesy of Drs. J.E. Koehler and J.W. Tappero. *NEJM* 337(1997):1876–1883.)

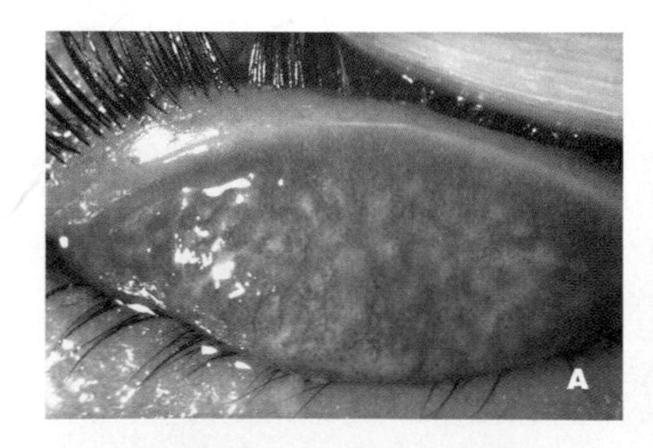


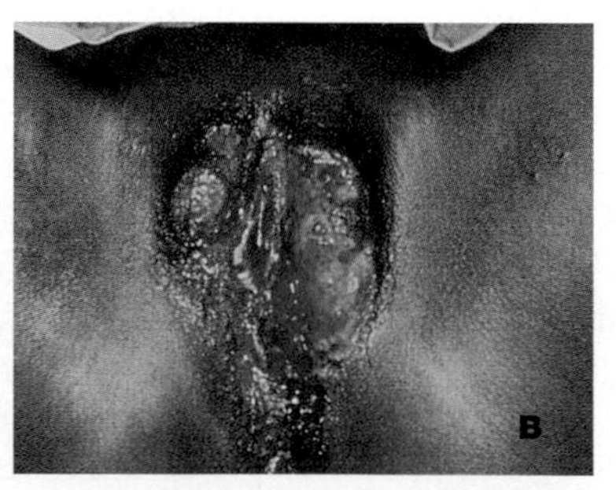


**Figure 156**
Clinical states caused by *Chlamydia trachomatis*. (a) Trachoma. The pronounced inflamed inner eyelid with many round, pale swelling (follicles) of the eye disease are shown. (Courtesy of Director-General and Programme Manager, Prevention of Blindness, World Health Organization.) (b) A clinical view of a lesion found on the genitalia of a female patient with lymphogranuloma venereum. This is one type of sexually transmitted disease caused by *C. trachomatis.*

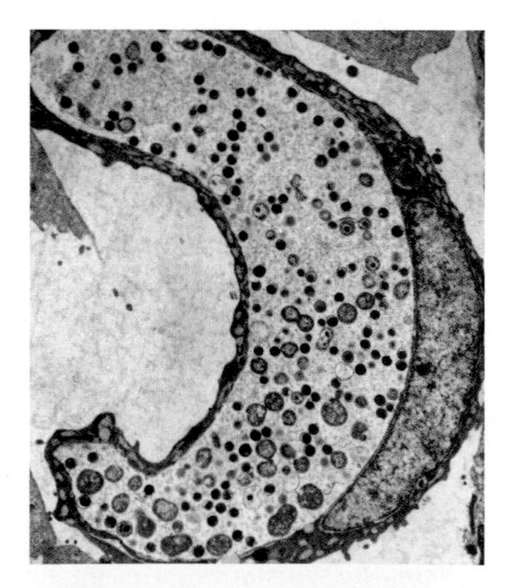

**Figure 157**
A transmission electron micrograph showing a membrane-bound vacuole in a cell infected with *Chlamydia trachomatis.* The vacuole contains both the larger reticulate bodies and the smaller, dark elementary bodies (4,664×). (From P.B. Wyrick and S.J. Richmond. *JAVMA* 195(1989):1509.)

## The Rickettsias and Chlamydias *(Continued)*

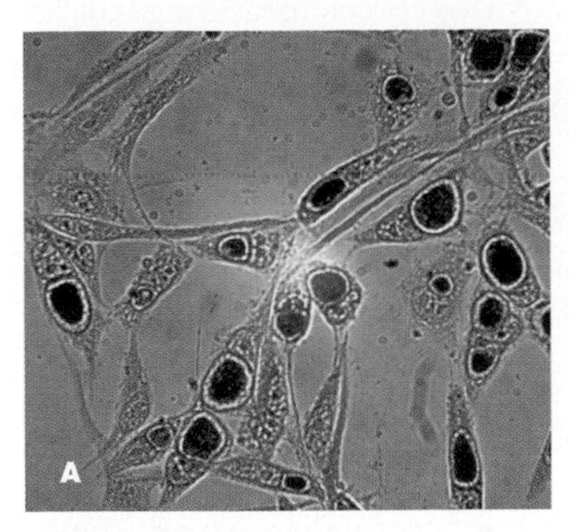


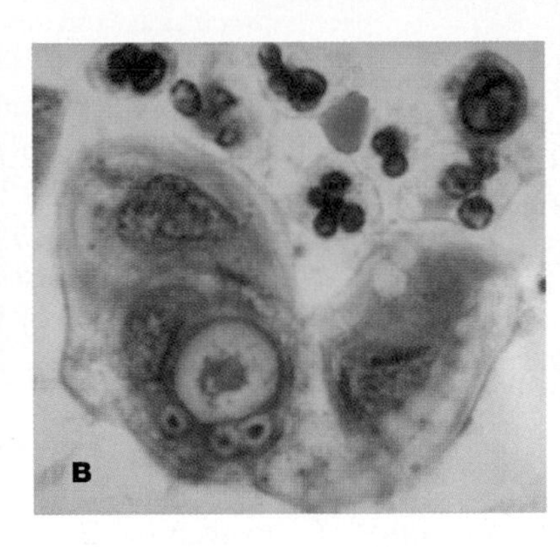


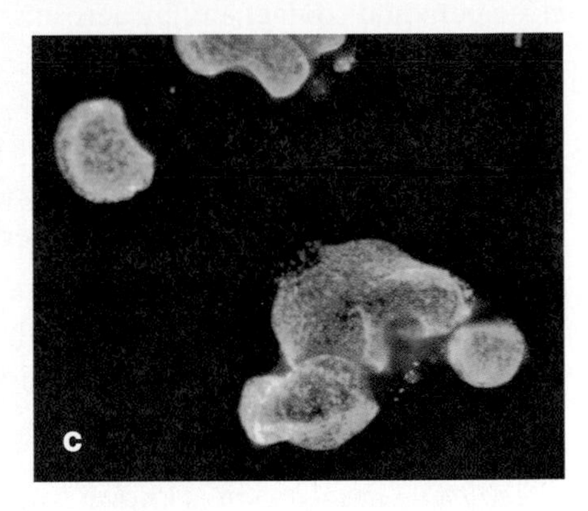


**Figure 158**
Microscopic features of *Chlamidia trachomatis*. (a) The organism in McCoy cells. (From R.L. Hodinka and P.B. Wyrick. *Inf. Immunol.* 56(1998):1456–63,.) (b) Intracytoplasmic inclusions from a specimen taken from a patient with lymphogranuloma venereum (1,200×). (c) *C. trachomatis* inclusions shown by fluorescent-antibody. (Courtesy of S.T. Knight and P.B. Wyrick.)

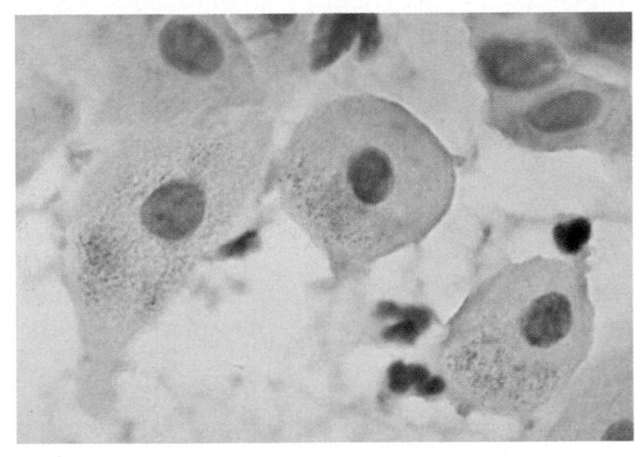

**Figure 159**
Chlamydial intracytoplasmic deposits revelaed by a nucleic acid probe technique. (Courtesy of Drs. D. Raisi, C. Ghirardini, and M. Portolani, University of Modena and Mirandala, Hospital Cytology Lab, Italy.)

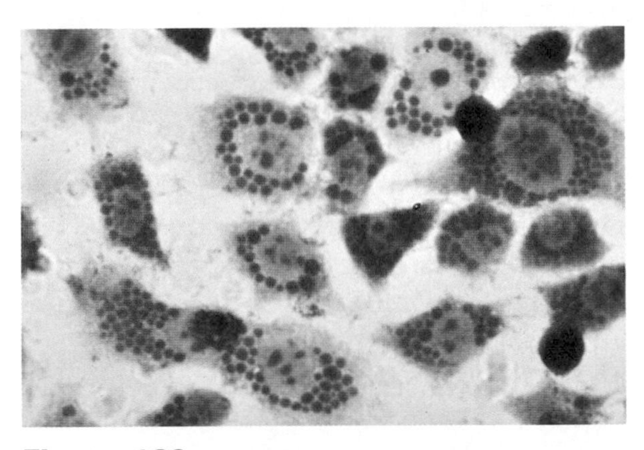

**Figure 160**
*Chlamydia psittaci* inclusions in McCoy cells. (From R.L. Hodinka and P.B. Wyrick, *Inf. Immunol.* 56(1998):1456–63.)

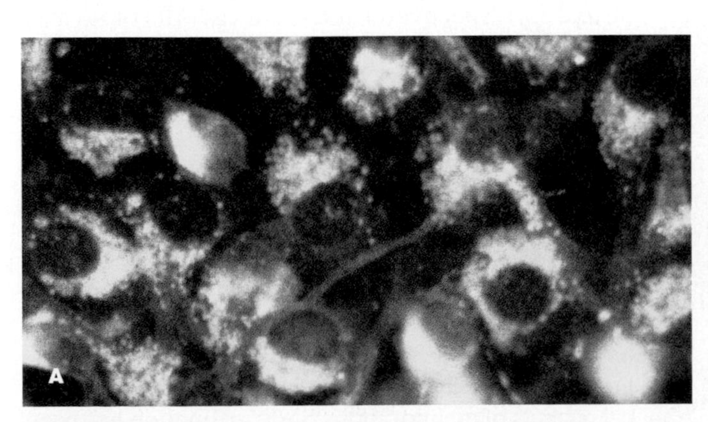


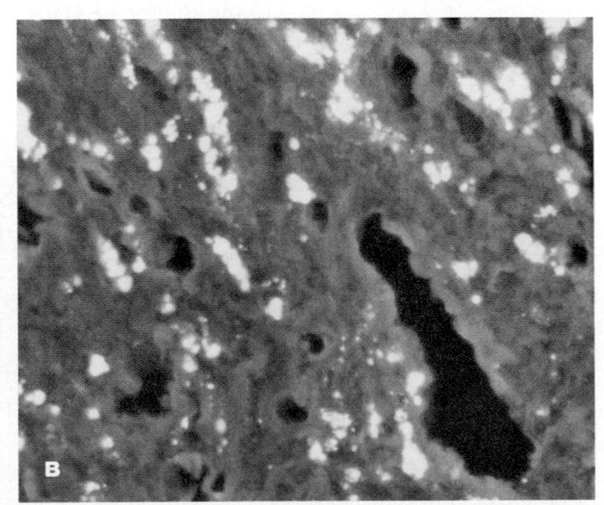


**Figure 161**
The presence of *Chlamydia psittaci* as shown by fluorescent antibody. (a) A tissue culture inoculated with a pharyngeal (throat) specimen. Chlamydial inclusions are shown. (b) Chlamydial antigen and inclusions (yellow glowing areas) in a mitral valve growth. (From D.S. Shapiro, S.C. Kenney, M. Johnson, C.H. Davis, S.T. Knight, and P.B. Wyrick. *NEJM* 326(1992):1192–95.)

**Transmission.** *Mycoplasma pneumoniae* is spread by close personal contact and by aerosols.

**Pathology and Clinical Features.** *Mycoplasma pneumoniae* infection is most prevalent in colder months and mainly affects children from ages 5 to 9 years. The incubation period is generally long and usually presents signs and symptoms such as a persistent cough, fever, and headache.

**Diagnosis.** Culture of *M. pneumoniae* from specimens and serological tests are mainly used for diagnosis.

### *Ureaplasma urealyticum*

*Ureaplasma urealyticum* is known to cause nongonococcal urethritis (NGU) in men and pelvic inflammatory disease (PID) and postabortal fever in women.

**Transmission.** *Ureaplasma urealyticum* is primarily spread through sexual contact.

**Pathology and Clinical Features.** Persons infected with *U. urealyticum* usually experience painful or difficult urination (dysuria), frequent urination, and an unusual urethral discharge.

## Gram-Positive Cocci

### *Aerococcus viridans*

*Aerococcus viridans* is a rare cause of human infection. It is a common airborne organism in various environments including hospitals and has been associated with nosocomial infections in immunosuppressed patients.

**Morphology and Cultural Properties.** *Aerococcus viridans* is a gram-positive coccus measuring 1.0–2.0 μm in diameter. Tetrads are commonly formed (Figure 164).

*Aerococcus viridans* is nonmotile and facultatively anaerobic. It grows best under reduced oxygen tension and at an optimal temperature of 30°C (Figure 165). The organism is catalase negative and does not liquefy gelatin or reduce nitrate. Acid without gas is produced from various carbohydrates. Unusual cultural features include the ability to grow at pH 9.6 and in 10% NaCl and in 40% bile.

### *Enterococcus faecalis*

*Enterococcus faecalis* is widely distributed in the environment, especially in the fecal matter of vertebrates. This organism is associated with urinary and biliary tract infections. In addition, it participates in mixed microbial infections involving the lungs, brain, abdomen, and pelvis.

**Transmission.** Transmission includes contaminated food or water, fomites, and aerosols.

**Morphology and Cultural Properties.** *Enterococcus faecalis* is a gram-positive coccus measuring 0.6–2.5 μm (Figure 166). Cells occur in pairs or in short chains. The organism is nonmotile, does not form capsules, and is facultatively anaerobic. Acid without gas is produced from a wide range of carbohydrates.

*Enterococcus faecalis* grows best at 37°C on most media (Figure 167) but can also grow at both 10° and 45°C, at a pH of 9.6, and in media containing 6.5% NaCl (see Figure 179b) or 40% bile (see Figure 179a).

**Pathology.** Enterococci are frequently associated with urinary tract infections and bacteriuria following urologic, intraabdominal, or hepatobiliary surgery. Enterococcal bacteremia also is often found in elderly patients with serious underlying medical problems. The prevalence of enterococcal infections has increased dramatically to the point that enterococci are now the second most common organisms recovered from patients with nosocomial infections.

**Diagnosis.** Identification involves isolation and biochemical tests of *E. faecalis*.

### *Micrococcus luteus*

*Micrococcus luteus* is primarily found on mammalian skin and in soil. It is also commonly isolated from the air and certain foods.

**Morphology and Cultural Properties.** *Micrococcus luteus* is a gram-positive coccus measuring 0.5–2.0 μm in diameter (Figure 168). Cells occur in pairs, tetrads, or irregular clusters but not in chains.

*Micrococcus luteus* is strictly aerobic and grows best at temperatures ranging from 25° to 37°C on simple media (Figure 169). Colonies are usually yellow. The organism is catalase and oxidase positive (see Figures 56 and 55, respectively) and produces little or no acid from carbohydrates.

### *Sarcina*

*Sarcina* species are widely distributed in the environment and commonly are isolated from mammalian intestinal tracts.

**Morphology and Cultural Properties.** *Sarcina* species are gram-positive cocci measuring 1.8–3.0 μm in diameter. Cells occur singly or in pairs, tetrads, or cubical packets of eight or more. Spore formation has been reported but is not usually seen.

*Sarcina* species are anaerobic and grow best at temperatures ranging from 30° to 37°C. Organisms are catalase negative and ferment carbohydrates.

## Mycoplasmas

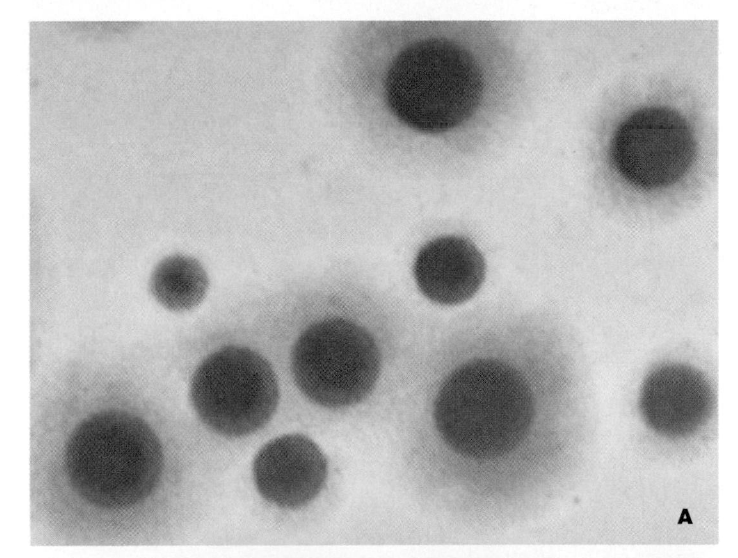


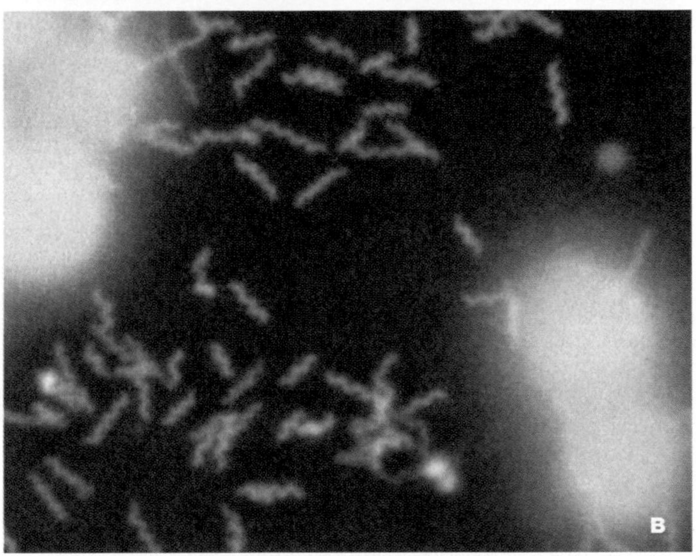


**Figure 162**
*Mycoplasma.* (a) Seven-day old stained *Mycoplasma fermentans* colonies on Dienes agar. Note the depressed centers surrounded by thin films of growth on the agar surfaces. Colonies range in size from 1 to 3 mm in diameter. (b) A fluorescent stained preparation of *Spiroplasma* species.

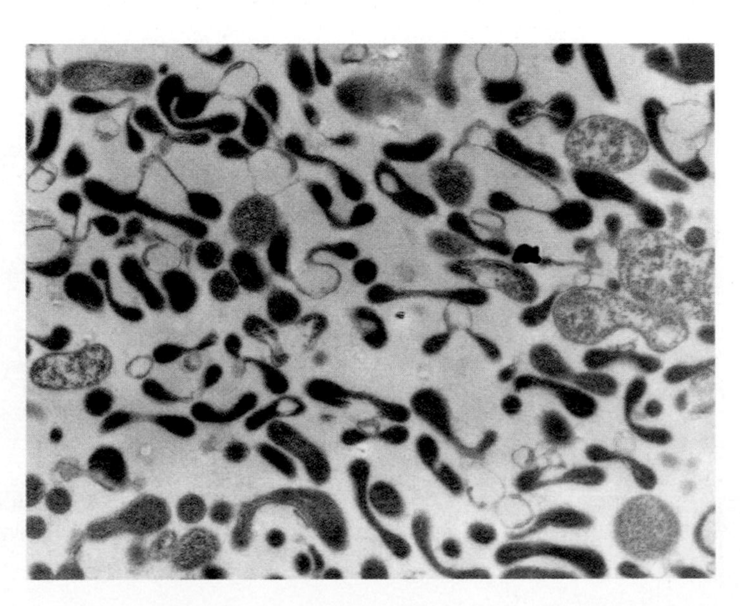

**Figure 163**
A transmission electron micrograph showing the wall-less nature of mycoplasmas.

## Gram-Positive Cocci

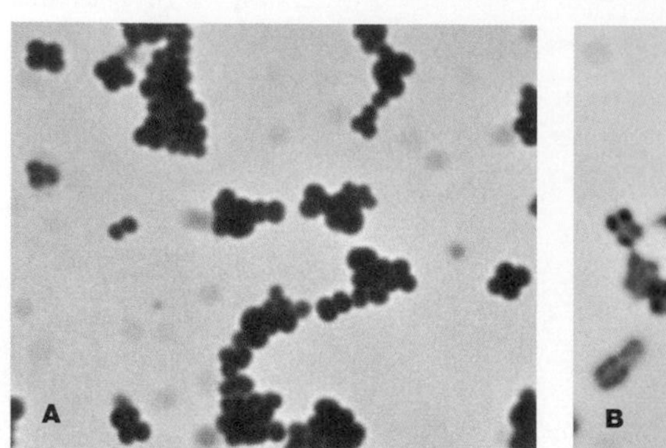

**Figure 164**
(A) Gram stain preparation of *Aerococcus.* Note the tetrad arrangements (1,000×). (b) A methylene blue stain preparation showing tetrads (1,000×).

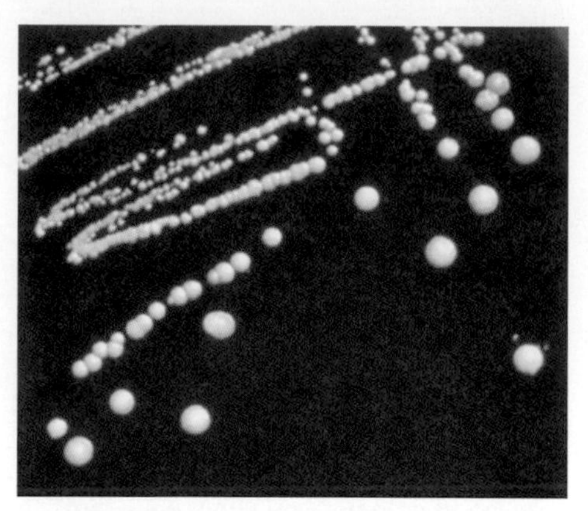

**Figure 165**
*Aerococcus* colonies on blood agar.

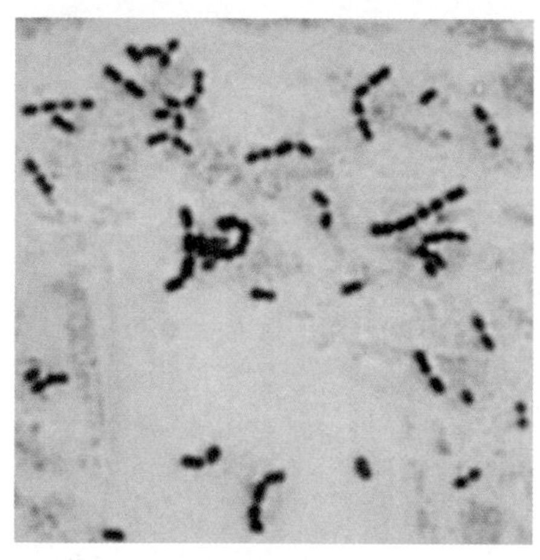

**Figure 166**
Gram stain of *Enterococcus faecalis.*

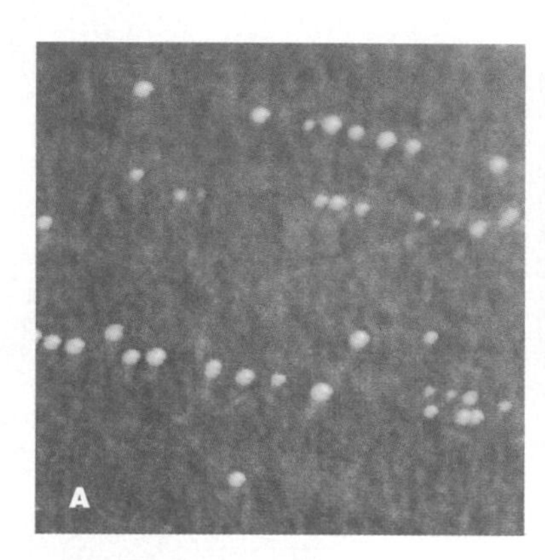

**Figure 167**
*Enterococcus faecalis* on culture media. (a) Colonies on nutrient agar. (b) Nonhemolytic colonies on blood agar.

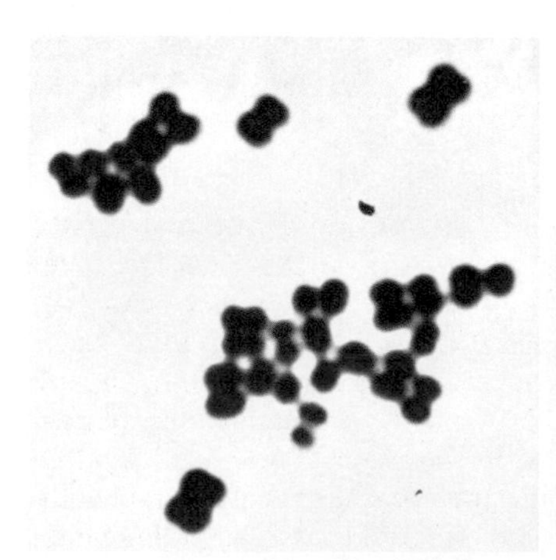

**Figure 168**
Gram stain preparation of *Micrococcus luteus* (1,000×).

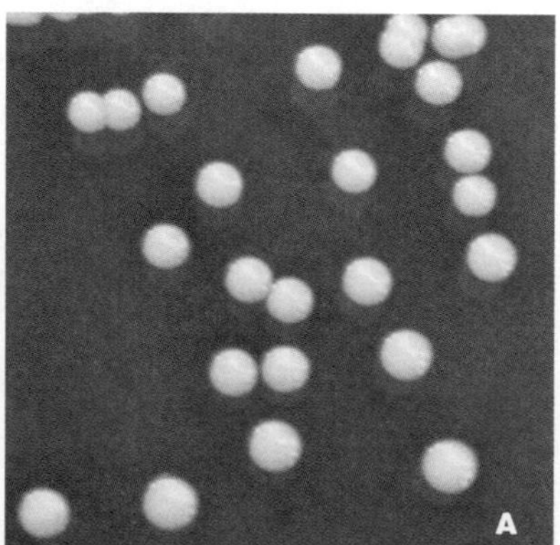

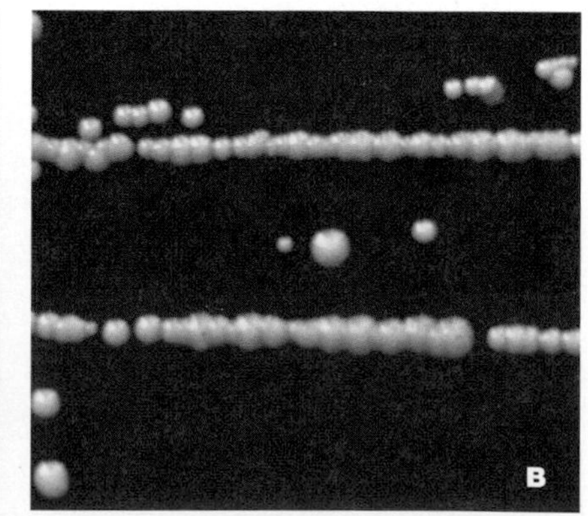

**Figure 169**
*Micrococcus luteus* on culture media. (a) Colonies on nutrient agar. (b) Nonhemolytic colonies on blood agar.

## *Staphylococcus*

Staphylococci are widespread in nature, though they are commonly found on the skin and in the mucous membranes of mammals and birds. Attention here will be given to only one species, *Staphylococcus aureus.*

**Transmission.** Staphylococci can be spread by a variety of ways, including aerosols, fomites, contact with infectious material from wounds, contaminated foods or water, and insects such as flies.

**Morphology and Cultural Properties.** Staphylococci are nonmotile, non-sporing, gram-positive cocci, 0.50–1.5 μm in diameter (Figure 171). Cells occur singly, in pairs, and in grapelike clusters.

*Staphylococcus aureus* is facultatively anaerobic and grows best at a temperature from 30° to 37°C. Colonies are usually opaque, smooth, circular, cream, and sometimes yellow (Figure 172) or yellow-orange. Blood agar colonies normally show beta-hemolysis (Figure 172a). The organism is catalase positive (see Figure 56) and oxidase negative (see Figure 55). Nitrate is often reduced to nitrite (see Figure 54).

**Pathology.** *Staphylococcus aureus* causes a bewildering array of infections including folliculitis (Figure 170a), superficial skin lesions (Figure 170b), and localized abscesses (Figure 170c); deep-seated infections involving bone (osteomyelitis), the central nervous system (meningitis), and the heart (endocarditis); toxic shock syndrome; pneumonia; and food poisoning. It, as well as *S. epidermidis,* are major causes of nosocomial infections of surgical wounds and in-dwelling medical devices such as urinary catheters. Many of the staphylococcal infections are associated with toxins secreted by the organisms.

**Diagnosis.** Several key characteristics of *S. aureus* can be used to identify the organism and to distinguish it from others. These include fermentation of mannitol on mannitol-salt agar (Figure 172b), DNAse production (see Figure 45), and a positive coagulase test (Figure 173). Coagulase is an enzyme that clots blood plasma. Commercially available molecular probes and identification systems, immunological tests, and bacterial viruses (phage typing) also are used for identification and related purposes.

## *Streptococcus*

The genus *Streptococcus* has great significance in medicine and industry. Various streptococci are important ecologically as members of the normal microbiota of humans and other animals. Some species are well known for the broad range of diseases they cause (Figures 174 and 175). Human diseases associated with the streptococci occur mainly in the respiratory tract or bloodstream or as skin infections. Industrially, several species are essential in various dairy processes and as indicators of pollution.

**Transmission.** The means of transmission include fomites, aerosols, and contaminated food and water.

**Morphology and Cultural Properties.** Streptococci are gram-positive, non-sporing, nonmotile cocci, measuring 0.50–2.0 μm in diameter (Figure 176). Cells occur in pairs or chains (Figure 176a). Some species are encapsulated (Figure 176b).

Streptococci are facultatively anaerobic and generally have a temperature growth range from 25° to 45°C. Optimal growth occurs at 37°C. These organisms are catalase negative and commonly attack media containing red bood cells (hemolysis). Streptococci can be divided into three groups by the type of hemolysis produced on blood agar: α hemolysis, producing green zones around colonies; β hemolysis, producing clear areas around colonies (Figure 177a); and γ hemolysis, producing no zones.

Streptococci are classified and placed into groups on the basis of colonial properties, hemolytic reactions, biochemical tests (Figures 178 and 179), and serological specificities based on antigenic differences in cell wall carbohydrates, proteins, and polysaccharide capsules. Examples of biochemical tests include bacitracin sensitivity, which differentiates group A streptococci (GAS) from other hemolytic streptococci (Figure 178a,b), bile solubility, and optochin sensitivity, which differentiates group D and enterococci from other viridans streptococci (Figure 178c and 179a).

The viridans group includes *S. salivarius* (Figure 177b), *S. mitis* (Figure 177c) and *S. mutans*, which are among the most common causes of infective endocarditis. *Streptococcus mutans* plays an important role in tooth decay (caries).

**Pathology.** An enormous number of streptococcal species have been identified over the years from a wide variety of human and other animal sources. Attention will be given only to a few representatives of the genus *Streptococcus*.

### *Group A*

Human disease is most commonly associated with group A streptococci (GAS). *Streptococcus pyogenes* and *S. pneumoniae* are members. *Streptococcus pyogenes* has long been recognized as an important cause of purulent diseases such as pharyngitis (strep throat), impetigo, erysipelas, and, less frequently, severe generalized diseases such as sepsis and toxic shock–like syndrome. Certain strains cause scarlet fever. In addition, immunologic-related diseases including acute glomerulonephritis, acute rheumatic fever, and rheumatic heart disease have been linked to complications of group A pharyngitis.

## Gram-Positive Cocci *(Continued)*

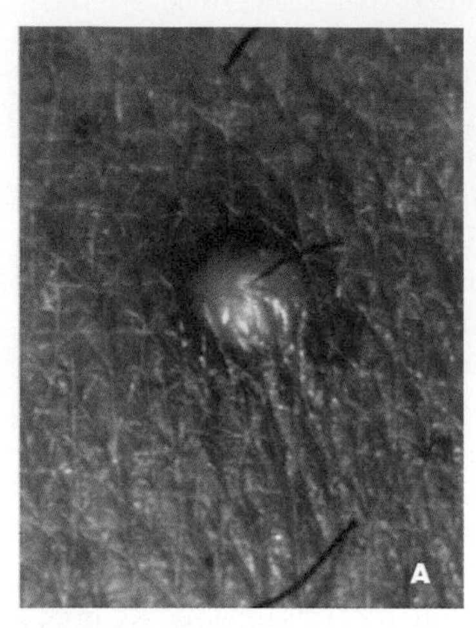
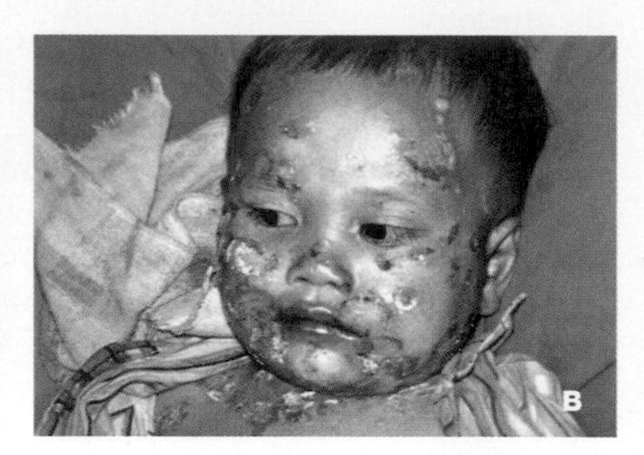
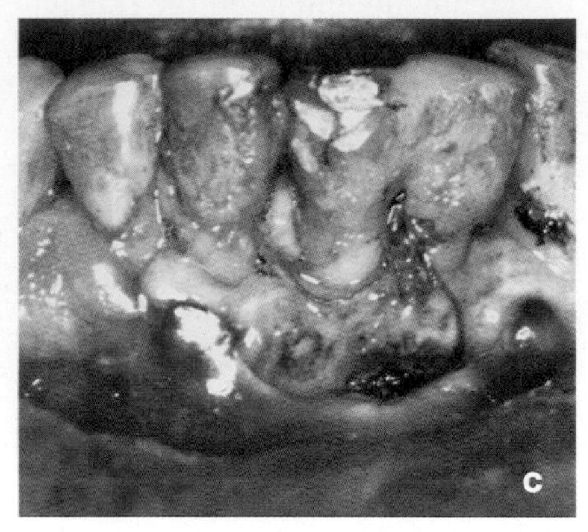

**Figure 170**
Clinical features of *Staphylococcus aureus* infections. (a) Folliculitis, inflammation of a hair follicle. (b) A child with impetigo. (c) Gum tissue injury (necrosis) associated with penicillin-resistant staphylococi. (From H. Helovuo, K. Kakkarainen, and K. Pannio. *Oral Microbiol. Immuno.* 8(1993):75–79.)

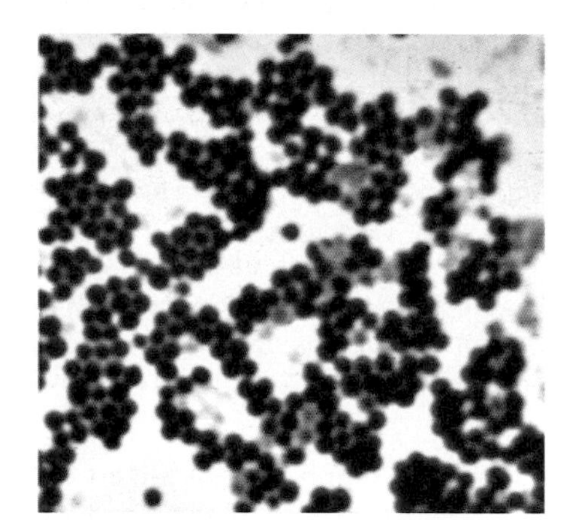

**Figure 171**
Gram stain of *Staphylococcus aureus.*

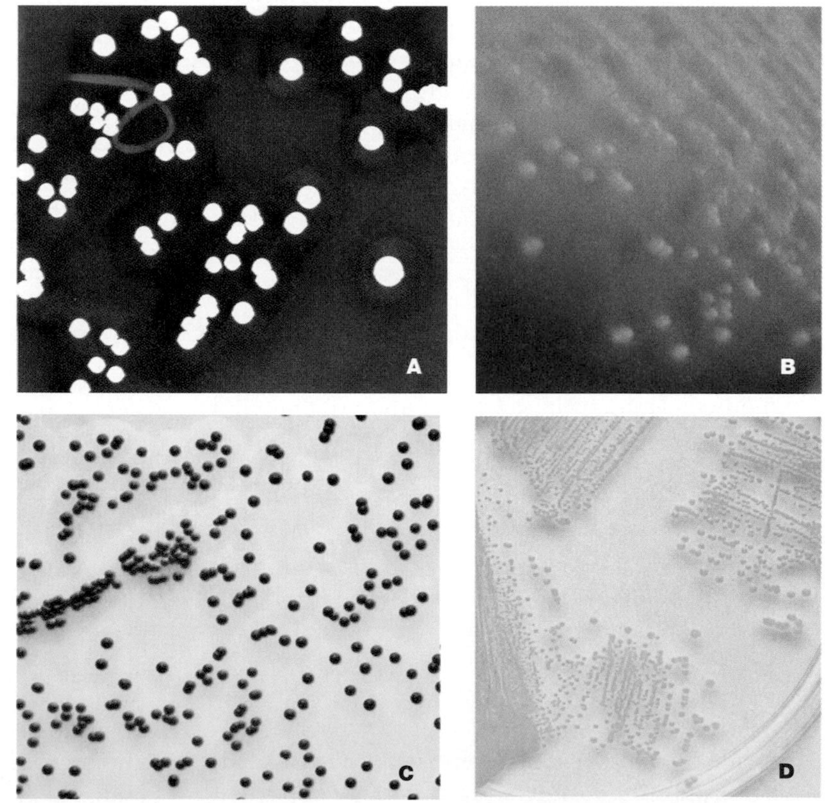

**Figure 172**
*Staphylococcus aureus* on culture media. (a) Beta-hemolytic colonies on blood agar. (b) Mannitol-salt agar reaction. (c) Growth on Baird Parker agar. (Courtesy of Difco Laboratories, Detroit, MI.) (d) Reaction on *Staphylococcus* medium 110°C. (Courtesy of Difco Laboratories, Detroit, MI.)

*Streptococcus pneumoniae* (Figures 176b and 178c) is the most common cause of community-acquired pneumonia and lobar pneumonia in the United States and of fatal bacterial pneumonia in developing countries. It also causes otitis media, sinusitis, meningitis, and bacteremia.

Pneumococci are frequently isolated from the nasopharynx of healthy people. Virtually all humans are colonized by pneumococci at some stage, and pneumococcal carriage rates are higher in young children and where people are living in crowded conditions. An important feature of *S. pneumoniae* is its capacity to produce a polysaccharide capsule, which is structurally distinct for each of the 84 currently known serotypes of the organism.

#### *Group B*

Group B streptococci (GBS) are leading causes of neonatal sepsis, pneumonia, and meningitis in the United States. Infection of newborns exhibiting early-onset neonatal disease occurs via vertical transmission during passage through the birth canal or by exposure to infected amniotic fluid.

**Diagnosis.** Diagnosis of streptococcal infections requires the isolation and culture of the likely agent from clinical specimens (Figure 174). Gram staining and biochemical and serological tests are of major value.

## Endospore-Forming Gram-Positive Rods and Cocci

### *Bacillus*

*Bacillus* species are widely distributed in nature, particularly in soil, from where they are spread in dust, in water, and on animal or plant materials. The majority of *Bacillus* species are nonpathogenic and rarely are associated with diseases in humans or lower animals. Exceptions include *B. anthracis,* the causative agent of anthrax, *B. cereus,* the causative agent of human food poisoning, meningitis and burn infections, and species such as *B. larvae, B. popilliae,* and *B. thuringiensis*, known pathogens of specific insect groups.

**Transmission.** *Bacillus* species are spread through aerosols and by contact with infected animals or animal products such as hides and various types of brushes made with or contaminated by infectious animal materials.

**Morphology and Cultural Properties.** *Bacillus* species are gram-positive, endospore-forming rods, 0.5–2.5 μm wide and 1.2–10.0 μm long. Cultures may become gram-negative with age. Only one endospore is formed in each cell (Figures 180 and 181). Endospores are generally oval or sometimes round and are resistant to unfavorable environmental conditions and related factors such as extreme heat, cold, radiation, the effects of drying, and disinfectants. *Bacillus anthracis* forms capsules.

*Bacillus* species are aerobic or facultatively anaerobic and exhibit a wide range of physiological abilities. Several species are motile. Most members grow on simple media (Figure 182) and over a broad temperature range. Some, such as *B. circulans,* form unusual colonies (Figure 182e).

### *Clostridium*

*Clostridium* species are widespread in the environment. They can be found in the soil as well as in the normal intestinal microbiota of humans and other mammals. The genus includes several saprophytes and pathogenic species. Pathogens are known to cause diseases such as botulism, tetanus, gas gangrene, and pseudomembranous colitis, a complication of antibiotic therapy (Figure 183).

**Transmission.** *Clostridium* species can be spread by a variety of ways including aerosols, insects, and contaminated foods and objects.

**Morphology and Cultural Properties.** Clostridia are gram-positive, commonly pleomorphic rods, 0.3–2.0 μm wide and 1.5–2.00 μm long (Figure 184). Cells occur in pairs or in short chains, with rounded or sometimes pointed ends. Endospores vary from oval to spherical forms and usually cause the cell to distend (Figure 185).

Clostridia are strictly (obligately) anaerobic to aerotolerant and usually are motile. Species are metabolically diverse and may be saccharolytic, proteolytic, neither, or both. Growth temperatures range from 10° to 65°C.

**Pathology.** The discussion of clostridial agents will be limited to only those primarily causing disease in humans.

**Diagnosis.** Several methods are available for the diagnosis of different clostridial infections. These include tissue Gram stains (if appropriate), bacteriological culture, tissue culture cytotoxicity, enzyme immunoassay, and polymerase chain technology.

#### *Clostridium botulinum*

*Clostridium botulinum* (Figure 186) is responsible for the paralyzing disease botulism. This species is divided into eight types (A, B, C alpha, C beta, D, E, F, G) on the basis of the specific toxins produced.

Three clinical forms of botulism may occur: foodborne, infant, and wound. Foodborne botulism occurs after ingestion of preformed toxin in contaminated food. Infant botulism results from the ingestion of *C. botulinum* spores, followed by the germination of the

## Gram-Positive Cocci *(Continued)*

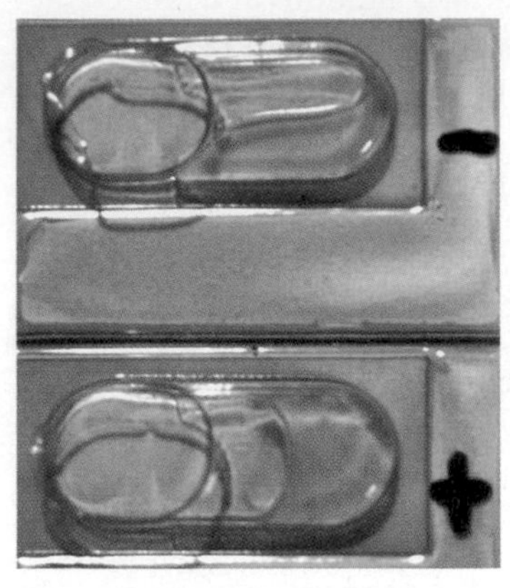


**Figure 173**
Coagulase reactions. A positive reaction is generally considered to be the best single indicator of potential pathogenicity. The formation of a plasma clot is a positive result; the absence of coagulation is a negative one.

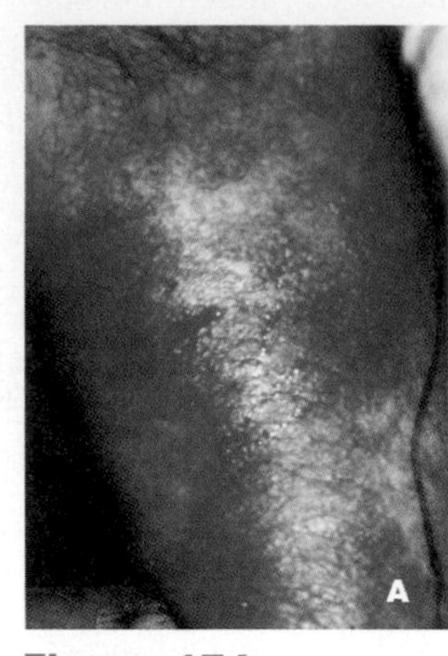


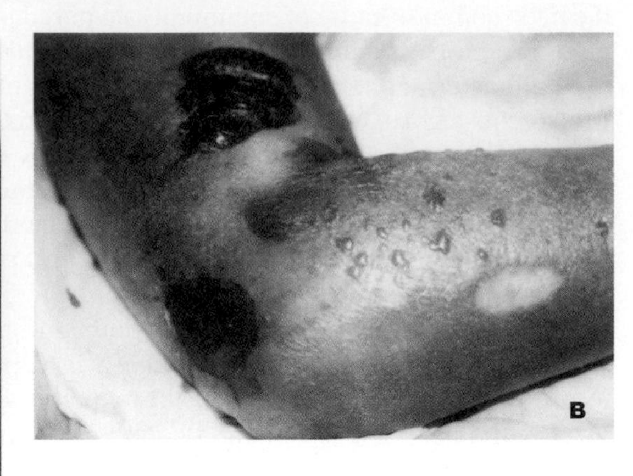


**Figure 174**
Examples of skin and tissue infections caused by group A streptococci (GAS). (a) Cellulitis of the arm, showing swelling and numerous thin, surface blisters oozing with bloody fluid. (b) Necrotizing fasciitis with blisters containing bloody fluid and a sloughing of skin from the elbow. (From B. Demers, A.E. Simor, H. Vellend, P.N. Schlievert, S. Byrne, F. Jamieson, S. Walmsley, and D.E. Low. *CID* 1(1993):792–800.)

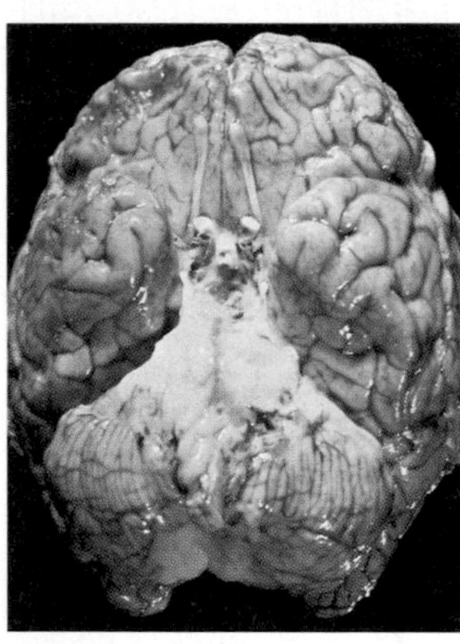

**Figure 175**
A view of the brain of a patient who died of *Streptococcus pneumoniae* meningitis. A thick, pussy accumulation can be seen covering a portion of the organ. (From J.A. Golden and D.N. Louis. *NEJM* 331(1994):34.)

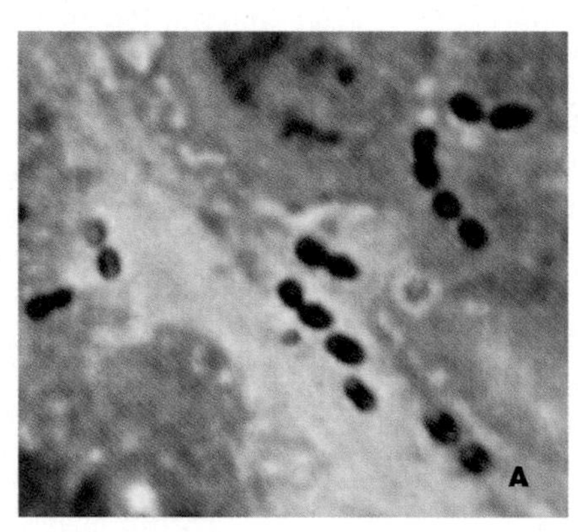


**Figure 176**
Microscopic views of streptococci. (a) A sputum specimen showing gram-positive diplococci and an occasional chain. (b) A capsule stain of *Streptococcus pneumoniae.*

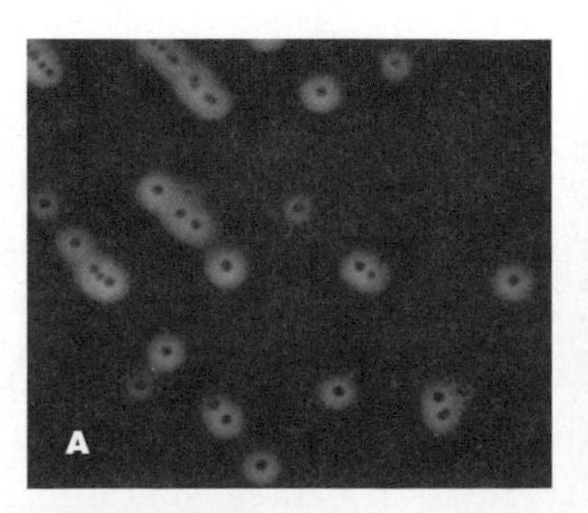


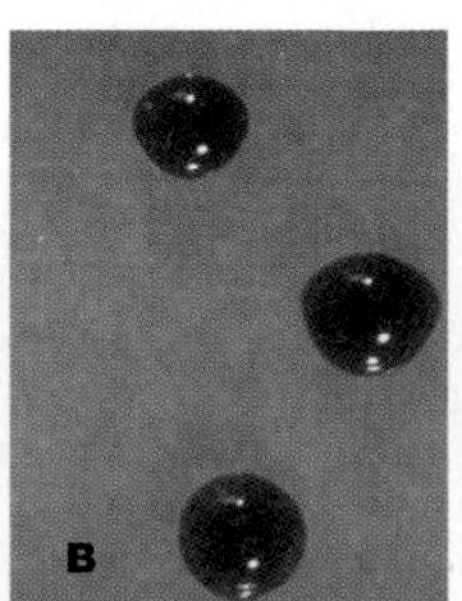


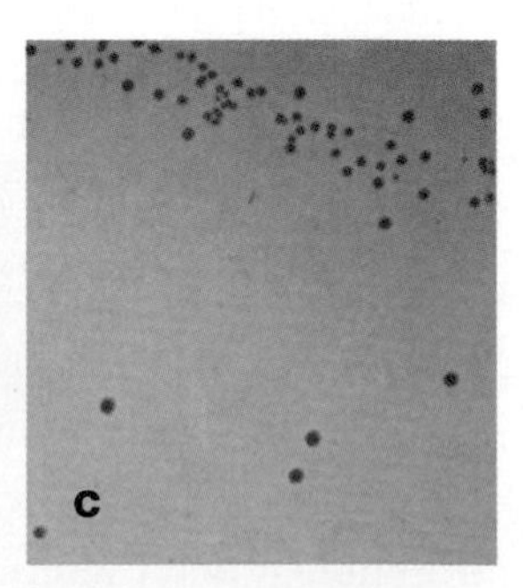


**Figure 177**
(a) Beta-hemolytic colonies of *Streptococcus pyogenes.* (b) Viridans streptococci *S. salivarius* colonies on *Mitis-Salivarius* agar. (c) *S. mitis,* another viridans streptococcus on the same medium.

## Gram-Positive Cocci *(Continued)*

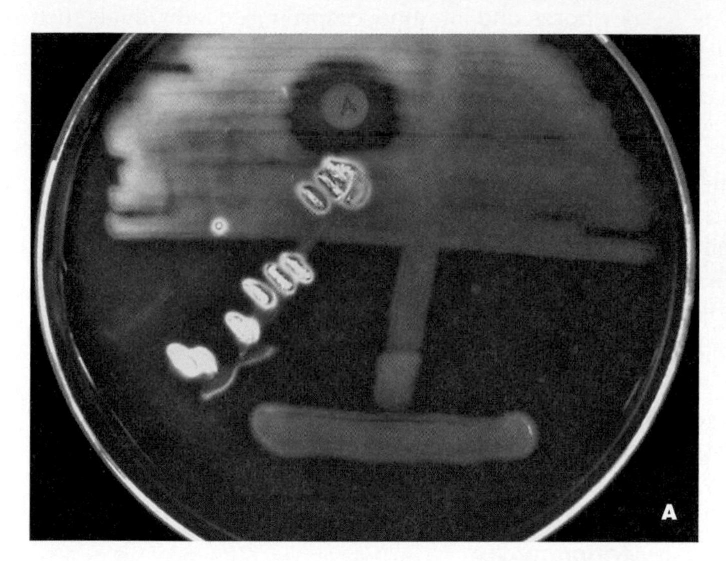


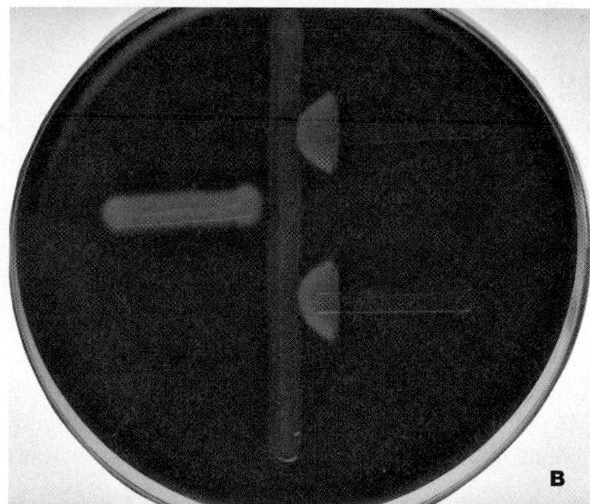

**Figure 178**
*Streptococcus* identification tests. (a) The typical characteristic growth inhibition by bacitracin. (The disk contains two units of the antibiotic.) Beta-hemolysis is evident from the agar stabs on the blood agar. The Christie, Atkins, and Munch-Peterson (CAMP) test is negative. (b) Positive CAMP reactions. The positive CAMP reaction is indicated by a triangle (arrowhead) perpendicular to the growth of a beta–lysin producing *Staphylococcus aureus* strain. (c) A positive growth inhibition optochin reaction. The *P* disk contains ethylhydrocupreine hydrochloride.

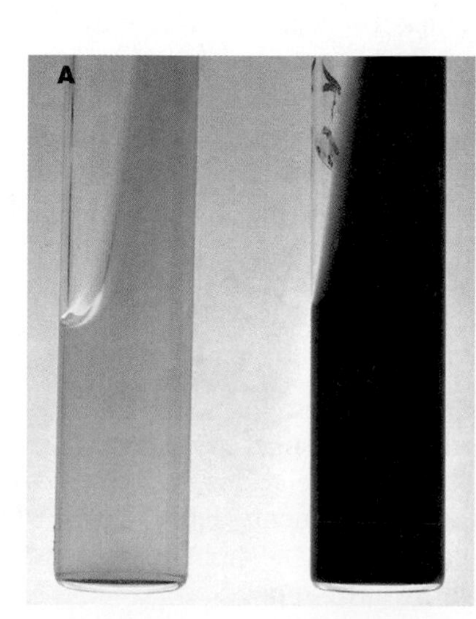

**Figure 179**
Tests for group B streptococci and certain enterococci. (a) The bile-esculin test. A positive reaction is a blackened slant. (b) Salt tolerance. A positive reaction is indicated when the brom-cresol purple indicator changes from purple to yellow.

spores, and toxin production in the infant's intestine. Wound botulism may occur after contamination of a deep wound by *C. botulinum*.

***Clostridium difficile***

*Clostridium difficile* causes most cases of antibiotic-associated pseudomembranous colitis. In this condition, raised whitish yellow or greenish yellow plaques form on the colon (Figure 183a). These plaques often combine with one another to form a pseudomembrane and cause substantial tissue destruction (Figure 183b), dehydration, and electrolyte imbalance.

***Clostridium perfringens***

Various clostridial species can cause wound infections, including gas gangrene. *Clostridium perfringens* is the most commonly associated species. It causes the infection after invading normal healthy muscle surrounding the wound. Fermentation of muscle carbohydrate produces gas in subcutaneous tissues that can be felt on touch.

***Clostridium tetani***

*Clostridium tetani* is the cause of tetanus, or lockjaw. When spores (Figure 185) are introduced into wounds by contaminated soil or foreign objects such as nails or glass splinters, they germinate into vegetative cells that subsequently produce exotoxins. The tetanus toxins cause muscular dysfunction and allow uncontrolled muscle stimulation.

## Regular, Non-Spore-Forming Gram-Positive Rods

### *Listeria*

*Listeria* species are widely distributed in the environment. Certain species are pathogenic.

***Listeria monocytogenes***

*Listeria monocytogenes* causes listeriosis and related conditions primarily in immunocompromised hosts.

**Transmission.** Listeriosis is primarily transmitted by contaminated food.

**Morphology and Cultural Properties.** *Listeria monocytogenes* is a gram-positive, non-spore-forming and nonencapsulated short rod, 0.4–0.5 μm in width and 0.5–2.0 μm in length (Figure 187). Cells have rounded ends and occur singly or in short chains.

The organism is a facultative anaerobe and is found intracellularly (Figure 187a). It is motile when grown at 20° to 25°C. However, its optimal temperature range is from 30° to 37°C. *Listeria* species are catalase positive and oxidase negative, and they ferment glucose.

**Pathology and Clinical Features.** Pregnant women, newborns, and immunocompromised individuals such as persons with AIDS are particularly susceptible to infections. The most common infections include bacteremia, spontaneous abortion, meningitis, and meningoencephalitis.

*Listeria monocytogenes* has created great interest because of its ability to reorganize host cell protein (actin) and to use this ability to spread to neighboring cells.

**Diagnosis.** Cultures of blood, cerebrospinal fluid, or wounds are used for diagnosis. Immunofluorescence and related methods also are of value.

## Irregular, Non-Spore-Forming Gram-Positive Rods

### *Actinomyces*

*Actinomyces* species are normal inhabitants of certain portions of the gastrointestinal tract of warm-blooded vertebrates. They are commonly found in the mouth and on mucous membranes.

***Actinomyces israelii***

Most cases of human actinomycosis (Figure 188) are caused by *A. israelii,* although other species also have been isolated from lesions.

**Transmission.** *Actinomyces* species are highly adjusted to mucosal surfaces and do not cause disease unless they are introduced into deeper tissues. Conditions contributing to establishing infection include those causing tissue injury, such as tooth extractions, or some other form of trauma to the mouth or jaw.

**Morphology and Cultural Properties.** *Actinomyces israelii* is a gram-positive, slender, straight, or slightly curved rod, 0.2–1.0 μm wide and 2.0–5.0 μm long. Organisms typically branch and may occur singly, in pairs, in V and Y arrangements, and side-by-side arrangements resembling a picket fence. Cells also may exhibit irregular staining and appear swollen.

*Actinomyces* species are nonmotile, nonsporing, and non-acid-fast. These organisms are facultative anaerobes and require $CO_2$ for good growth. Optimal growth temperature ranges between 35° and 37°C. Mature colonies appear after 7 to 14 days of incubation and are usually rough and crumbly in texture.

*Actinomyces* are catalase positive and indole negative. Acid but no gas is produced from carbohydrates.

**Pathology and Clinical Features.** Once an infection starts, microbial growth occurs as small colonies referred to as sulfur granules because of their color. Such colonies consist of branching cellular filaments (Figure 188).

Infections are most common in the mouth or jaw and usually give the face a swollen appearance.

## Endospore-Forming Gram-Positive Rods and Cocci

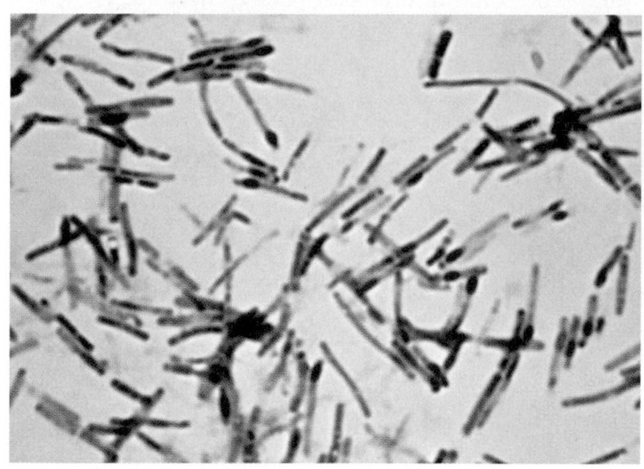

**Figure 180**
Spore stained preparation of *Bacillus subtilis.* Carbol fuchsin and methylene blue were used as the primary and secondary stains, respectively. Spores appear as red swollen areas in rods (1,000×).

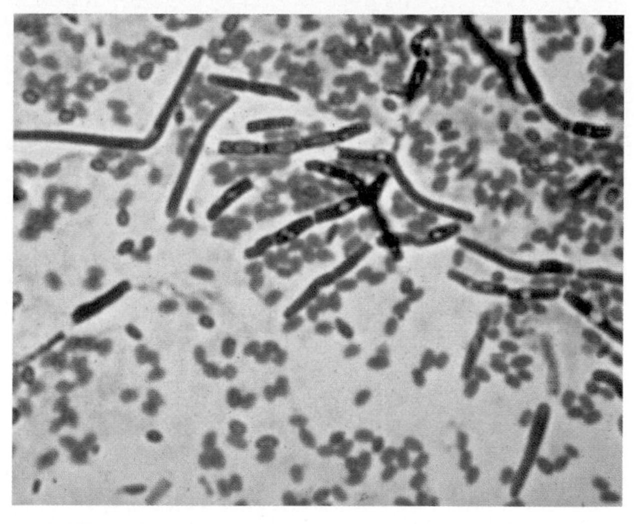

**Figure 181**
Spore stain of *B. megaterium.* Spores are green, and vegetative cells are red (1,000×).

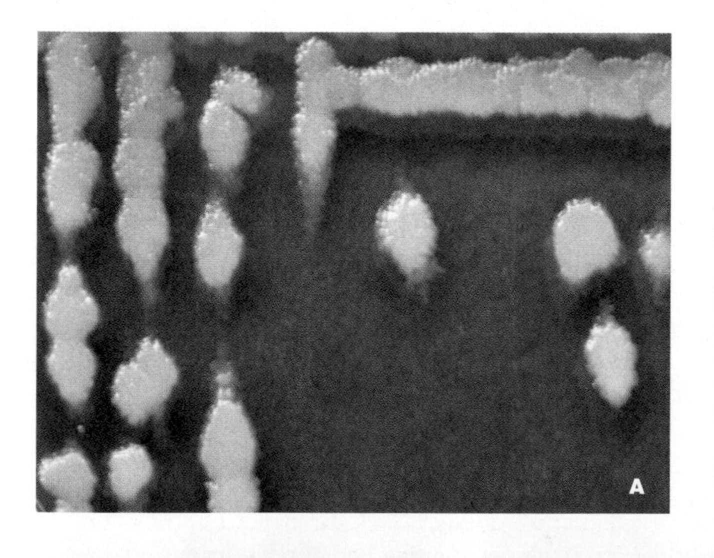


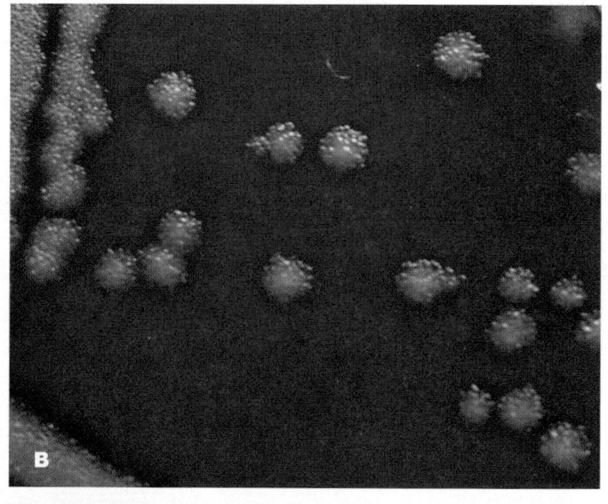


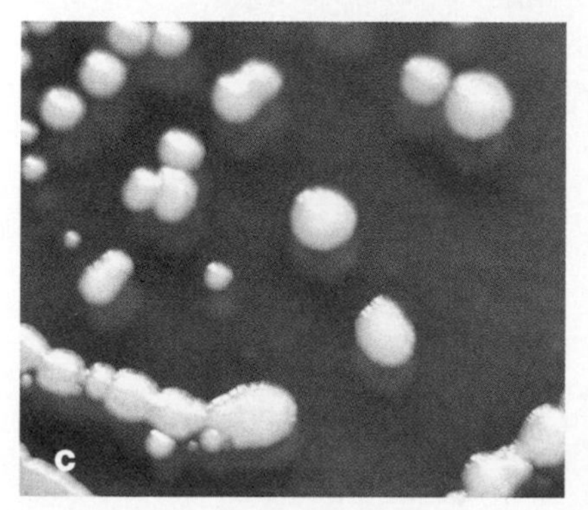


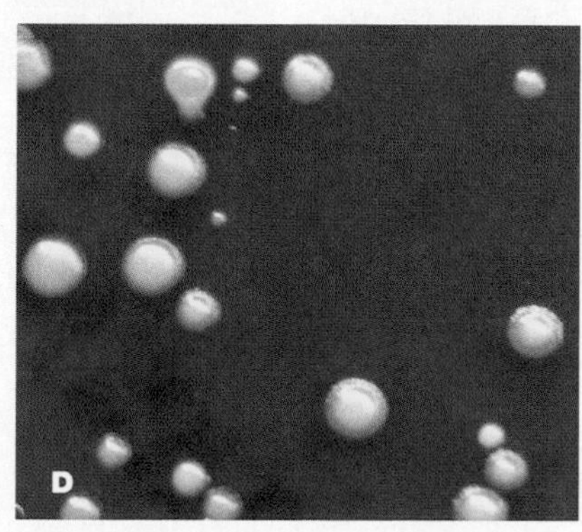


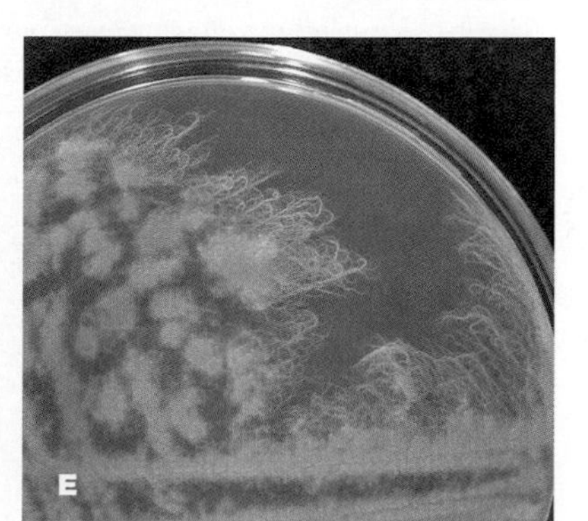


**Figure 182**
*Bacillus* species on nutrient media. (a) *B. subtilis* on nutrient agar. (b) *B. subtilis* on blood agar. (c) *B. megaterium* on nutrient agar. (d) *B. megaterium* on blood agar. (e) The unusual filamentous growth of *B. circulans.*

## Endospore-Forming Gram-Positive Rods and Cocci *(Continued)*

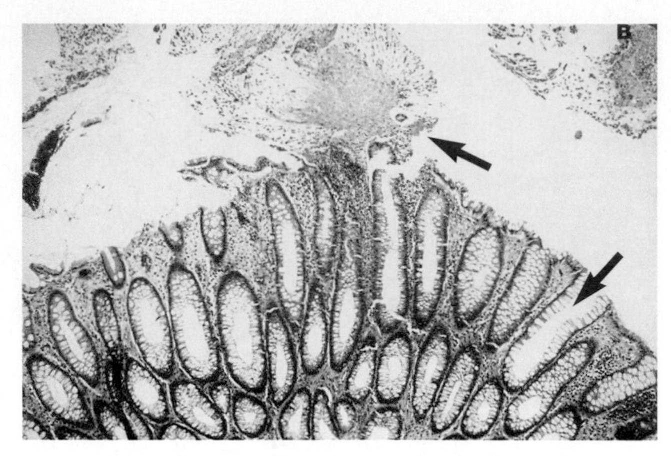


**Figure 183**
Pseudomembranous colitis. (a) A large intestine specimen showing yellow plaques (enlarged areas). (b) A microscopic view of a specimen showing local destruction (lower arrow) and pseudomembrane material and destroyed tissue (upper arrow). (From C.P. Kelley, C. Pothoulakis, and J. T. LaMont. *NEJM* 330(1994):257.)

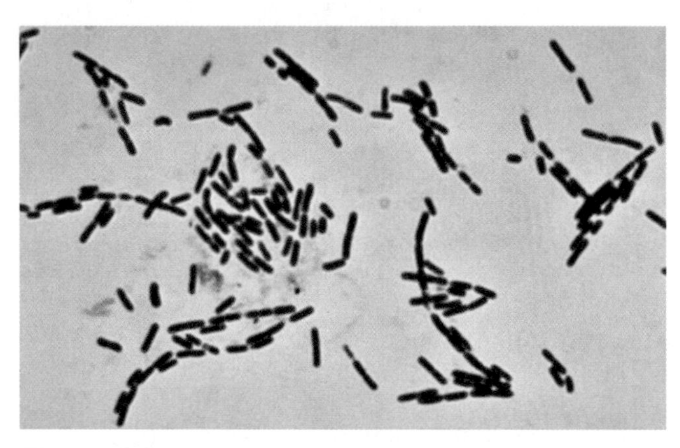

**Figure 184**
Gram-positive rods of *Clostridium* species.

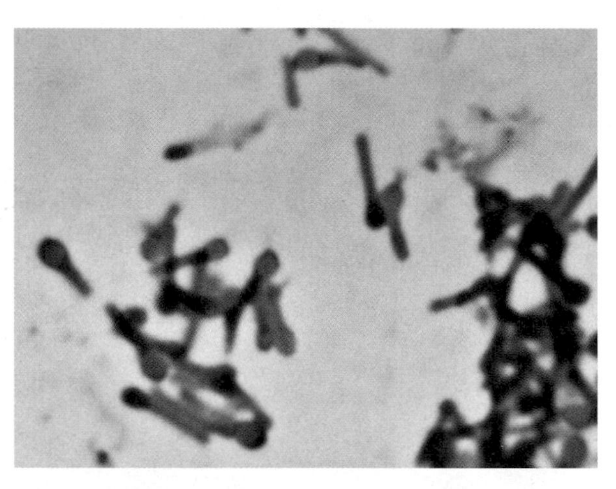

**Figure 185**
Microscopic view of *Clostridium tetani* spores. The red spores appear at the ends of cells.

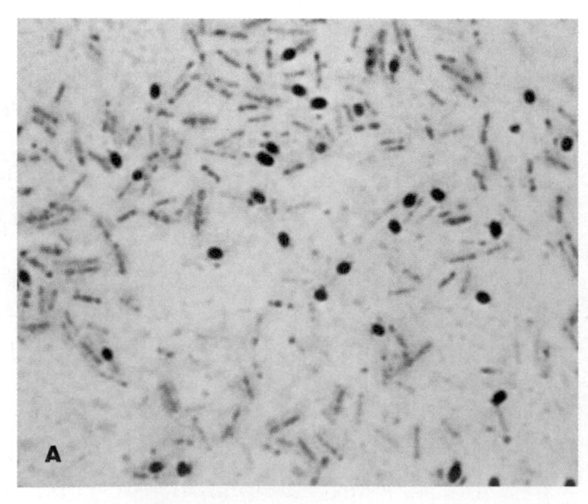


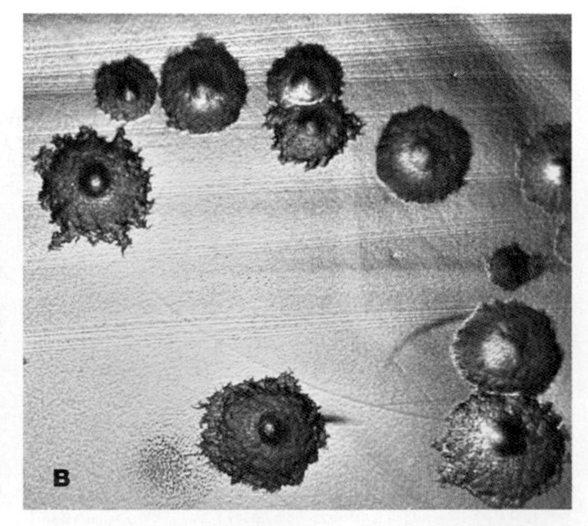


**Figure 186**
*Clostridium botulinum.* (a) Spore stain. (b) Colonies.

**Diagnosis.** Tissue specimens (biopsies) are taken for culture and staining. Demonstrating the presence of the sulfur granule in tissue is of value. A longer than usual incubation period is necessary for *Actinomyces* species.

### *Corynebacterium*

Corynebacteria are found in the environment on plants and as members of the microbiota of humans, lower animals, and plants. Several are commonly isolated in clinical laboratories.

#### *Corynebacterium diphtheriae*

*Corynebacterium diphtheriae* is the major pathogen of the genus and is the causative agent of diphtheria.

**Transmission.** *Corynebacterium diphtheriae* is spread by aerosols, direct contact with skin infections (Figure 189), and to some extent by fomites.

**Morphology and Cultural Properties.** Corynebacteria are gram-positive, straight or slightly curved, slender rods 0.3–0.8 µm wide, and 1.5–8.0 µm long. Cells are usually found singly or in pairs and are arranged in a V or palisade formation. Some cells also can exhibit clubbed ends. Metachromatic granules or stored energy material (polymetaphosphate) can be demonstrated by special staining methods (Figure 190).

Corynebacteria are nonmotile, facultative anaerobes. Organisms are catalase positive, and they often reduce nitrate and tellurite. Most species produce acid but no gas from glucose and certain other carbohydrates. The optimal growth temperature is 35°C. Colonies of *C. diphtheriae* on tellurite-containing media appear gray or black (Figure 191).

**Pathology and Clinical Features.** Diphtheria is the most common disease caused by *C. diphtheriae*. The disease includes both local infection of the upper respiratory tract and the systemic effects of the organism's exotoxin. The signs and symptoms generally associated with the disease are the formation on the mucous membrane of a gray white membrane consisting of fibrin, white blood cells, and dead cell remains that may extend from the throat area to the larynx and trachea; sore throat; general discomfort; headache; and nausea. Death can occur from respiratory blockage or destruction of heart tissue by the toxin.

Another form of the disease is cutaneous diphtheria (Figure 189).

**Diagnosis.** The initial diagnosis of diphtheria is based on clinical signs and symptoms. Definitive diagnosis is based on the isolation and culture of *C. diphtheriae* from the infection site and demonstration of its toxin production. Selective media containing potassium tellurite are used for isolation (Figure 191).

## The Mycobacteria

### *Mycobacterium*

The mycobacteria are widely distributed in soil and water. Several species are pathogenic and can cause disease involving various vertebrates (Figure 192). The most commonly encountered pathogen is *Mycobacterium tuberculosis* (Figures 193 and 194), followed by *M. avium* complex (Figures 199 and 200), *M. kansasii* (Figures 197 and 198), and *M. scrofulaceum*. Unclassified, atypical, or nontuberculous organisms are frequently referred to as mycobacteria other than tubercle (MOTT) bacilli.

**Transmission.** *Mycobacterium tuberculosis*, *M. avium* complex (MAC), *M. kansasii,* and others causing respiratory disease are spread primarily through the inhalation of airborne droplet nuclei. Nosocomial transmission from patients or specimens is of major concern to health care workers.

Contaminated food and fomites also can spread mycobacteria.

**Morphology and Cultural Properties.** Mycobacteria are straight or slightly curved acid-fast rods, 0.20–0.7 µm wide and 1.0–10.0 µm long. Cells are poorly stained by the Gram method and are weakly gram-positive.

Mycobacteria are nonmotile, non-spore-forming, and aerobic. Organisms grow slowly, with visible colonies appearing within 2 to 60 days depending on the species. Colonies of some species are often pigmented, especially when exposed to light (Figure 198). Mycobacteria are catalase positive.

#### *Mycobacterium tuberculosis complex*

On the basis of cultural and biochemical properties, *M. tuberculosis*, *M. bovis*, *M. africanum,* and *M. microtii* are members of the *M. tuberculosis* complex. Other species are grouped together as **nontuberculosis mycobacteria**. These include *M. kansasii* and *M. avium intracellulare* (Figures 198–200) which cause respiratory disease similar to tuberculosis, and *M. ulcerans* and *M. marinum,* which cause skin and soft tissue infections (Figure 192).

**Pathlogy and Clinical Features.** Tuberculosis primarily affects the lower respiratory system. A chronic productive cough, fever, night sweats, and weight loss are typical of the disease. Tuberculous mycobacteria are able to avoid destruction by the host's defenses and form the primary lesion known as the **tubercle**. Organisms spread to regional lymph nodes, enter the blood, and reseed the lungs. Tissue injury results from cell-mediated hypersensitivity.

## Regular, Non-Spore-Forming Gram-Positive Rods

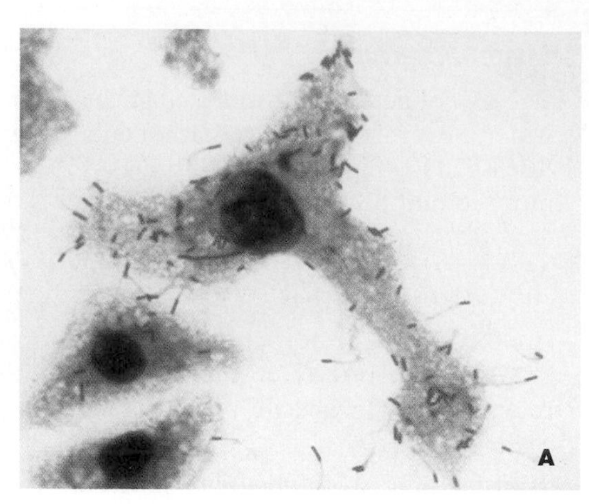


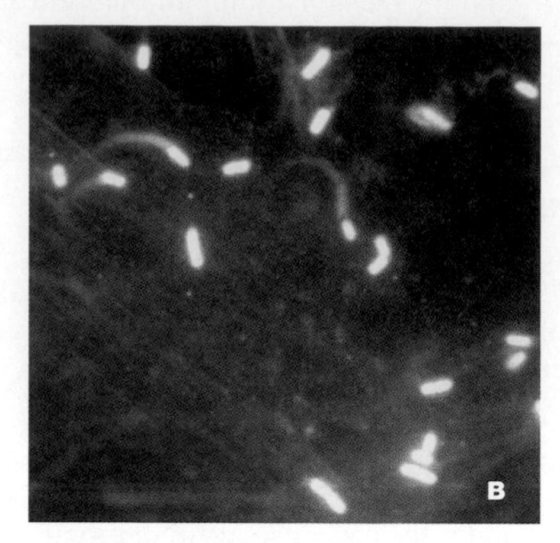


**Figure 187**
Microscopic views of *Listeria monocytogenes.* (a) Cells infected with *L. monocytogenes.* (Courtesy of Drs. S. Jones and D.A. Portroy.) (b) *L. monocytogenes* as shown by immunofluorescence.(From E. Gouin, P. Dehaoux, J. Mengaud, C. Kocks, and P. Cossart. *Inf. Immunol.* 3(1995):2729–37.)

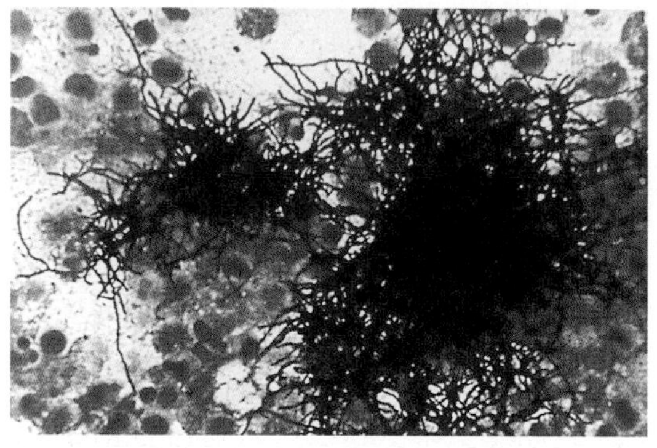

**Figure 188**
A Gram-stained specimen from a patient with actinomycosis. The lesion, which included sulfur granules contained *Actinomyces israelii.* (Reprinted by permission of *Infections in Medicine* 9(1992):13. SCP Communications.)

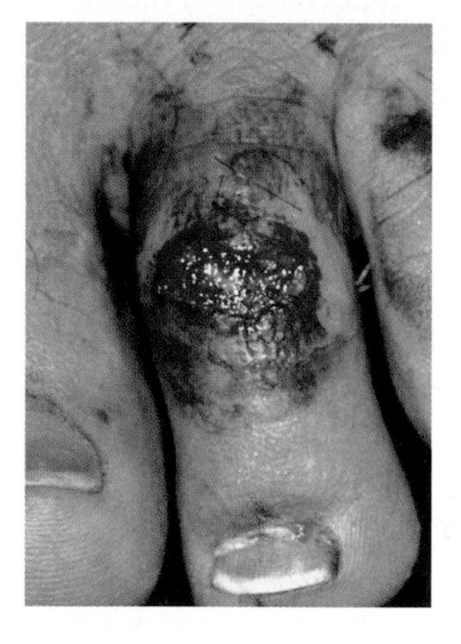

**Figure 189**
The most common form of cutaneous diphtheria, the ulcer or ecthyma diphthericum. (From W. Hofler. *Internat, J. Dermatol.* 30(1991):845.)

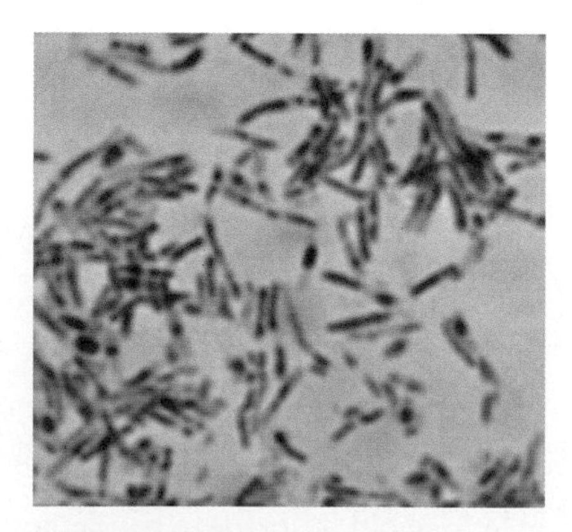

**Figure 190**
Metachromatic granules (enlarged and darkly stained areas) of *Corynebacterium diphtheriae.* (1,000×).

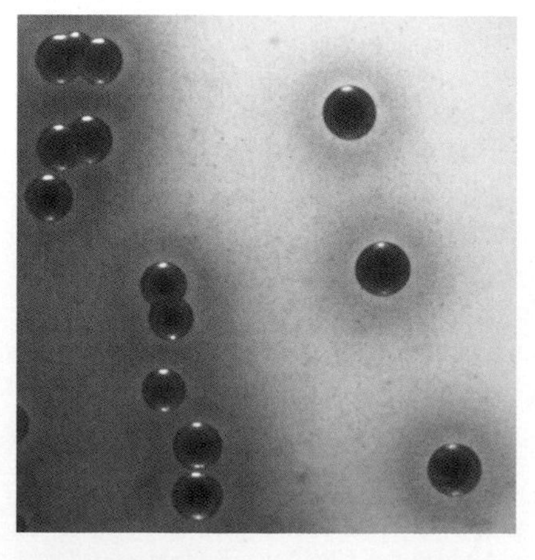

**Figure 191**
Ccolonies of *Corynebacterium diphtheriae* on a tellurite-containing medium.

**Diagnosis.** Active disease is diagnosed on clinical grounds, an abnormal chest radiograph, acid-fast organisms in sputum or other specimens (Figure 193), and isolation of the organism. Various approaches are used to isolate or to detect mycobacteria, including the BBL Septi-Check AFB system (Figure 195) and the BBL *Mycobacterium* Growth Indicator Tube (Figure 196).

***Mycobacterium avium-intracellulare (M. avium complex, MAC)***

The majority of disease states caused by *M. avium-intracellulare* (Figure 200) and other MOTT bacilli appear as chronic pulmonary disease (Figure 199), local lymph node inflammation, and diseases of skin, soft tissue, bones, and joints. Overwhelming MOTT-disseminated disease also occurs in patients with AIDS. MAC causes the third most common opportunistic bacterial diseases affecting adults with HIV infections.

***Mycobacterium leprae***

*Mycobacterium leprae* is the causative agent of leprosy (Hansen's disease), a chronic tumor-associated (granulomatous) disease of the peripheral nerves and mucous membranes (mucosa) of the nose.

**Transmission.** Transmission of *M. leprae* is believed to occur most commonly by contamination of the nasal mucosa or skin lesions with infectious secretions. The reservoir of the disease is the infected human.

**Cultural Properties.** Growth of *M. leprae* in artificial media or in tissue cultures is generally limited to only a few generations. The organism can be successfully grown in the foot pads of normal mice, irradiated thymectomized mice, and the armadillo, which may also be infected by natural means.

**Pathology and Clinical Features.** Two major forms of leprosy are recognized; **tuberculoid** and **lepromatous**. Intermediate forms also occur. The incubation period for the disease, which in general is very long, is measured in years.

Individuals with tuberculoid leprosy (Figure 201) can exhibit large flattened patches with raised or elevated red edges and dry, pale, hairless centers on any body surface. Loss of sensation on the skin also may develop as a result of invasion of peripheral sensory nerves.

Individuals with lepramatous leprosy exhibit extensive skin involvement, with thickening of looser skin parts of the lips, forehead, and ears. The classic lion face is typical of such infections. Extensive penetration of *M. leprae* into body tissue can cause severe body damage resulting in the loss of facial bones, fingers, and toes.

**Diagnosis.** Tuberculoid leprosy is diagnosed on clinical features. The lepromatous form involves demonstrating *M. leprae* in stained infected tissues (Figure 202).

***Mycobacterium smegmatis***

*M. smegmatis* (Figure 203) is an acid-fast rod that can be isolated as a common contaminant from urine specimens taken from males. Finding this bacterium in urine does not necessarily indicate an infection.

### *Bacillus of Calmette and Guérin (BCG)*

The bacillus of Calmette and Guérin (Figure 204) is the basis for the only available live vaccine used as a preventative measure against tuberculosis. The vaccine developed by Albert Calmette and Camile Guérin was originally derived from a *Mycobacterium bovis* strain that was weakened by repeated subculturing. *M. bovis* is the cause of bovine tuberculosis. Once a serious health threat, this disease is now rarely encountered.

## Nocardia

### *Nocardia*

The nocardia are widely distributed in soil. Some species are pathogenic opportunists for humans and lower animals. *Nocardia asteroides* and *N. brasiliensis* are of medical importance. These organisms cause pulmonary infection (*N. asteroides*), and skin and subcutaneous infections (*N. brasiliensis*).

**Transmission.** Pulmonary nocardiosis is acquired by inhalation of the disease agent, which is present in dust and soil and on contaminated mucosal surfaces. Many individuals with the disease have poorly functioning immune systems.

Skin and related infections generally result from direct inoculation of *Nocardia* by a thorn or a wood sliver.

**Morphology and Cultural Properties.** *Nocardia* species are gram-positive rods or are coccoid in shape and measure 0.50–1.20 μm in diameter. Cells are partially acid-fast (Figure 205) and give a beaded appearance.

*Nocardia* are aerobic and grow well on most enriched media, forming colonies within 2 to 7 days and exhibiting a variety of colors including tan, gray, orange, purple, and white (Figure 206).

**Diagnosis.** Laboratory diagnosis includes direct smears from infected sites, culture, and biochemical tests.

## The Mycobacteria

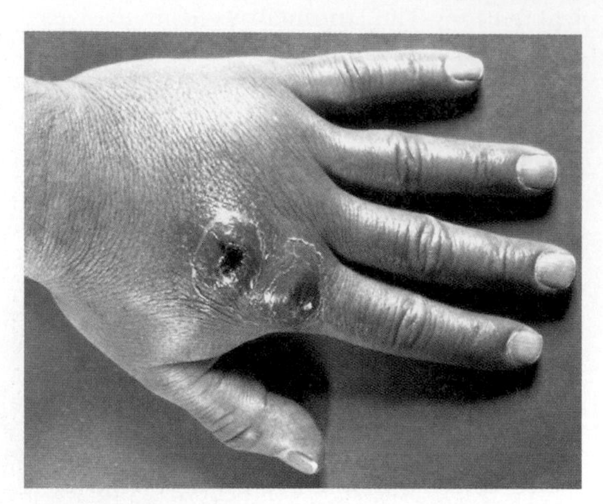

**Figure 192**
Skin ulcers and disease caused by *Mycobacterium marinum.*

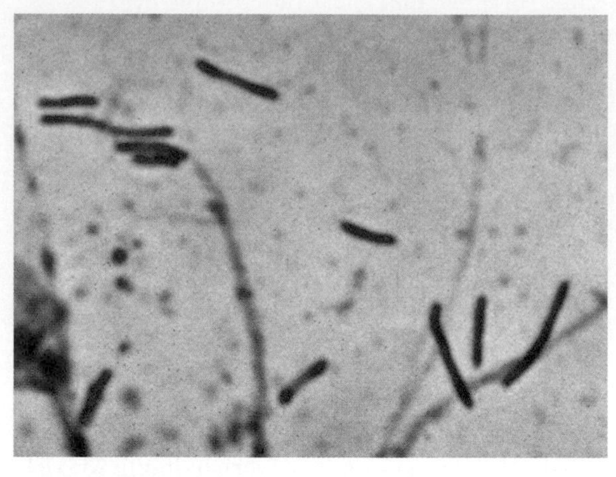

**Figure 193**
The acid-fast rods of *Mycobacterium tuberculosis.*

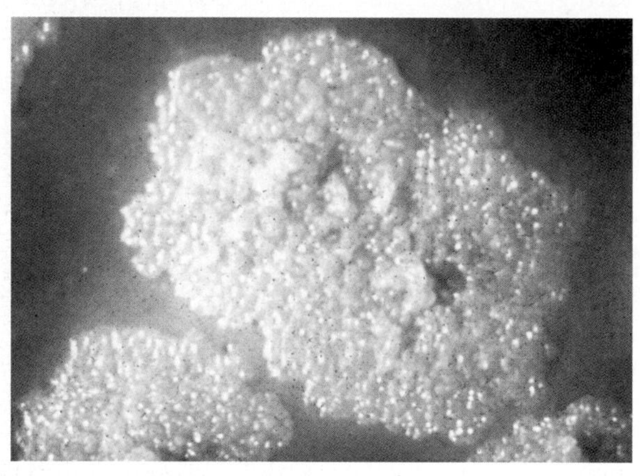

**Figure 194**
Colonies of *Mycobacterium tuberculosis.*

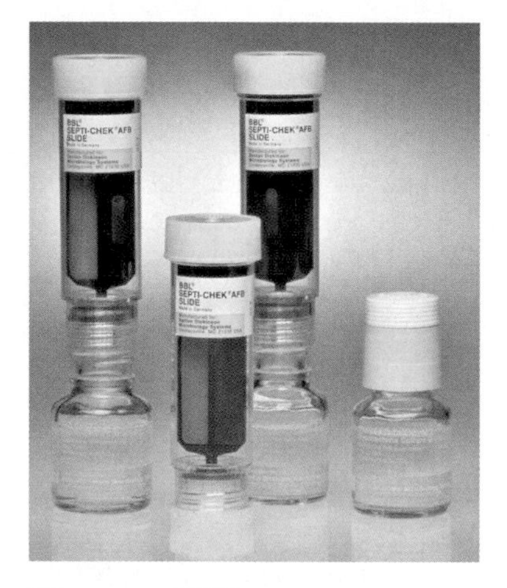


**Figure 195**
BBL™ Septi-Chek AFB *Mycobacterium* culture/subculture system. (Courtesy of Becton Dickinson Microbiology Systems.)

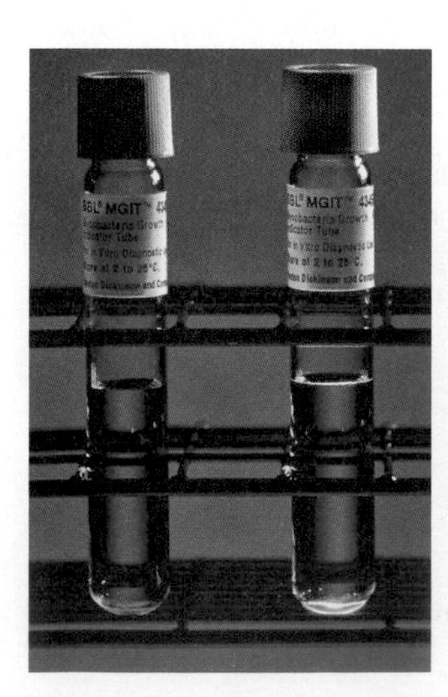


**Figure 196**
The BBL™ *Mycobacterium* Growth Indicator Tube (MGIT)™. Growth of mycobacteria is indicated by an orange fluorescent glow upon exposure to a long-wave ultraviolet light source. (Courtesy of Becton Dickinson Microbiology Systems.)

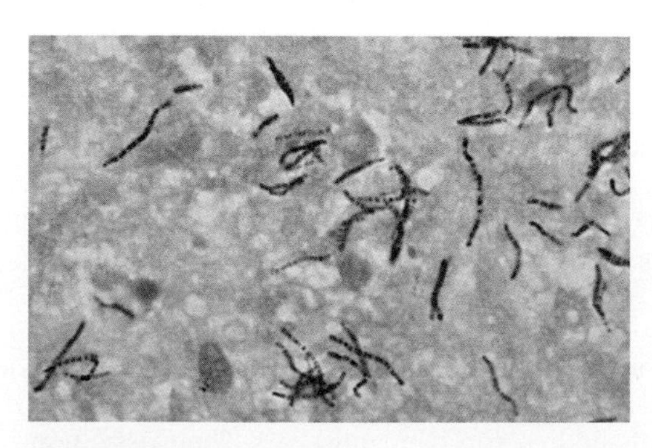

**Figure 197**
A Fite-stained lymph node specimen containing *Mycobacterium kansasii.* The rods are long and beaded (1,000×). (From F.S. Jannotta and M.K. Sidaway. *Arch. Path. Lab. Med.* 113(1989):1120.)

## The Mycobacteria *(Continued)*

**Figure 198**
*Mycobacterium kansasii* colonies.

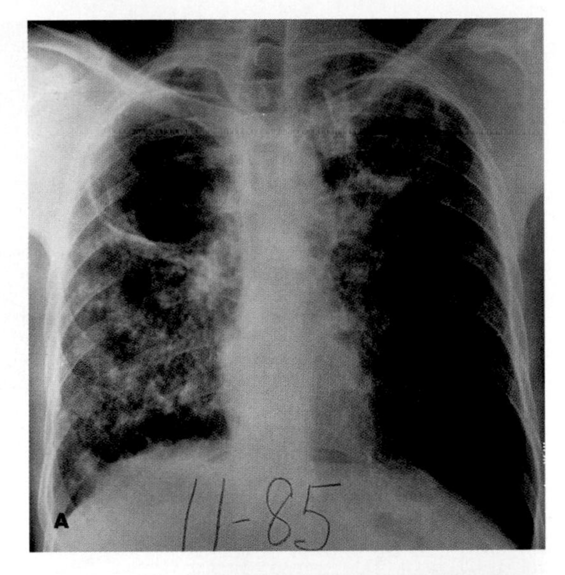


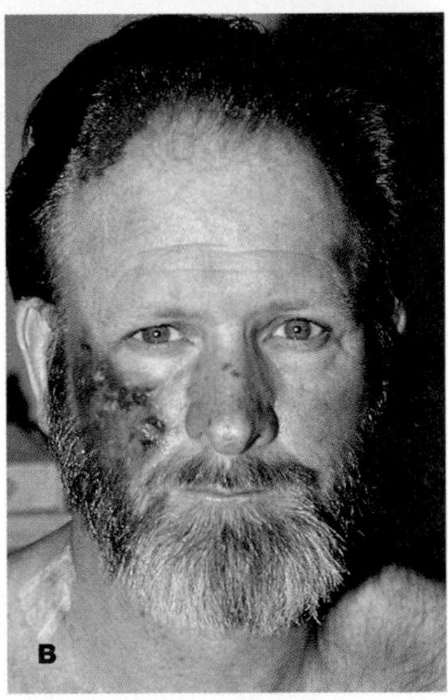


**Figure 199**
*Mycobaterium avium* complex (MAC) clinical states. (a) An x-ray showing extreme involvement of the lungs (large dense area).(From E. Wolinsky. *CID* 15(1992):1–12.) (b) A human immunodeficiency virus–infected person with a nontuberculous, atypical mycobacterium skin infection caused by MAC. (From S.T. Nedorost, B. Elewski, J.W. Tomford, and C. Camisa. *Internat. J. Dermatol.* 30(1991):491–97.)

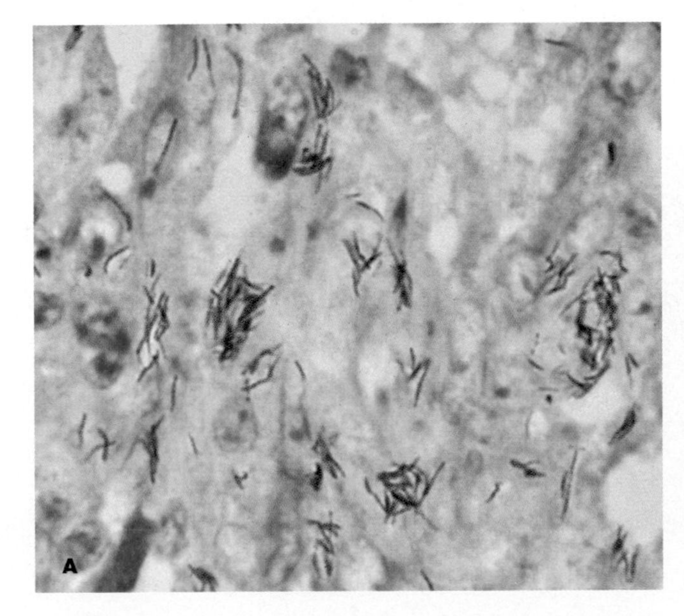


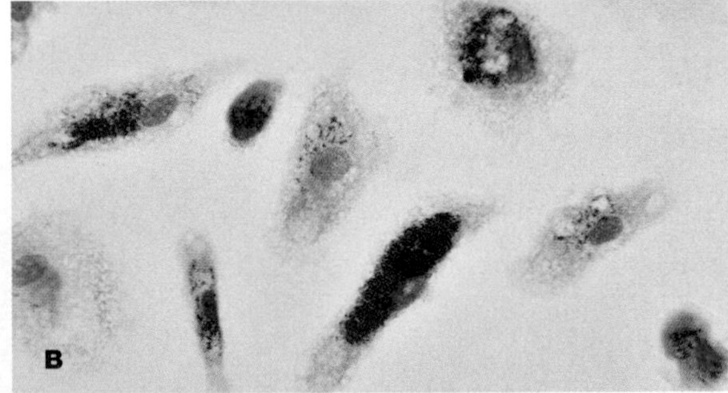


**Figure 200**
Microscopic views of *Mycobacterium avium-intracellulare* (MAC). (a) A Fite-stained lymph node specimen. The rods are short and beaded (1,000×). (From F.S. Jannotta and M.K. Sidaway. *Arch. Path. Lab. Med.* 113(1988):1120.) (b) A skin specimen showing intracellular acid-fast rods. (From S.E. Hoffner, G. Kallenius, and S.B. Svenson. *Res. Microbiol.* 143(1992):391–98.)

## The Mycobacteria *(Continued)*

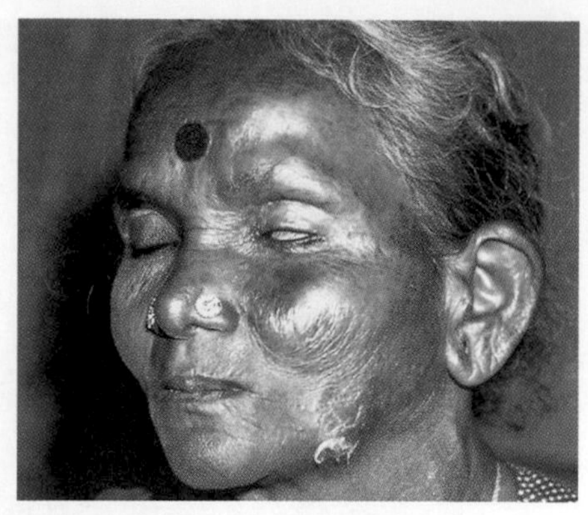

**Figure 201**
A case of early leprosy. (From M. Hogeweg, M. *Trop. Doctor* Suppl. 1(1992):15–21.)

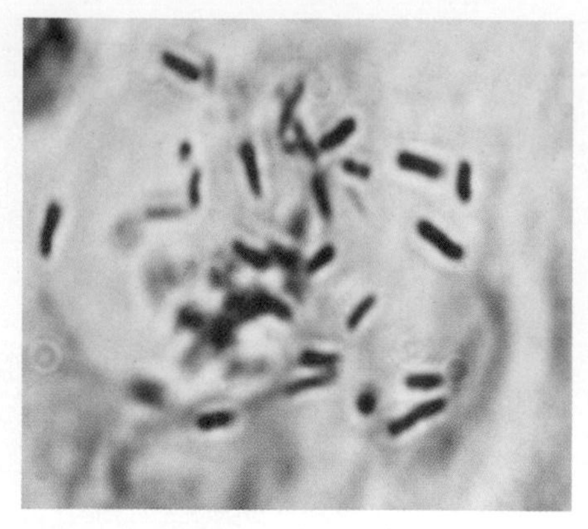

**Figure 202**
The acid-fast reaction of *Mycobacterium leprae.*

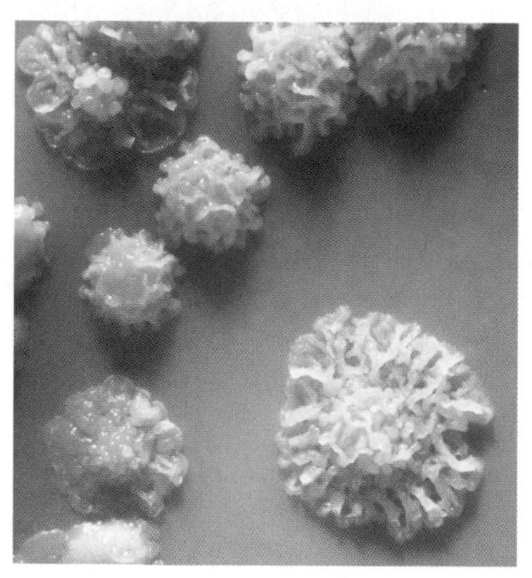

**Figure 203**
Colonies of *Mycobacterium smegmatis.*

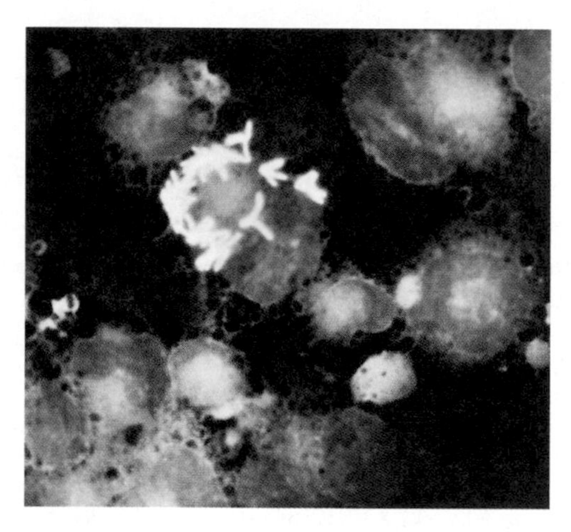

**Figure 204**
Bacillus of Calmette and Guérin (BCG) as shown by immunofluorescence. (From F.C. Bange, A.M. Brown, and W.R. Jacobs Jr. *Inf. Immunol.* 64(1996):1794–99.)

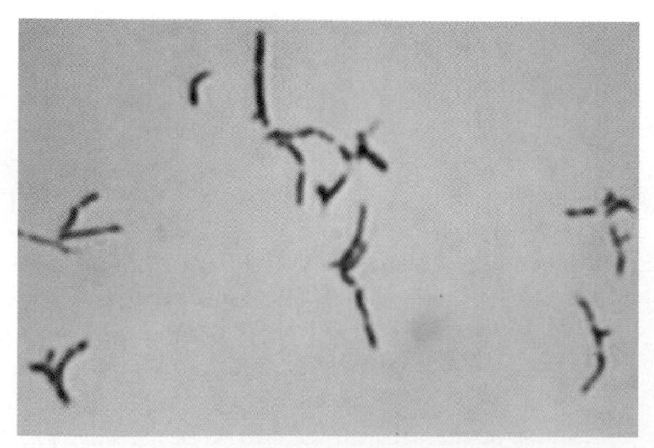

**Figure 205**
Acid-fast rods of *Nocardia asteroides* (1,000×).

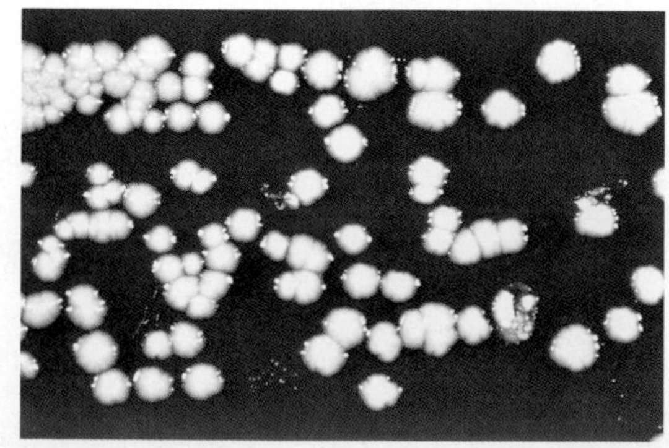

**Figure 206**
*Nocardia asteroides* colonies on blood agar.

SECTION 4

# Mycology

*There are more things in heaven and earth, Horatio,*
*Than are dreamt of in your philosophy.*

*—William Shakespeare, Hamlet*

Mycology (the study of fungi) had modest beginnings in the eighteenth century, often as a Sunday hobby for physicians. Today, fungal forms of life are still of interest to physicians, but a number of other specialists have been added to those concerned with fungi. These include the geneticist, genetic engineer, molecular biologist, plant pathologist, ecologist, and commercial microbiologist. This list reflects the widespread distribution of fungi, their usefulness as research organisms, and their involvement in many aspects of everyday life. Here is a brief exploratory look at this group of microorganisms.

## Properties of Fungi

Like bacteria, the fungi are extremely diverse. Unlike bacteria, however, they have a eukaryotic form of cellular organization. They are also much larger and contain the organelles typical of eukaryotic cells. In contrast to algae (the next group to be considered), the fungi lack chlorophyll, and, consequently, despite the fact that several are green, do not carry out photosynthesis. Fungi also have cell walls, which generally contain **chitin**, a polysaccharide that is found in the skeletons of insects, the shells of crabs, and related forms of life.

Although the fungi are a large and diverse group of eukaryotic microorganisms, three fungal groups have major practical importance: the *microscopic* **molds** (Figure 207a) and **yeasts** (Figure 207b) and the *macroscopic* **mushrooms** (Figure 207c,d).

Fungi can be found in a wide variety of habitats. Most are found in soil or on dead plant matter. Here, these types play crucial roles in the mineralization of organic carbon in nature. Equipped with some of the most powerful digestive enzymes known, decomposing fungi are capable of reducing wood, fiber, and foods to their basic chemicals with staggering efficiency. Like animals, fungi are heterotrophic and must obtain preformed organic substances from their environment. Some fungi are aquatic, living primarily in fresh water, and a few marine forms are known.

Fungi also are recognized for their disease-producing capabilities. Plants and animals including humans are susceptible. These pathogenic fungi have the enzymes necessary to obtain nutrients directly from the living host.

### Molds and Yeasts

Fungi are identified in the laboratory on the basis of their vegetative and reproductive structures (Figures 208 to 211). The most commonly seen fungi in the laboratory exist in one or both of two forms: molds and yeasts.

Molds, or filamentous fungi, form tubelike filaments called **hyphae** (singular, *hypa*). Some hyphae have cross-walls, or **septa** to separate cells (Figure 208a), others lack septa and are called **coenocytic,** or **nonseptate**. Hyphae begin as single filaments and then branch repeatedly, forming a network known as a **mycelium** (plural, *mycelia*). Many are dry, cottony, raised masses of branching hyphae (Figures 221 through 225). Three basic types of mycelia are recognized: vegetative, aerial, and reproductive. The rhizoids of a vegetative mycelium penetrate the surface of nutrient materials and absorbs nutrients (Figure 208b); an aerial mycelium grows above surfaces of nutrients, such as an agar medium; and a reproductive mycelium gives rise to and bears the reproductive structures called **spores** (Figures 208a and 209).

Asexual reproduction may involve the formation of **conidia** (singular, *conidium*), or spores (Figure 208). Asexual spores, known as **sporangiospores** (Figure 208c), form within a saclike structure known as a **sporangium**. They are the result of cytoplasmic division in the sporangium. The sporangia are typically formed on specialized hyphae known as **sporangiophores**. A distinctive domelike internal structure, the **columella**,

# MYCOLOGY
## Properties of Fungi

**Figure 207**
Examples of fungi. (a) The cottony mycelium (colony) of the mold *Phialiophora repens* on potato dextrose agar. (From M. Hironga and associates. *J. Clin. Microbiol.* 27(1989):394–99.) (b) Colonies of Brewer's yeast, *Saccharomyces cerevisiae.* (c) A group of common mushrooms, *Lepiota cristata,* growing on a lawn. (d) The shelf mushroom, *Polyporus versicolor,* growing on dead wood.

is formed during sporangium development. Identification of fungi belonging to the phylum Zygomycetes (Table 12) is based on the presence of these reproductive structures and the rootlike rhizoids (Figure 208b). Several fungi are able to form thick-walled, resistant spores called chlamydoconidia (**chlamydospores**) (Figure 240b). Table 13 briefly describes the features of conidia.

Conidia are nonmotile reproductive structures and typically are produced on aerial hyphae (Figure 209). Conidia can be single-celled or can appear as multicompartment forms separated by internal crosswalls. The crosswalled type is known as **macroconidia** (Figures 209d and 231c). A variety of conidia are formed by the fungi. Their development, arrangement, and microscopic appearance are of major importance to rapid and precise identification. Spores and conidia often are pigmented and resistant to drying. They serve to spread fungi to new habitats. Their presence gives a rather dry, dusty appearance to mycelial surfaces.

Some molds also produce sexual spores, which result from sexual reproduction (Figure 210). Fungal species belonging to different taxonomic groups produce different sexual spore types (see Table 12).

**Table 12** Classification and General Properties of Fungi

| Group | Common Name | Hyphae | Asexual Structures | Type of Sexual Spore | Typical Representatives |
|---|---|---|---|---|---|
| Ascomycetes (Ascomycota) | Sac fungi | Septate | Conidia, blastoconidia | Ascospore | *Neurospora, Saccharomyces, Morchella* (morels) |
| Basidiomycetes (Basidomycota) | Club fungi, mushrooms | Septate | Basidiospores | Basidiospore | *Amanita* (poisonous mushroom), *Agaricus* (edible mushroom) |
| Zygomycetes (Zygomycota) | Bread molds | Coenocytic | Spores, conidia | Zygospore | *Mucor, Rhizopus* (common bread mold) |
| Oomycetes (Chytridiomycota) | Water molds | Coenocytic | Spores, conidia | Oospore | *Allomyces* |
| Deuteromycetes (Fungi Imperfecti) | Fungi imperfecti | Septate | Conidia, blastoconidia, phialoconidia | None | *Penicillium,* Aspergillus, Candida |

**Table 13** Examples of Conidia

| Conidium Type | Brief Description |
|---|---|
| Arthroconidium (arthrospores) | Produced by fragmentation of fertile hyphae |
| Blastoconidium (blastospores) | Formed by outgrowth of new cells |
| Chlamydoconidium (chlamydospores) | Formed by enlarging and developing thick walls; produced during unfavorable conditions |
| Macroconidium | Large multiseptate spores |
| Microconidium | Small spores |
| Phialoconidium | Formed at the tip of a flask-shaped specialized conidiophore, the phialide |

Yeasts generally form, smooth, moist colonies (Figure 207b). Microscopically, they are usually spherical or oval (Figure 208d). Yeasts reproduce by **budding,** a process by which a new cell (**blastoconidium,** or bud) forms as a small outgrowth of the parent cell. The bud gradually grows and then breaks away. Yeasts do not form filaments or mycelia. At times, yeasts may not separate from one another and form a chain called a **pseudomycelium**. Some yeasts also can undergo sexual reproduction, resulting in the formation of a fertilized cell called a **zygote**.

### Dimorphism

Fungi that are able to grow as two different forms are referred to as being **dimorphic**. This ability is usually temperature-dependent, but it may be influenced by other factors as well. For example, *Blastomyces dermatitidis* (Figure 211), the fungus that causes blastomycosis, an infection of the lungs and other body tissues, grows as a yeast at 37°C and as a mold at 23° to 25°C (room temperature).

### Classification

A number of fungal properties are used for purposes of identification and classification. These include the type of hyphae and spores produced, habitats, biochemical reactions and cultural characteristics. Table 12 compares the major phyla of fungi.

## Culture and Microscopic Techniques

Although, with few exceptions, the direct microscopic examination of specimens rapidly yields important information, the identification of a fungus depends on culture. Several types of media are available for a variety of purposes. As is the case for bacteria, media may be enriched, selective, or differential. Sabouraud's dextrose agar is a commonly used selective agar for fungal cultivation. Other media containing different ingredients such as bird seed (Figure 212a) and dyes (Figure 212b) are used for the isolation and identification of fungi.

### Direct Microscopic Examination

Because fungi tend to grow more slowly than other microorganisms or are more difficult to isolate, significant attention is placed on direct microscopic examination and related techniques. Several methods are available for direct microscopic examination.

## Properties of Fungi *(Continued)*

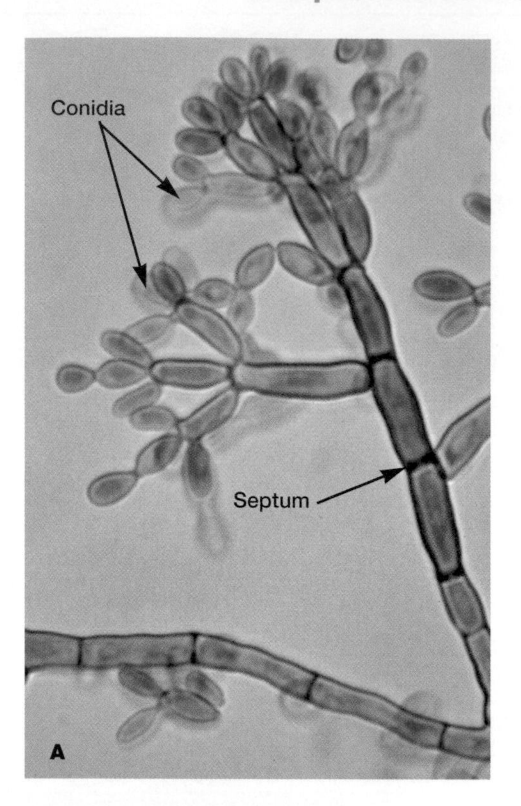


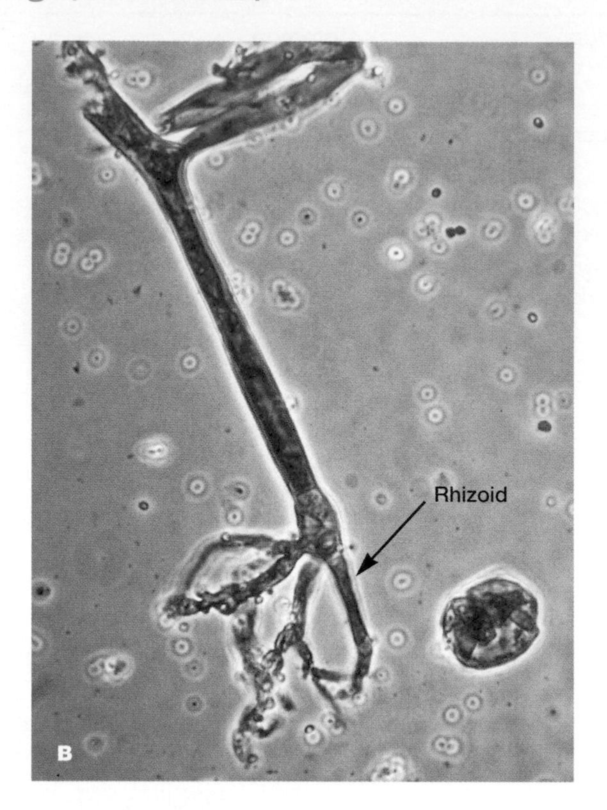


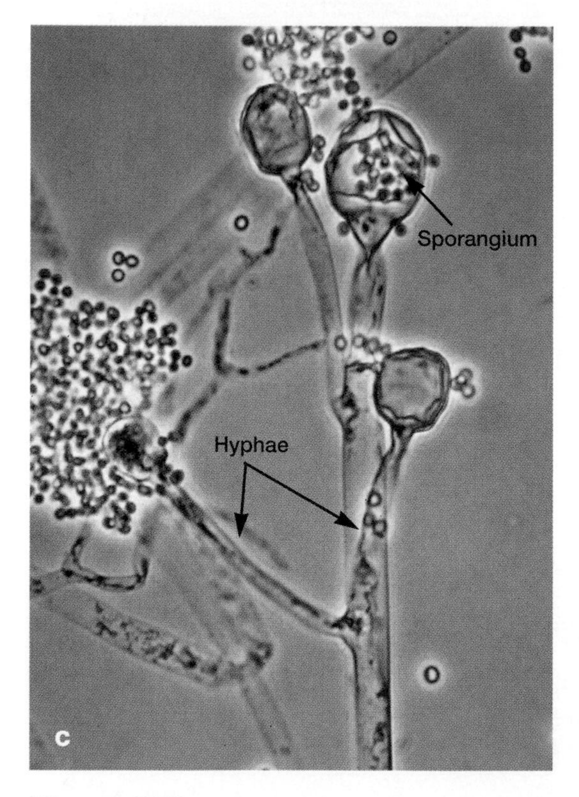


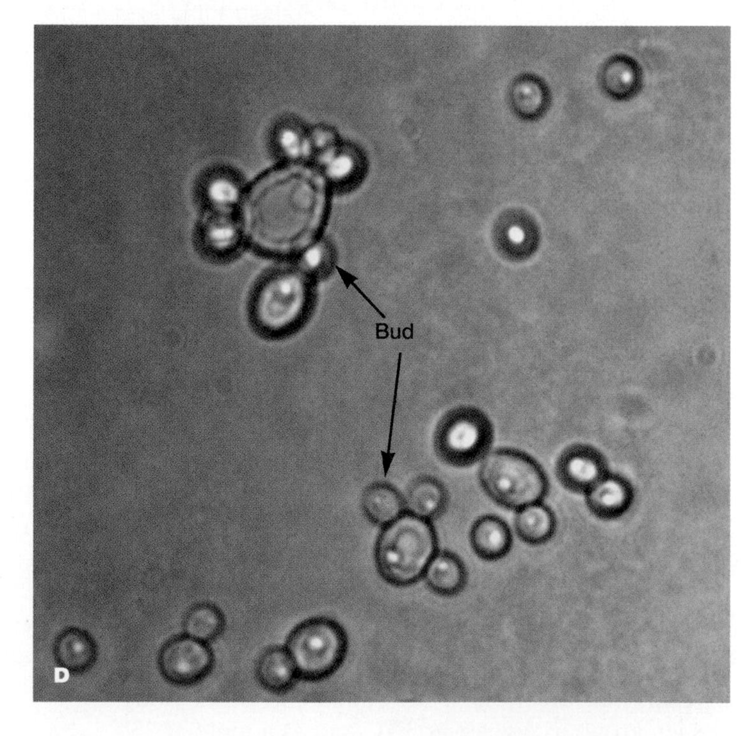


**Figure 208**
Microscopic views of fungi showing typical structures and certain differences between molds and yeast. (a) A microscopic view of a fungus showing septate hyphae and a round to oval type of spore called a *conidium*. (Courtesy of Dr. T. Matsumoto.) (b) The rootlike rhizoids. (c) The saclike sporangia and sporangiospores. Nonseptate hyphae also can be seen in these micrographs. (From G. St. Germain, A. Robert, M. Ishak, C. Tremblay, and S. Claveau. *Clin. Inf. Dis.* 16(1993):640–45.) (d) The yeast *Saccharomyces cerevisiae,* showing parent cells and buds.

## Properties of Fungi *(Continued)*

**Figure 209**

Microscopic views of conidia-producing fungi. (a) *Cladosporium carrionii.* The conidiophores produce long, branching chains of smooth-walled, oval, somewhat pointed conidia. (b) *Curvularia geniculata.* Conidia are large, usually contain four cells, and may appear curved owing to the swelling of a central cell. (c) *Exophiala spinifera.* Conidia are oval and gather in clusters at the end and sides of conidiophores and at points along filamentous hyphae. (Courtesy of Dr. T. Matsumoto.) (d) *Alternaria* species. Conidia are large and brown and have both transverse and longitudinal crosswalls. They are single or in chains.

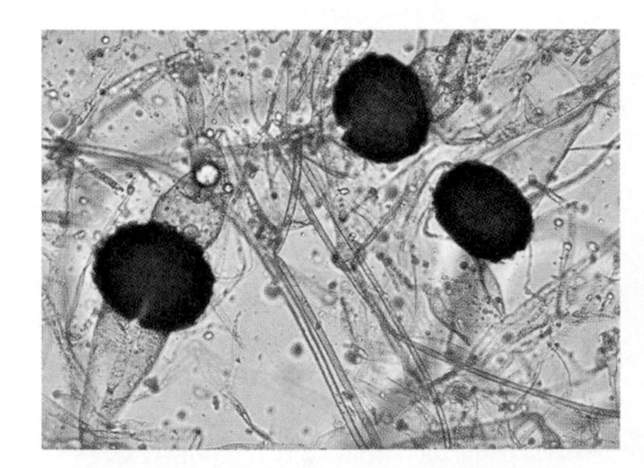

**Figure 210**

Representative thick black zygospores of *Rhizopus* species. Note the two hyphal branches connecting to the zygospores.

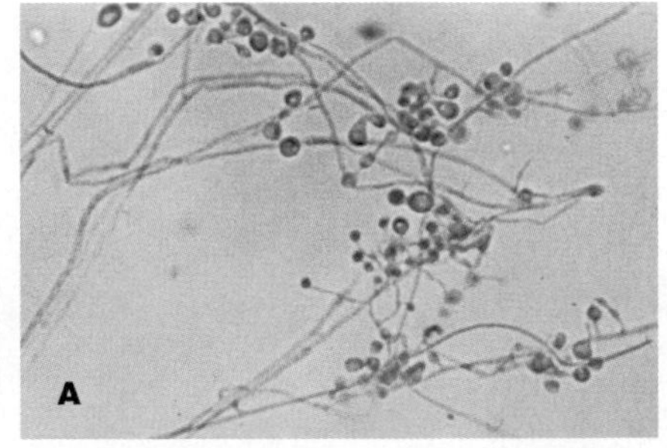


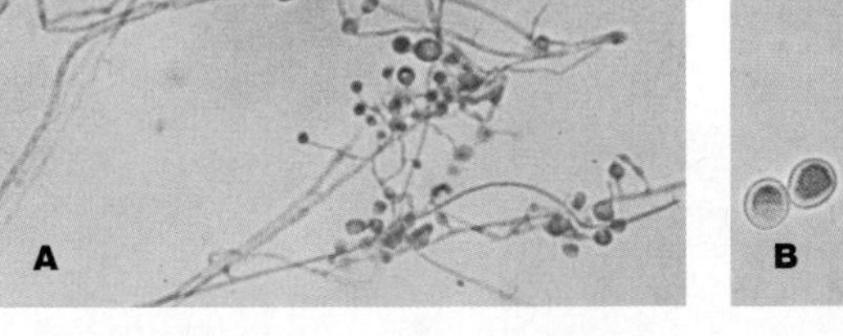


**Figure 211**

Dimorphism. (a) A temporary wet mount with lactophenol cotton blue stain preparation. *Blastomyces dermatitidis* in its mycelial phase growing at 25°C. Single conidia are directly attached to hyphae or are on small conidiophores (400×). (b) Yeast phase of the fungus growing at 37°C. Single-budding cells can be seen (400×). (From S.E. Hoy. and T.-Y. Chuang. *Cutis.* 48(1991):193–96.)

1. **Temporary wet mount**. Direct examination of a specimen taken from a mycelium can be achieved by placing it in a good-sized drop of lactophenol cotton blue, a mounting fluid containing both disinfectant (phenol) and stain (cotton blue). Lactic acid acts as a clearing agent and aids in preserving fungal structures (Figures 211a and 213a).
2. **Potassium hydroxide (KOH) preparation**. This technique involves placing a specimen into a drop of KOH and then mixing and gently heating it. After cooling, the preparation is examined. The KOH digests protein and other types of tissue debris, thereby making it easier to see fungal structures (Figure 213b).
3. **India ink preparation**. India ink is used to highlight certain structures such as capsules (Figure 214). The India ink does not stain the structure but provides a dark background against which the capsule can be seen.

A number of techniques also are available to demonstrate and to study fungi in tissues. These include the acid-fast stain, modified Gram stain (Figure 216), and specific histopathologic stains such as:

1. **Gomori-methenamine-silver nitrate stain**. This staining procedure sharply outlines fungi in black due to the depositing of silver on their cell walls (Figure 215). The internal parts of hyphae stain deep rose to black and the background appears green.
2. **Periodic acid-Schiff stain**. The hydroxyl groups in the carbohydrates of fungal cell walls are oxidized by the periodic acid to form aldehydes. The aldehydes in turn combine with basic fuchsin dye to produce a pink-purple color (Figure 235b).

Immunofluorescence microscopy and related techniques also are used to detect fungi directly in tissues and body fluids.

## Culture and Microscopic Techniques

**Figure 212**
Culture Techniques. (a) On the right, orange to brown colonies of *Filobasidiella* (*Cryptococcus*) *neoformans* on *Guizotia abyssinica*–creatinine agar are shown. On the left, nonpigmented colonies of the same organism growing on Sabouraud's glucose agar. (Courtesy of Dr. F. Staib, Former Chief, Mycology Unit, Robert Koch Institute.) (b) Colonies of *F. neoformans* (dark blue) and *Candida albicans* (light blue) on Sabouraud's glucose agar containing trypan blue.

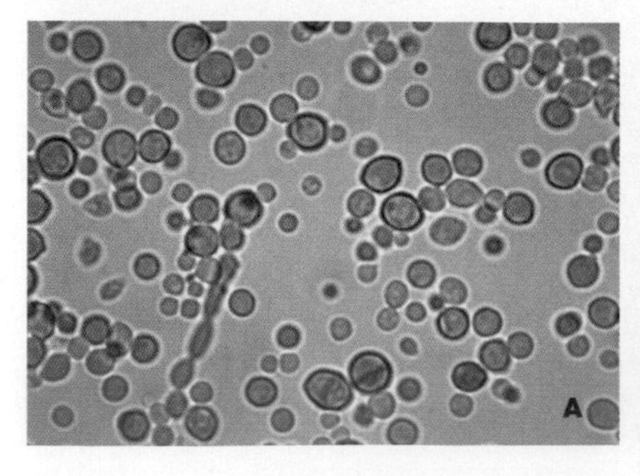


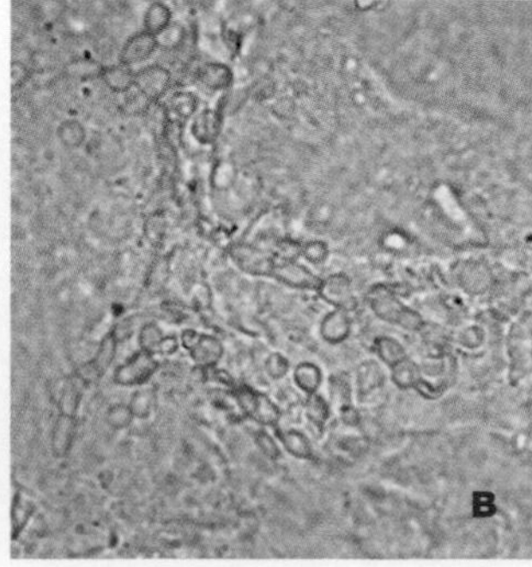


**Figure 213**
Microscopic techniques. (a) Temporary wet mount of a culture showing yeastlike cells and septate hyphae. (Courtesy of Drs. T. Matsumoto and T. Matsuda.) (b) A potassium hydroxide preparation of a nail specimen showing filamentous hyphae and spherical cells. (Courtesy of Drs. T. Matsumoto and T. Matsuda.)

## Culture and Microscopic Techniques *(Continued)*

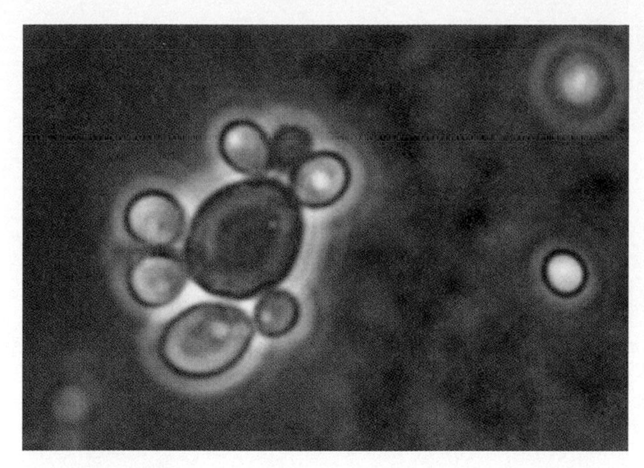

**Figure 214**
An India ink preparation of *Filobasidiella (Cryptococcus) neoformans.*

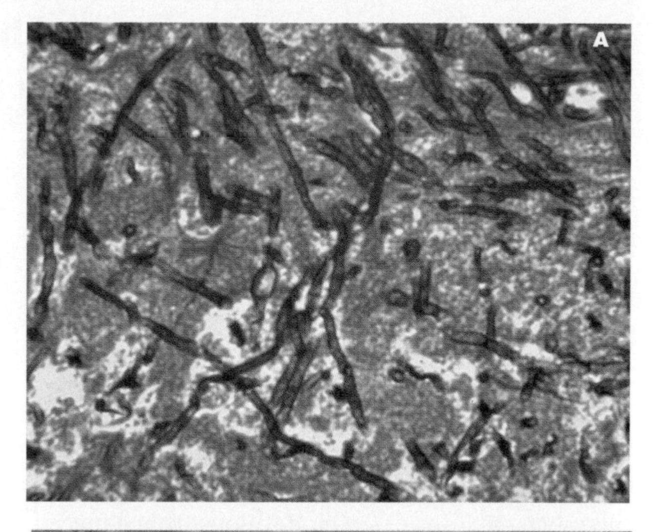


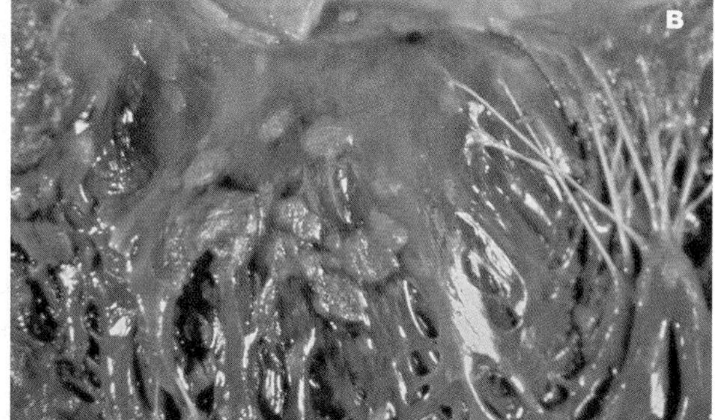


**Figure 215**
(a) A Gomori-methenamine-silver nitrate stained preparation of heart tissue showing the presence of hyphal elements. (b) A fungus-caused infective endocarditis. Specimens were taken from the irregular yellow-tan growths on the left ventricle. (Courtesy Dr. F. Welty, Harvard Medical School).

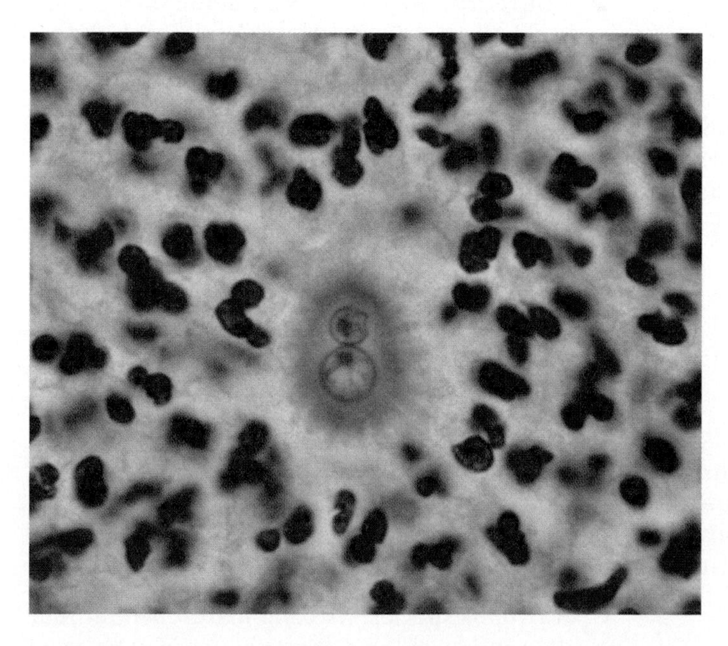

**Figure 216**
A Brown-Hopps Gram-stained preparation showing a budding yeast cell of *Sporotrichum (Sporothrix) schenckii* in a skin specimen (750×). (From R.C. Neafie and A.M. Marty. *Clin. Microbiol. Revs.* 6(1993):34–56.)

# Representative Fungi, Their Distinctive Properties, and Selected Disease States

## Macroscopic Views of Fungi

### Mushrooms

Mushrooms (Figures 217 and 218) represent a large group of filamentous fungi that typically form large structures called **fruiting bodies,** or **basidiocarps.** The most familiar of these visible forms include the gilled mushrooms and the fleshy fungi (Figure 217e).

The mushroom represents the reproductive phase of the fungus that formed it. It develops from a mycelium that is hidden from view in soil or another substrate. A mushroom is a specialized body that ensures the maximum production of spores (**basidiospores**). There are many variations of the basic mushroom structure. However, in general, a mushroom consists of a **cap,** or **pileus**, radiating layers of **lamellae** (gills) found on the underside of the cap and representing the spore-producing region, and a **stipe,** or **stalk,** on which the cap sits. Many mushrooms may have one or two layers of tissue, known as **veils**, on young fruiting bodies. Patches of veil material may be found on the cap and on portions of the stipe. A persistent ring of tissue on the stipe is called an **annulus**. Figure 217a shows the characteristic parts of a mushroom.

The identification of a particular mushroom involves noting the presence or absence of these parts and other properties including the size, shape, cap and stipe, color, surface appearance and texture, odor, color changes of structures following injury, habitat, and spore color. Determination of spore color is one of the most essential steps in mushroom identification. A spore print (Figure 220) is the technique usually used to determine spore characteristics as well as certain features of gills.

### Puffballs

Puffballs (earth-balls) are fleshy (later dry) fruiting bodies found in rich soil and decaying wood. The inside of the developing puffballs contains numerous small chambers containing club-shaped structures that produce large numbers of basidiospores.

### Morels

True morels (Figure 219) represent a group of tasty edible mushrooms, many of which are highly prized by connoisseurs of fine food. They are often found in forests around the world.

## Common Molds and Yeasts

A number of fungi are commonly studied in microbiology and related courses. Several of them are also considered as contaminants, but under certain circumstances they may be labeled as pathogens. Fungi such as *Rhizopus nigricans* (Figure 221), *Penicillium* species (Figure 223), *Fusarium* species (Figure 225a), and *Scopulariopsis* species (Figure 225c), though generally harmless to healthy persons, cause disease in individuals whose immune systems have been weakened by severe diseases such as AIDS or by immunosuppressive therapy. Some authorities refer to such fungi as **opportunists**.

The following descriptions of mycelia (both macroscopic and microscopic) are based on fungal growth on Sabouraud's dextrose agar.

1. ***Alternaria* species**. Mycelial surface is grayish white at first and later becomes greenish black or brown (Figure 224). Hyphae are septate and dark. Microscopic structures include conidiophores and brown macroconidia (Figure 209d).
2. ***Aspergillus* species**. Mycelial color, which varies widely, from white to shades of green, black, yellow, brown, and gray, usually is determined by the color of conidia. Microscopic structures include conidia (phialoconidia), conidiophores, vesicles, and small vessel-shaped structures (phialides) on which conidia are formed. (See Figure 222).
3. ***Fusarium* species**. Mycelia are cottony, often producing a diffusible pink, purple, or yellow pigment (Figure 225a).
4. ***Penicillium* species**. Mycelia are greenish blue, grow rapidly, and appear with a powdery or velvet texture (Figure 223). Some species secrete yellow surface droplets. Microscopically, the fungus appear as a broom or brush form (Figure 223b). Conidia (phialoconidia), conidiophores, and phialides also are seen.
5. ***Rhizopus* species**. Mycelia mature within four days and quickly cover media surfaces. Mycelial appearance is cottony and is white at first and then becomes dotted with black dots (sporangia). Microscopic structures include rhizoids, sporangiophores, sporangia, columellae, and sporangiospores (Figure 221b).

## Macroscopic Views of Fungi

**Figure 217**

Examples of macroscopic common fungi. (a) The structures of a mature mushroom (*Agaricus* species). The specific parts shown include the cap (pileus) attached to the stalk (stipe), a ring (annulus) around the stalk, and the lower surface of a cap showing the radiating strips of tissue called gills (lamellae). (b) The oyster mushroom (*Pleurotus ostreatus*) and its small buttons. (c) The bush or hen of the woods mushroom, *Polyporus frondosus.* (d) Cup-shaped mushroom, *Peziza* species. (e) An example of a fleshy mushroom.

## Macroscopic Views of Fungi *(Continued)*

**Figure 218**
Poisonous mushrooms. (a) *Amanita* species, a highly poisonous species. (b) Common earth-ball, *Scleroderma citrinum.*

**Figure 219**
*Morchella esculenta,* the morel, a highly edible variety.

**Figure 220**
A spore print.

6. ***Rhodoturula* species.** Mycelia develop rapidly and are soft and pink to coral in color (Figure 225b). Microscopically, *Rhodoturula* appear as budding cells that are round or oval.
7. ***Saccharomyces cerevisiae.*** Colonies appear as smooth, moist, and white to cream colored growths. *Saccharomyces* appear as budding yeast cells.
8. ***Scopulariopsis* species.** Mycelia range in color from light tan to dull gray (Figure 225c). Microscopic structures include conidiophores and chains of pear-shaped, thick-walled conidia.

## Funguslike Protists

Slime molds are examples of funguslike protists. They are eukaryotic and have properties similar to both fungi and protozoa. Slime molds can be divided into two groups, the **cellular slime molds,** which are composed of single ameba-like cells, and the **acellular slime molds,** which consist of **plasmodia**, naked protoplasm masses of indefinite size and shape (Figure 226).

## Spectrum of Mycoses

Most fungi that infect humans and other animals and cause disease are categorized by the tissue or organ levels that are the primary sites of involvement. These categories include **superficial, cutaneous, subcutaneous,** and **systemic mycoses**.

### Agents of Superficial and Cutaneous Mycoses (Dermatophytoses)

Superficial fungal infections (mycoses) involve only the outermost layers of the skin or parts of hair shafts. *Phaeoannellomyces werneckii* causes such skin infections, and *Trichosporon beigeli* and *Piedraia hortae* cause the hair shaft infections known as **white piedra** and **black piedra**, respectively. Potassium hydroxide preparations are helpful to diagnosis.

The cutaneous mycoses, or dermatophytoses, are infections of the hair, nail, or skin on a living host (Figures 227 and 228) and are caused by species of the three genera *Epidermophyton, Microsporum,* and *Trichophyton* (Figures 229 through 234). These fungi attack and are limited to the keratinized tissues. They are referred to as **dermatophytes**. Occasionally, the terms *tinea* and *ringworm* are used to indicate the diseases (dermatophytoses) caused. Table 14 lists several examples. In general, *Microsporum* species attack hair and skin but not nails. *Trichophyton* species attack hair, skin, and nails. *Epidermophyton floccosum* infects skin and occasionally nails. The dermatophytoses are generally acquired through direct contact with infected persons or other animals and fomites.

Endothrix and ectothrix are the two known forms of hair invasion. In the **endothrix** infection, hyphae grow down the hair follicle and penetrate the hair shaft (Figure 228). The **ectothrix** infection is characterized by spores, arthroconidia (formed by frequent fragmenting in the area surrounding the hair), forming a sheathlike covering, and hyphae growing around the hair shaft and eventually destroying it.

Diagnosis of a dermatophytosis includes KOH preparations of appropriate specimens and culture. Mycelial characteristics and microscopic examination of cultures are also used for fungal identification (Figures 229 through 234).

### Agents of Subcutaneous Mycoses

Subcutaneous mycoses affect the deeper layers of the skin (Figure 235). The causative fungi gain access to subcutaneous tissues through some form of injury. Examples of such diseases include chromoblastomycosis and sporotrichosis (Figure 236). The organisms causing these two diseases differ from other fungi that cause subcutaneous mycoses by being dimorphic.

#### *Chromoblastomycosis*

This mycosis begins after the causative agent is implanted into the skin. Lesions develop slowly and gradually increase in size to form a subcutaneous nodule or tumor (Figure 235a).

Fungi that cause chromoblastomycosis include *Fonsecaea pedrosoi* (Figure 235c–e), *Cladosporium carrionii* (see Figure 209a), and *Exophiala spinifera* (see Figure 209c). These organisms appear in tissue as thick-walled spherical cells that are called **muriform cells** or **Medlar bodies** (Figure 235b).

#### *Sporotrichosis*

This disease also is caused through tissue injury. The classic form of the disease includes the formation of numerous nodules, abscesses, and ulcers that develop along the lymphatics that drain the site of entry. The disease generally does not extend beyond this area.

The causative agent *Sporotrichum* (Sporothrix) *schenckii,* grows as a mold in cultures at 25°C and as yeastlike cells in tissues (Figure 236). These properties are of value in diagnosis.

### Agents of Systemic Mycoses

Several fungi are known to cause systemic mycoses. Only three are discussed here: *Blastomyces dermatitidis* (see Figure 211), *Coccidioides immitis* (Figure

## Common Molds and Yeasts

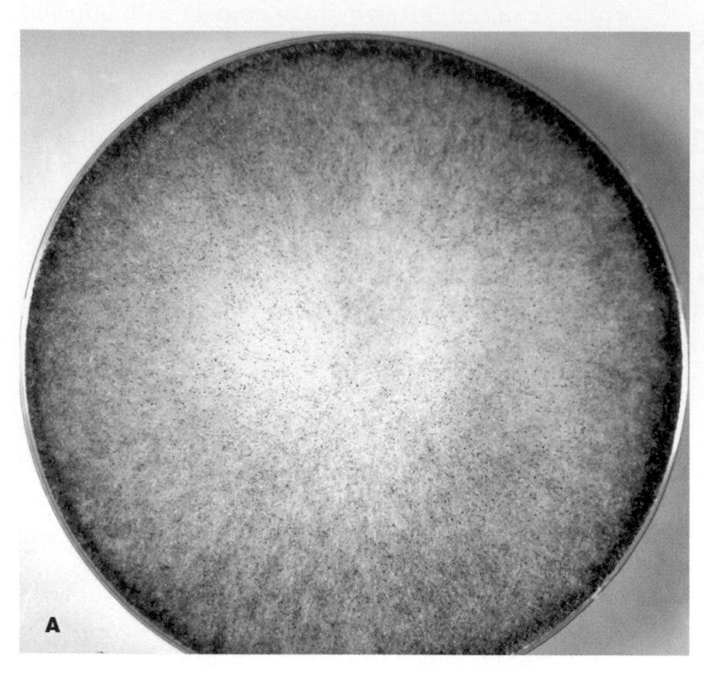


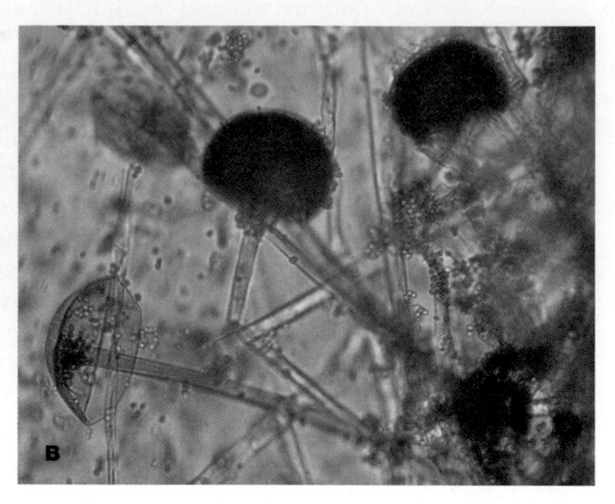


**Figure 221**
The bread mold *Rhizopus nigricans.* (a) The mycelium grown on Sabouraud's dextrose agar. Note the different pattern of growth. The black dotlike structures are sporangia (sacs of sporangiospores). (b) A microscopic view. The black sporangia (S) filled with spores, columella (C) and free spores can be seen.

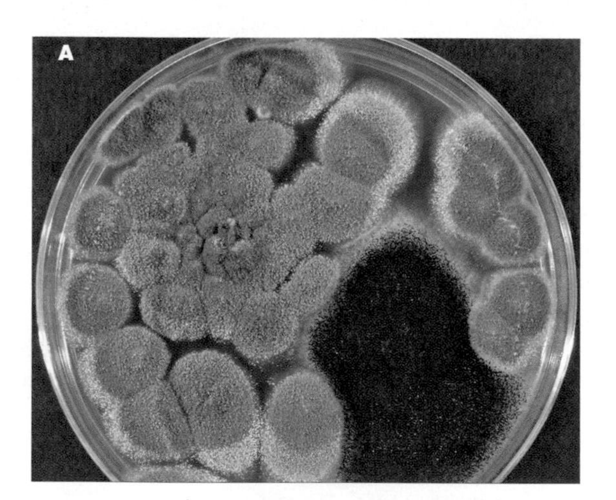


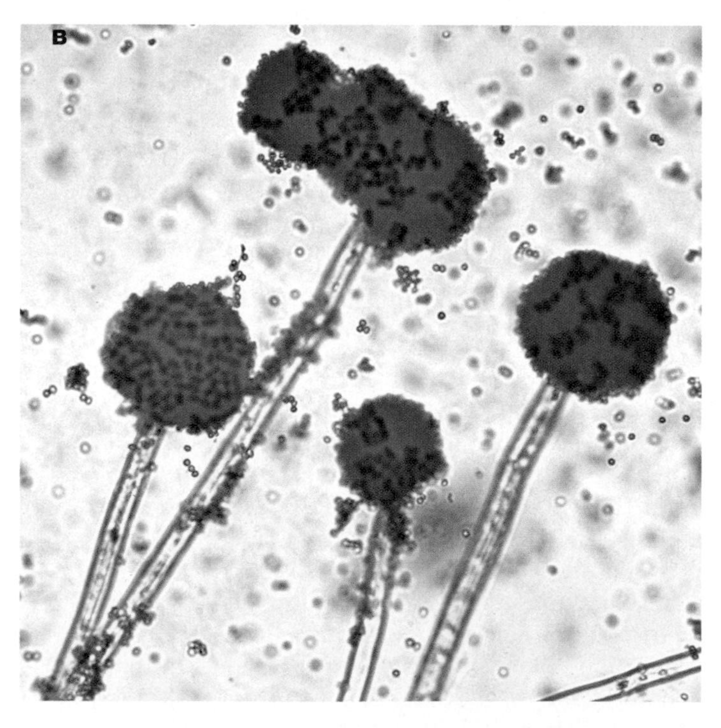


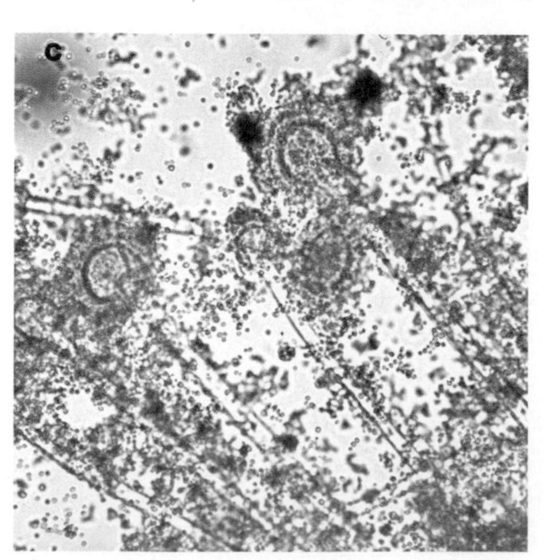


**Figure 222**
*Aspergillus* species. (a) Two different mycelia, *A. niger* (black) and *A. flavus* (greenish) are shown. Microscopic views of *A. niger* (b) and *A. flavus* (c) showing a conidiophore (lower supporting structure), the swollen vesicle connecting to the phialide area, and the oval conidia.

## Common Molds and Yeasts *(Continued)*

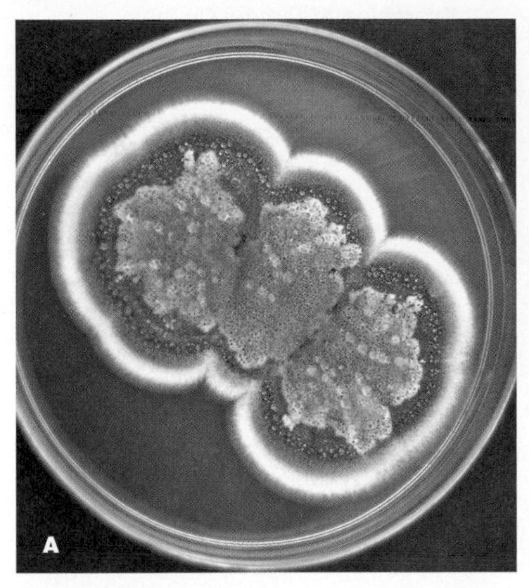


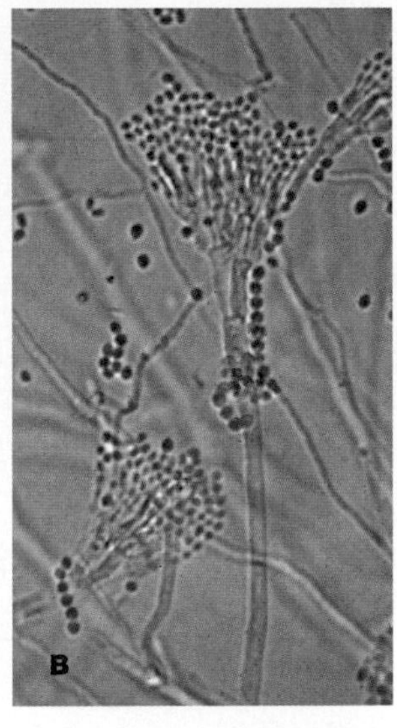


**Figure 223**
*Penicillium* species. (a) Mycelium. (b) Microscopic view showing lower supporting conidiophore, connecting to a phialide, and the oval conidia.

**Figure 224**
*Alternaria* species mycelium. See Figure 209d for the spores of the mold.

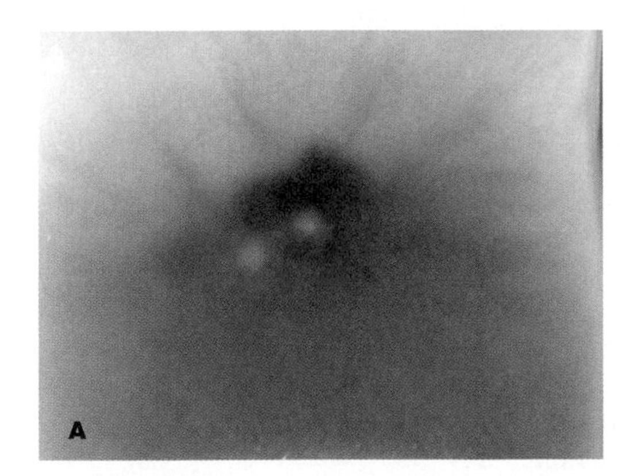


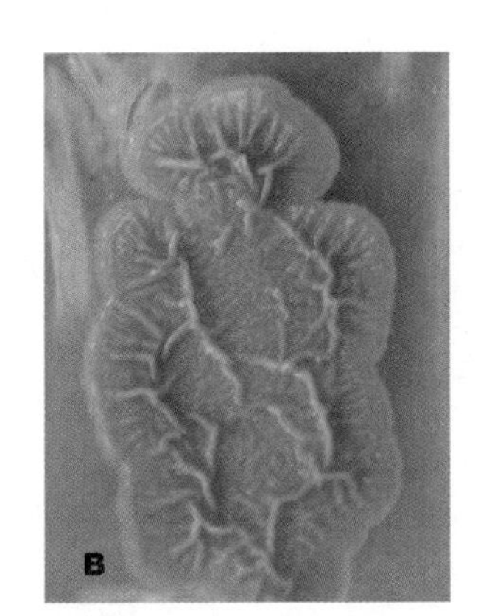


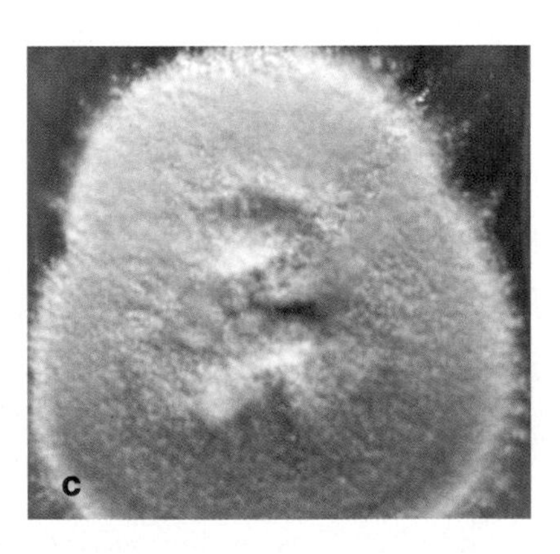


**Figure 225**
Common contaminants, all of which can cause infections in individuals with weakened immune systems. (a) *Fusarium* species. This mold forms a fluffy, cottony mycelium, which may be accompanied by a diffusible, pink, purple, or yellow pigment. (b) *Rhodoturula* species. This genus includes nonfermenting yeasts usually forming mucoid, yellow to red colonies. (c) *Scopulariopsis* species. This mold forms white to light beige or dull grayish mycelia.

## Funguslike Protists

**Figure 226**
The scrambled egg–appearing slime mold *Fuligo septica.* Both well-formed yellow plasmodia and filamentous newly forming (whitish) plasmodia are shown.

Table 14 Representative Dermatophytoses

| Disease | Description | Examples of Fungal Causes |
|---|---|---|
| Tinea barbae (ringworm of the beard) | Circular patches on bearded area; some loss of hair | *Trichophyton mentagrophytes T. rubrum, Microsporum canis* |
| Tinea capitis (ringworm of the scalp) | Involvement of the scalp and hair; crusty, scaly lesions on scalp | *Mycrosporum canis, Trichophyton tonsorans* |
| Tinea corporis (ringworm of the body) | Ringlike lesions with a central scaly area | *Trichophyton rubrum, T. mentagrophytes, Microsporum canis* |
| Tinea cruris (ringworm of the groin) | Ringlike, sometimes leathery, patches and itching in skin folds of pubic region | *Epidermophyton floccosum, Trichophyton rubrum, T. mentagrophytes* |
| Tinea pedis (ringworm of the feet) | Fluid-filled lesions, skin cracks, peeling, itching between toes | *Epidermophyton floccosum, Trichophyton mentagrophytes,T. rubrum* |
| Tinea unguium (ringworm of the nails | Hardening and discoloration of the nails | *Epidermophyton floccosum, Trichophyton mentagrophytes, T. rubrum* |

237), and *Histoplasma capsulatum* (Figure 238). Infection due to these fungi is usually acquired by inhalation of conidia from an environmental source.

### *Blastomyces dermatitidis*

Blastomycosis is a chronic, tumor-forming infection of the lungs that may spread to other tissues. The causative agent, *Blastomyces dermatitidis* is found worldwide and can be found in soil and wood.

**Transmission.** *Blastomyces dermatitidis* is inhaled. However, it is not spread from person to person or from lower animals to humans.

**Morphology and Cultural Properties.** *Blastomyces dermatitidis* is a dimorphic fungus. It appears as a mold (Figure 211a) at room temperature (25°C) and as a yeast in body tissues and on cultures grown at 27°C (Figure 211b)

**Pathology and Clinical Features.** Most infections are without signs or symptoms. Symptomatic disease usually appears as a mild respiratory infection, which usually is accompanied by fever, weight loss, general weakness, and a productive cough. If the fungus spreads, tissues such as skin, bones, joints, prostate gland, and testes are most commonly involved.

**Diagnosis.** The most rapid means of diagnosis is the direct demonstration of yeast cells in KOH preparations. Culture on routine mycologic media is also used but may take as long as 4 weeks.

### *Coccidioides immitis*

Coccidioidomycosis, or valley fever, is the oldest of the major systemic mycoses on the basis of reported cases. The disease is also the most geographically restricted of this group of diseases because the causative agent, *Coccidioides immitis,* grows only in soil in semiarid climates (dry areas with high summer temperatures and mild winters).

**Transmission.** Infection usually results from the inhalation of arthroconidia (Figure 237g). However, introduction of this fungus into other sites also can result in a self-limited infection (Figure 237a,b). Domesticated, zoo, and wild animals are susceptible to infection.

**Morphology and Cultural Properties.** *Coccidioides immitis* is a dimorphic fungus. On routine culture at room temperature and at 37°C it grows as a mold (Figure 237e). In tissue, instead of a yeast phase, the fungus produces large, distinctive, round-walled structures, **spherules,** containing up to hundreds of endospores (Figure 237f). This structure is unique among pathogenic fungi.

**Pathology and Clinical Features.** In the self-limited form of the disease, general symptoms include cough, chest pain, loss of appetite, general discomfort, and fatigue. More than half of infections show no symptoms, or the disease is so mild it is not detected. Disseminated coccidioidomycosis may involve cutaneous and subcutaneous tissues (Figure 237d), the meninges, and visceral organs.

**Diagnosis.** The most rapid means of laboratory diagnosis is the direct KOH demonstration of the thick-walled spherules in sputum and other specimens. Culture from sputum and skin and other tissues and serological and skin tests also are useful.

### *Histoplasma capsulatum*

Histoplasmosis is a disease of worldwide distribution. The causative agent, *Histoplasma capsulatum,* can be found in soil, chicken or pigeon coops, and bat caves.

## Superficial and Cutaneous Mycoses (Dermatophytoses)

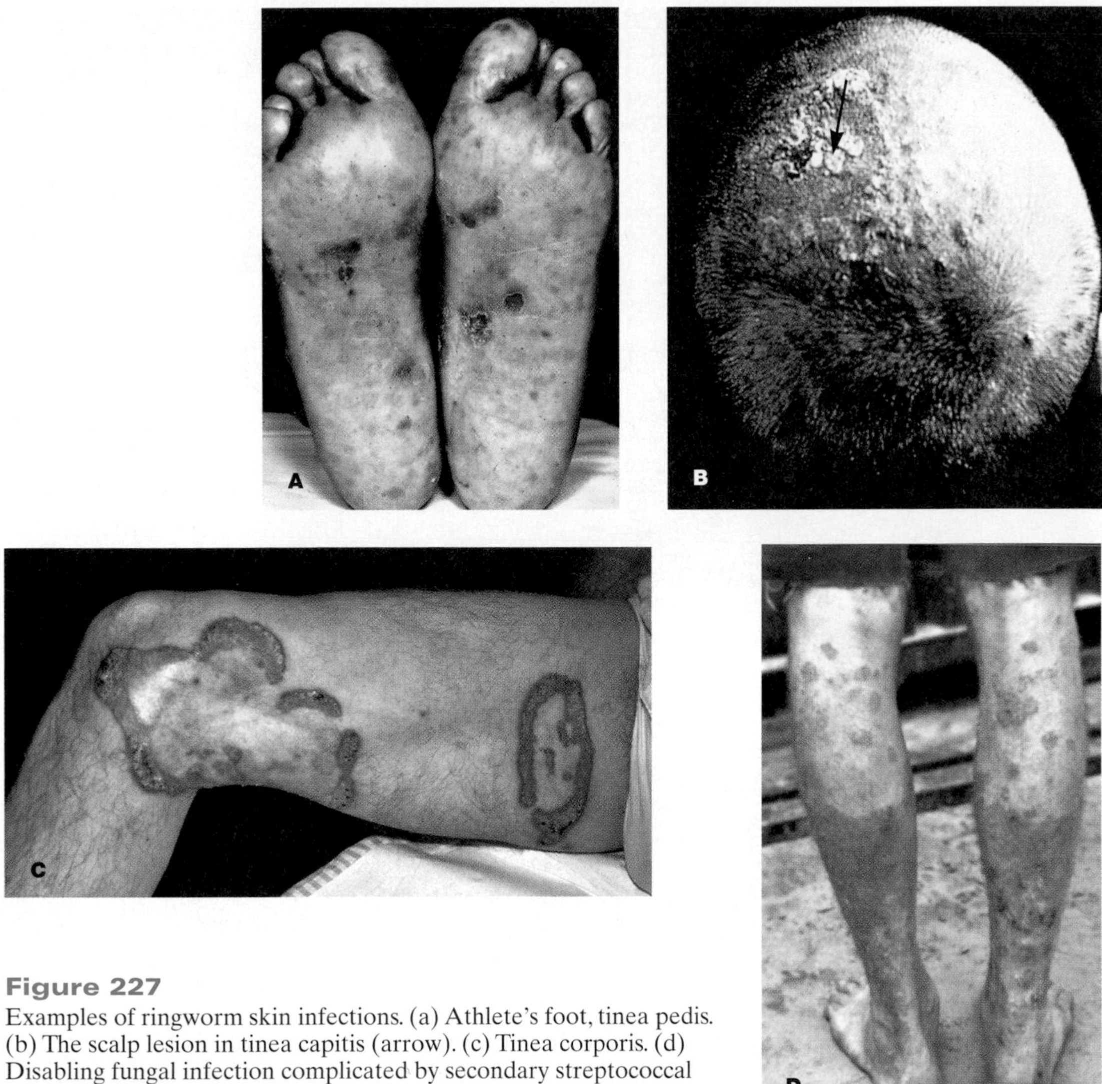


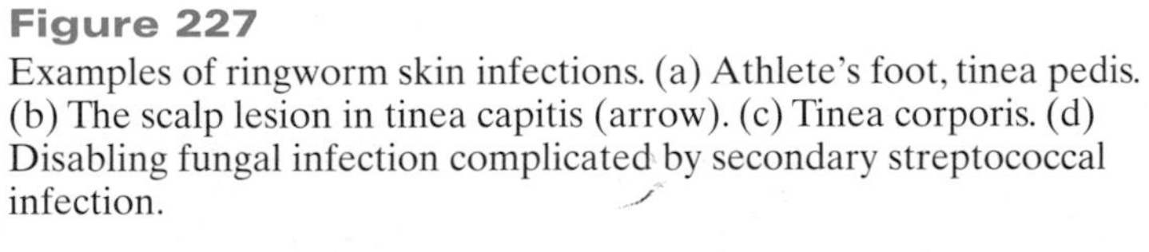

**Figure 227**
Examples of ringworm skin infections. (a) Athlete's foot, tinea pedis. (b) The scalp lesion in tinea capitis (arrow). (c) Tinea corporis. (d) Disabling fungal infection complicated by secondary streptococcal infection.

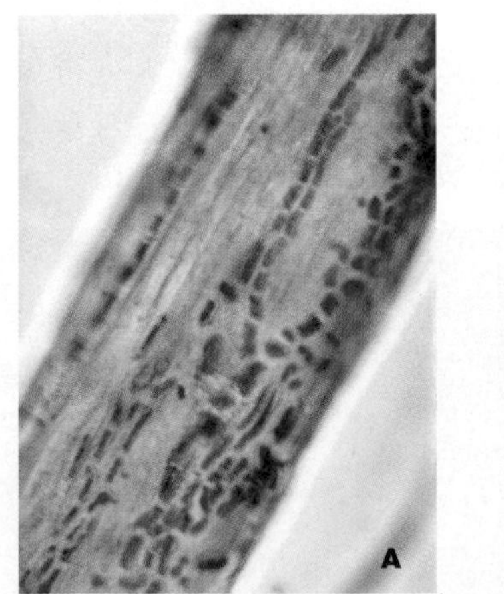


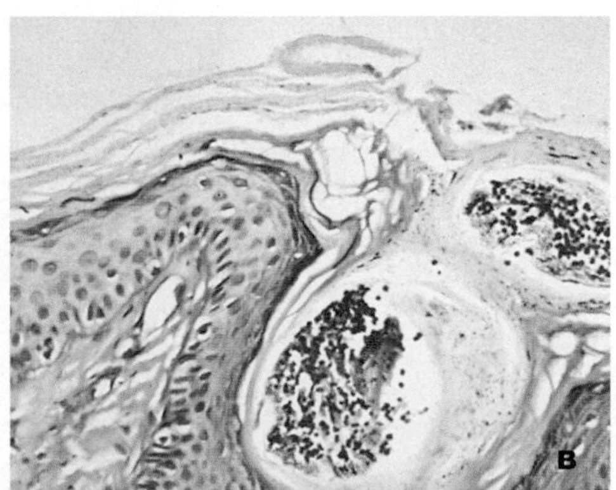


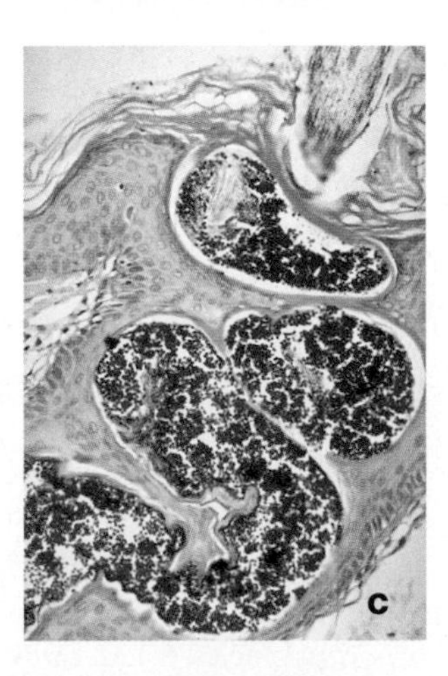


**Figure 228**
Infection of the hair. (a) An endothrix infection. Arthroconidia can be seen in the infected hair. (b) Spore-ridden hair shafts forming darkly staining plugs in the surface layer of the scalp (stratum corneum). (c) Spores almost completely replacing the contents of a hair. (From J. Lee, Y.-Y., and M.-L. Hsu. *J. Cutan. Pathol.* 19(1992):54–58.)

## Superficial and Cutaneous Mycoses (Dermatophytoses) *(Continued)*

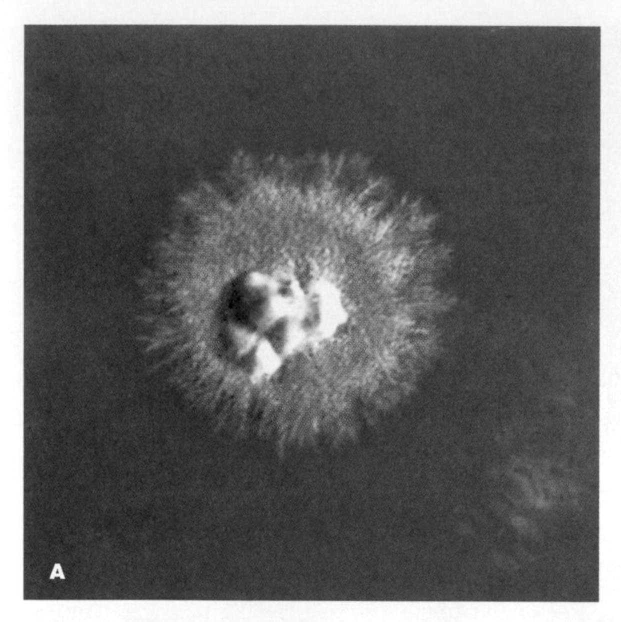

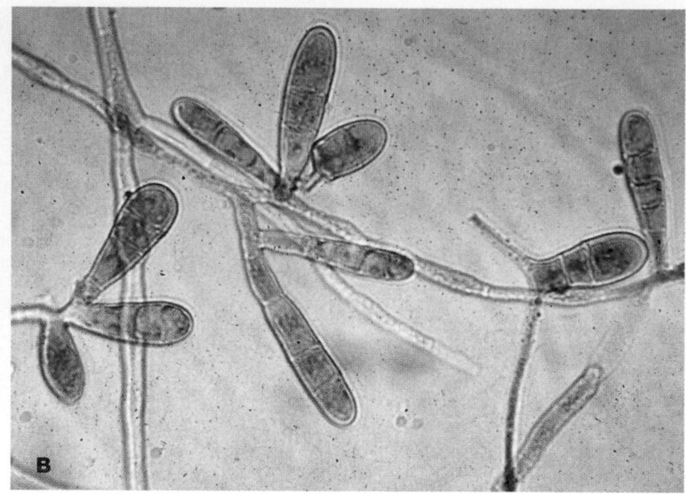

**Figure 229**
*Epidermophyton floccosum.* (a) Mycelium on Sabouraud's dextrose agar. (b) Microscopic view showing macroconidia.

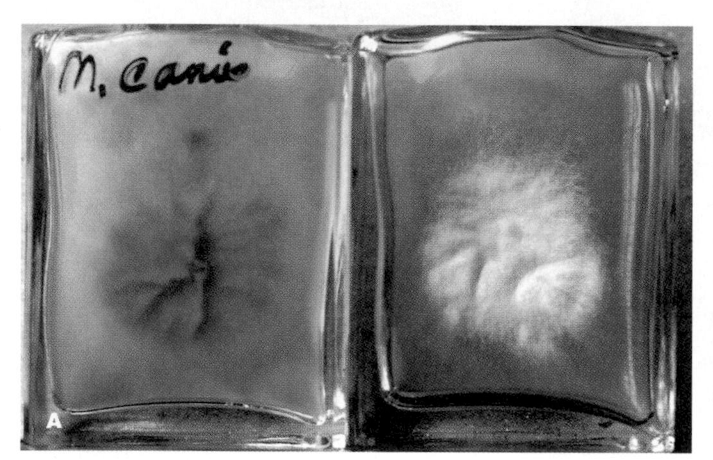


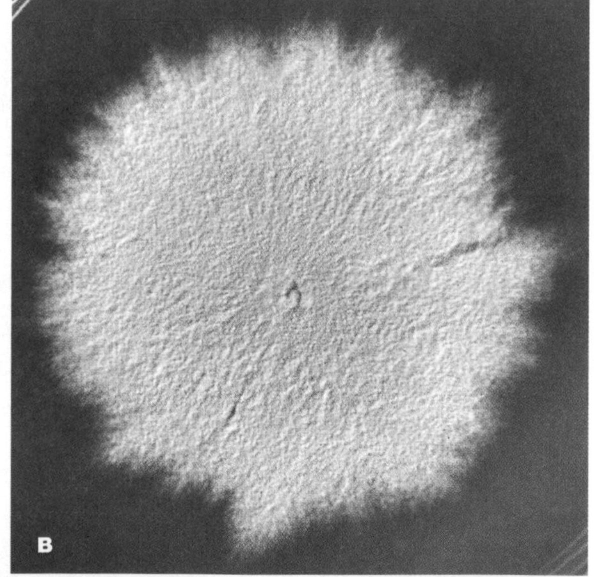

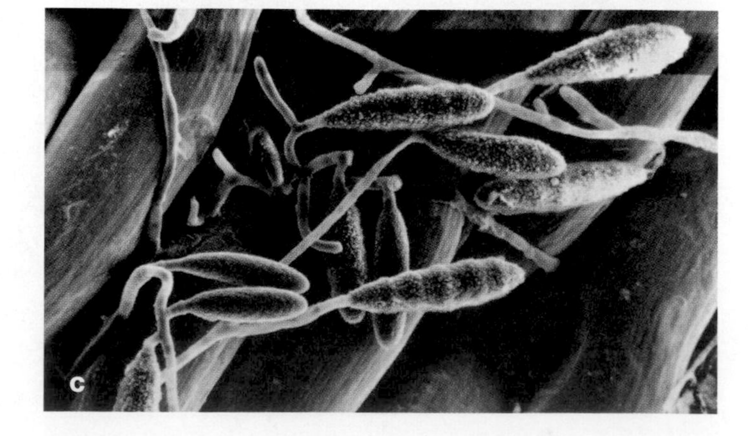

**Figure 230**
*Microsporum* species. (a) *M. canis* mycelial surfaces. (b) *M. gypseum* mycelium. (c) Scanning electron micrograph of *M. gypseum* macroconidia (800×). (Courtesy of N. Contet-Audonneau and M. Miegeville.)

## Superficial and Cutaneous Mycoses (Dermatophytoses) *(Continued)*

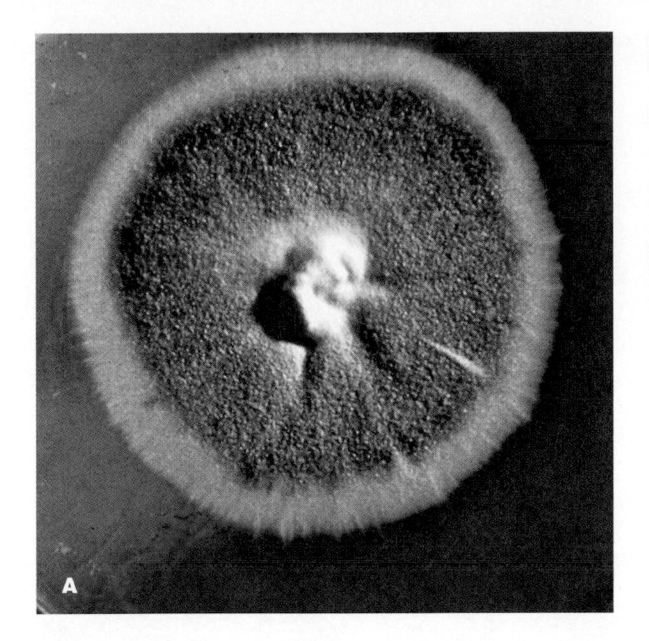

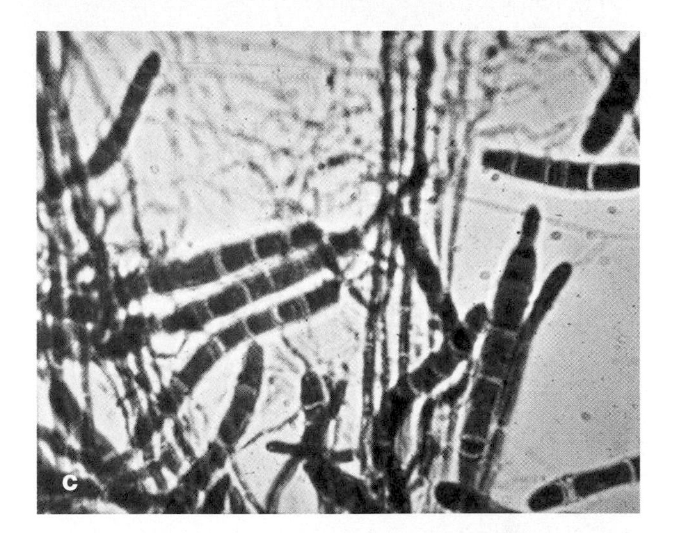

**Figure 231**
*Trichophyton rubrum.* (a) Mycelium on Sabouraud's dextrose agar. (b) The red to purplish pigment produced by the mold can be seen on the undersurface. (c) *T. rubrum* macroconidia.

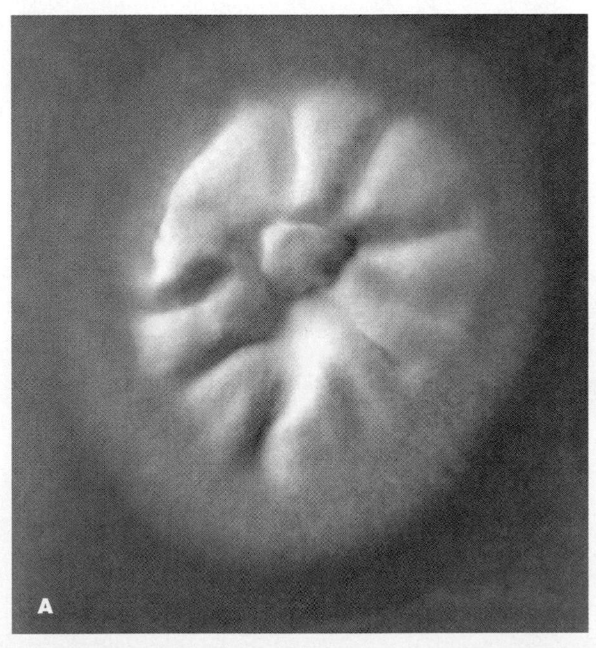

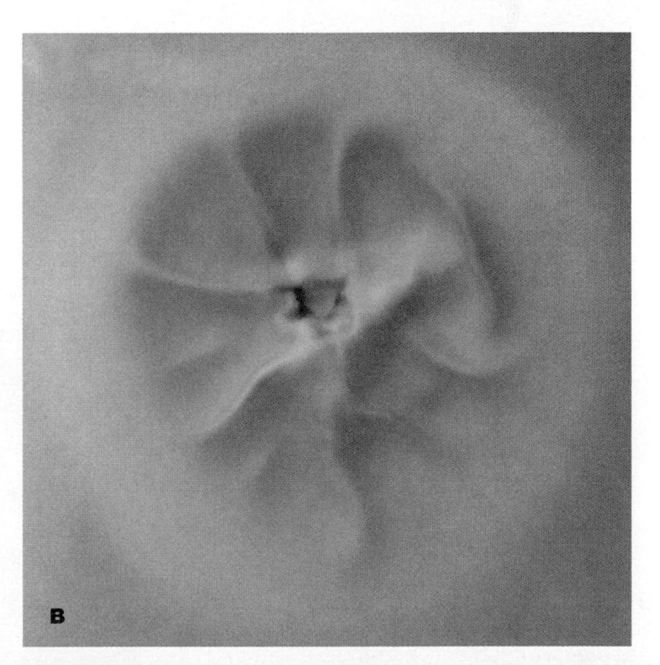

**Figure 232**
*Trichophyton tonsorans.* (a) Top view of mycelium. (b) Undersurface of mycelium.

## Superficial and Cutaneous Mycoses (Dermatophytoses) *(Continued)*

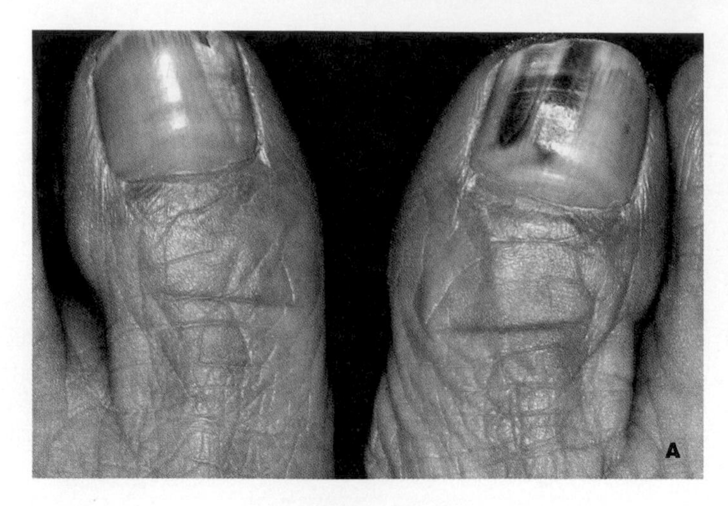

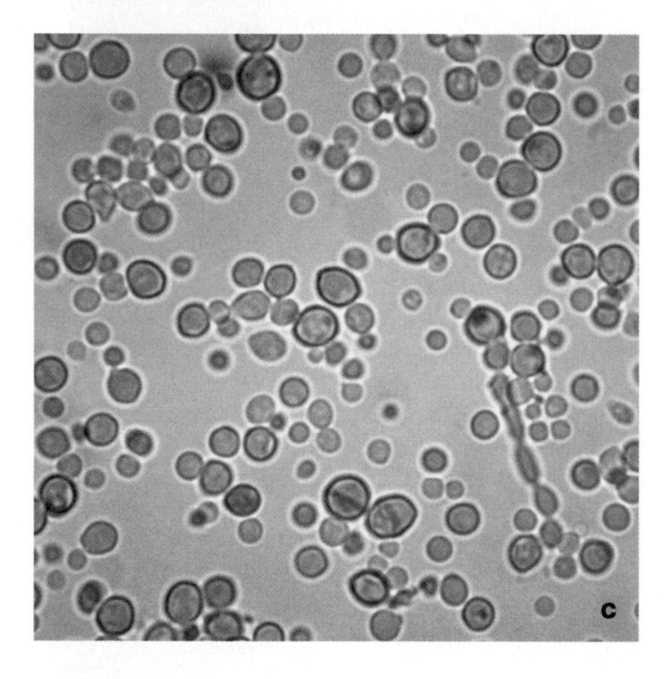


**Figure 233**
Nail infection, a case of fungal melanonychia (black pigmentation of the nail). (a) Black discoloration of toenails. (b) *Wangiella dermatitidis* mycelium, the causative agent. (c) Hyphae and conidia (600×). (From T. Matsumoto, T. Matsuda, et al. *Clin. Exp. Dermatol.* 17(1992):83–86.)

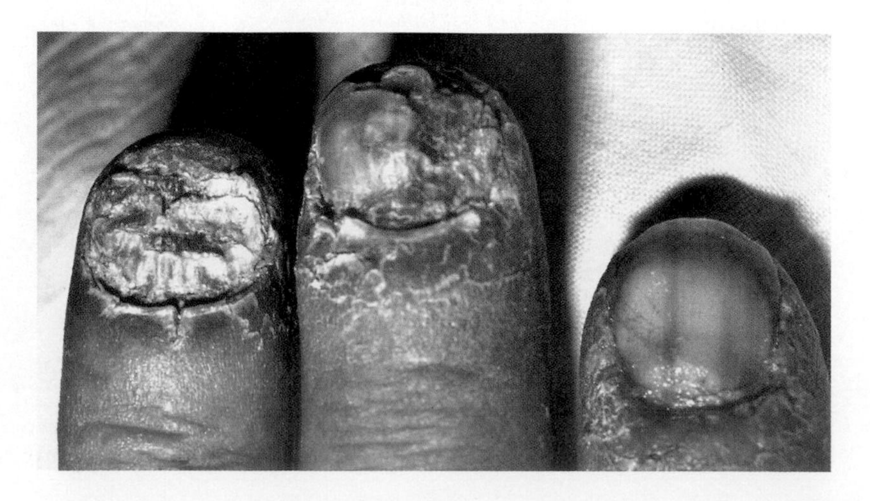

**Figure 234**
*Trichophyton rubrum* nail infections. A severe case of onychomycosis in a 37–year-old man with AIDS.

(From N.S. Prose, K.G. Abson, and R.K. Scher. *Internat. J. Dermatol.* 31(1992):453.)

## Subcutaneous Mycoses

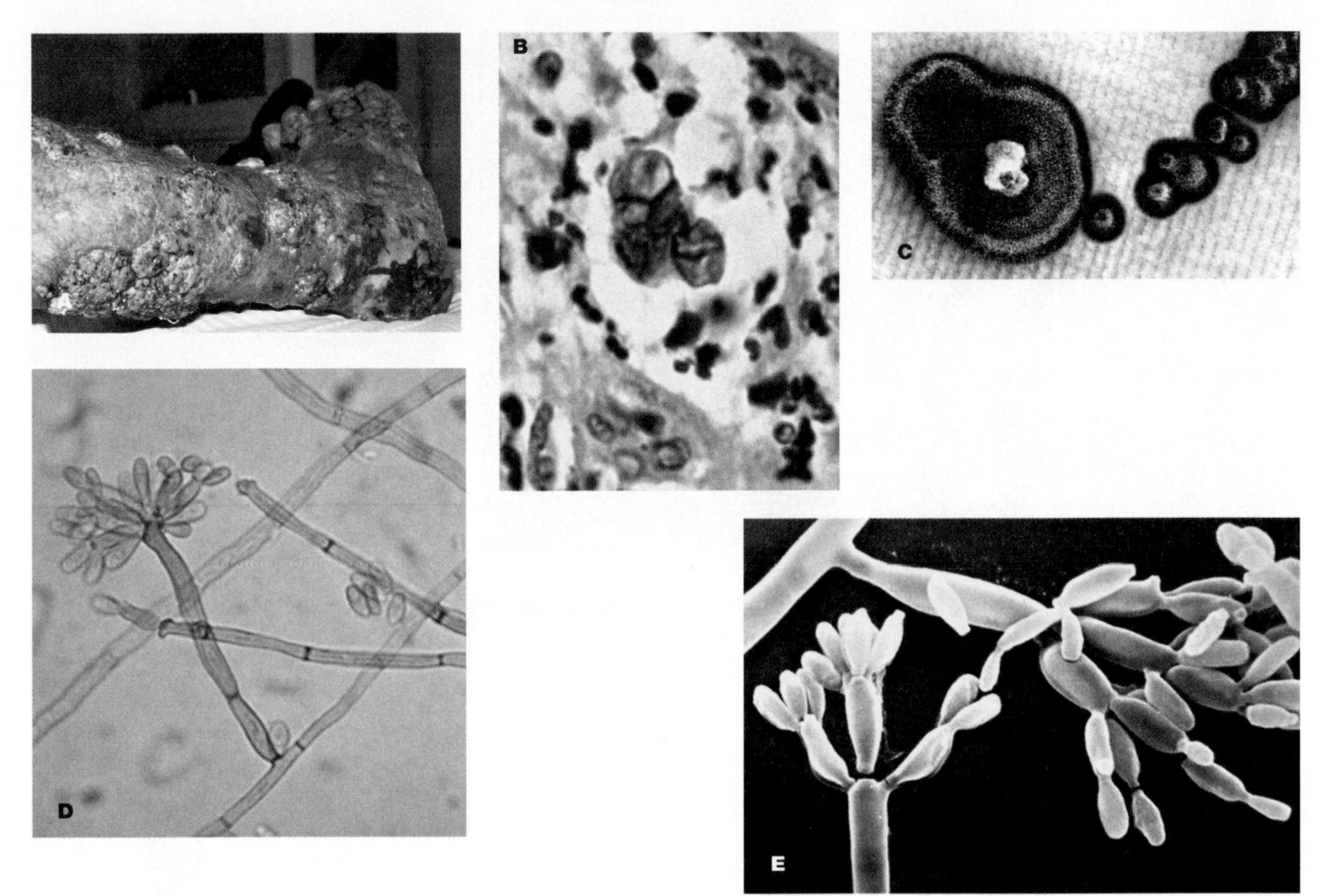


**Figure 235**
Chromoblastomycosis. (a) Clinical appearance. (b) A skin biopsy showing muriform cells (an intermediate form between a yeast and a mold) stained by the periodic acid-schiff technique. (c) *Fonsecaea pedrosoi* on Sabouraud's dextrose agar. (d) Microscopic view of *F. pedrosoi* showing septate hyphae and the typical fonsecaea-type conidia formation. (e) A scanning electron micrograph of *F. pedrosoi* conidia (oval cells). (From F. Queiroz-Telles, et al. *Internat. J. Dermat.* 31(1992):805–12.)

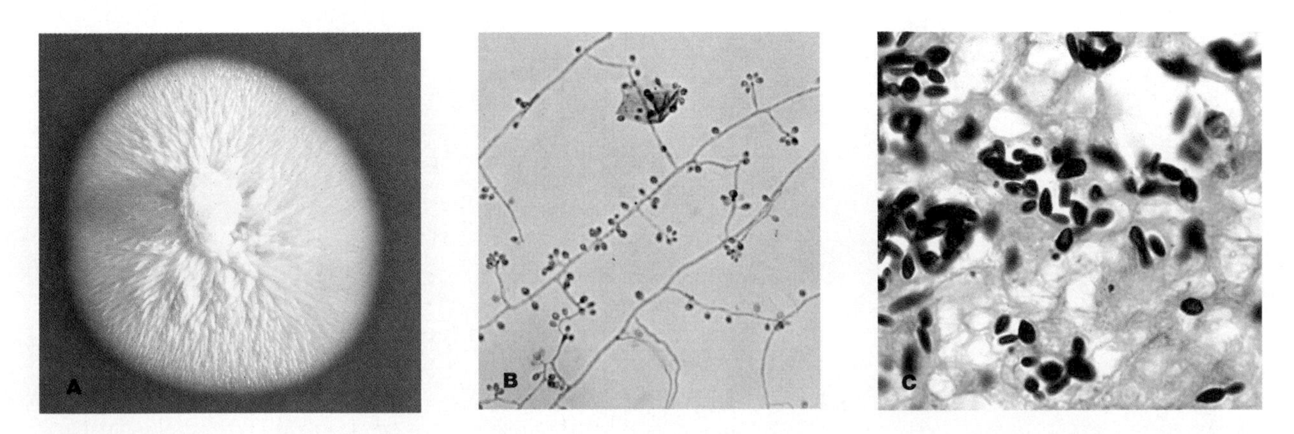


**Figure 236**
*Sporotrichum (Sporothrix) schenckii.* (a) Mycelium on Sabouraud's dextrose agar. (b) Microscopic view of cells grown at 25°C and stained with lactophenol cotton blue. Note the flowerlike arrangement of the oval conidia (400×). (From S.E. How and T.-Y. Chuang. *Cutis* 48(1991):193–96.) (c) Yeast phase in a Gomori-methenamine-silver nitrate stained tissue specimen (750×). (From R.C. Neafie and A.M. Marty. *Clin. Microbiol. Revs.* 6(1993):34–56.)

## Systemic Mycoses

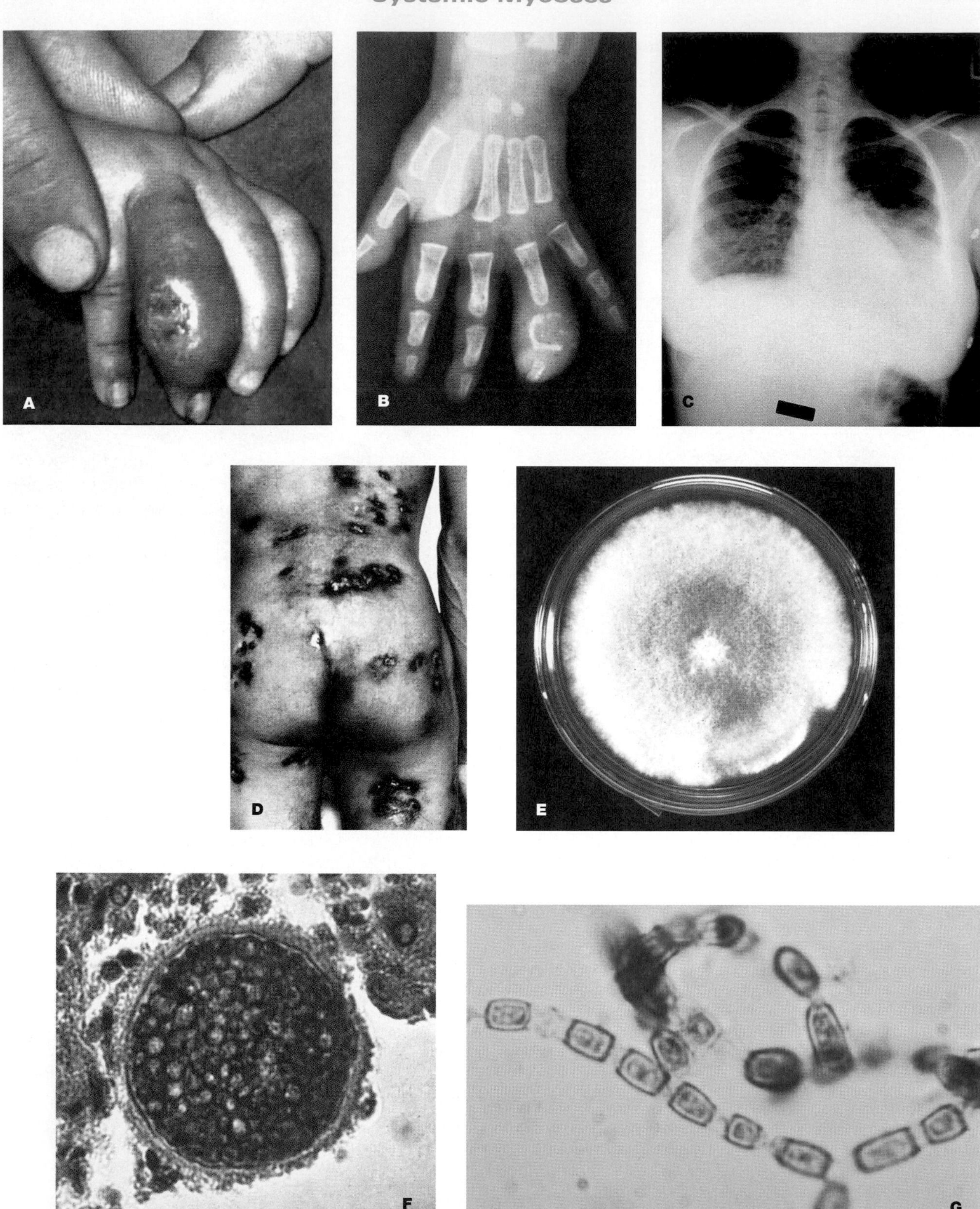


**Figure 237**
*Coccidioides immitis.* (a) Finger infection. (b) X-ray of infected finger. (c) A chest x-ray showing an infiltrate in the lower left lobe of a patient with the disease. (From S.A. Westphal and G.A. Sarosi, *CID* 18(1994):974–78.) (d) Surface lesions in a case of disseminated disease. (e) Mycelium on Sabouraud's dextrose agar. (f) A spherule containing endospores in tissue. (g) Stained arthroconidia from a culture grown at 25°C.

**Transmission.** *Histoplasma capsulatum* is acquired through inhalation of conidia from an aerosol. Infected bats also are possible sources of infection. Person-to-person transmission does not occur.

**Morphology and Cultural Properties.** *Histoplasma capsulatum* is a dimorphic fungus that appears as a yeast in tissue and in cultures grown at 37°C. Cultures incubated at 22° to 25°C exhibit mold characteristics with the formation of a diagnostic structure, the **tuberculate macroconidium** (Figure 238c). *Histoplasma capsulatum* grows on standard media used for fungi, but it may take several weeks to produce a recognizable mycelium.

**Pathology and Clinical Features.** Histoplasmosis is an infection of the reticuloendothelial system, where *H. capsulatum* invades, grows, and reproduces in macrophages and giant cells. Most cases of the disease are asymptomatic or show only fever and cough for a few days or weeks. More severe cases may exhibit chills, chest pain, and general discomfort. Progressive disease develops along a path similar to pulmonary tuberculosis. Signs and symptoms include lung cavities, sputum production, night sweats, and weight loss.

Disseminated histoplasmosis may involve several organs, including the skin, central nervous system, lungs, gastrointestinal tract, and adrenal glands (Figure 238a,b).

**Diagnosis.** Direct microscopic examination of sputum in pulmonary histoplasmosis is rarely helpful. Culture of blood and biopsy specimens from a reticuloendothelial organ such as the spleen or lymph nodes are more likely to be successful. Staining of tissue specimens with methenamine silver (Figure 215) is also helpful to diagnosis.

## Agents of Opportunistic Mycoses

Certain microorganisms regularly cause infection and disease when they enter a nonimmune host. Such organisms are known as primary pathogens. In contrast to these infectious disease agents, there are also opportunists. These organisms rarely cause disease in healthy humans. However, if an individual's defense system has been compromised or weakened by a burn or instrumentation, opportunists can cause serious and often fatal disease. Opportunists take advantage of the opportunity provided by an immunodeficient host's weakened state. Opportunistic infections (OIs) cause disease with higher frequency or higher severity, or both, among HIV-infected persons than among the general population, presumably because of immunosuppression. More than 100 microorganisms, including bacteria, fungi, protozoa, and viruses cause OIs. Such infections are associated with considerable morbidity and mortality. Three examples of fungal opportunists will be described.

### *Candida albicans*

*Candida albicans*, a yeast, is one of the most frequently found fungal opportunists as well as one of the most common causes of several serious fungal diseases (Figure 239).

**Transmission.** *Candida albicans* is a normal inhabitant of the oral cavity, the lower gastrointestinal tract, and the female genitalia. Most infections are caused by an organism in an individual's own microbiota. However, infections can also result from direct mucosal contact with lesions in others, as in cases of sexual contact or by the introduction of *C. albicans* with invasive procedures.

**Morphology and Cultural Properties.** *Candida albicans* grows in either of two basic forms. One is a yeast form with blastoconidia (Figure 208d) and hyphal elements, which include pseudohyphae and true hyphae. The second form grows on specialized media. This form develops thick-walled **chlamydoconidia,** or chlamydospores (Figure 240b), which distinguish it from other *Candida* species. The organism grows well on most standard media (Figure 240a).

**Pathology and Clinical Features.** *Candida* infections involve the skin or mucous membranes, or spread to body organs (Figure 239c). In skin infections, *Candida* invades the skin and nails, producing effects similar to those of dermatophytes (Figure 239d). In mucous membrane infections of the mouth, the yeast may produce a condition known as **thrush**. Involvement of vaginal mucosal surfaces, a frequent site of *Candida* infection, is known as **vulvovaginitis**. Individuals with immunologic defects, such as persons with AIDS, may experience various mucocutaneous types of candidiasis (Figure 239b).

**Diagnosis.** Superfical *C. albicans* infections provide adequate material with which to diagnose. Microscopic examination of KOH preparations or Gram stains will show numerous budding yeast cells (Figure 240c). Cultures of various types of specimens, including direct biopsy and bronchoalveolar lavage (BAL), are of value. Biochemical tests also are used to distinguish among *Candida* species.

### *Filobasidiella (Cryptococcus) neoformans*

*Filobasidiella (Cryptococcus) neoformans* is the causative agent of cryptococcosis. The incidence of this disease has increased substantially in recent years because of the large number of AIDS cases in the world. One-half of the patients with cryptococcosis are infected with HIV.

## Systemic Mycoses *(Continued)*

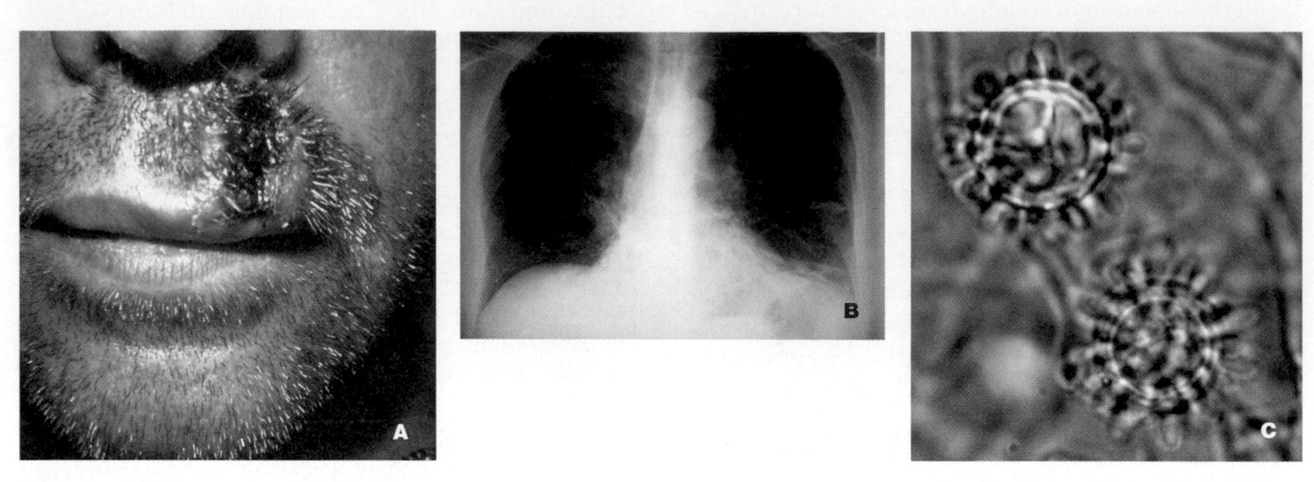

**Figure 238**

Disseminated *Histoplasma capsulatum.* (a) Skin infection. (b) Lung involvement. (c) Microscopic view from a 25°C culture grown on Sabouraud's dextrose agar. Note the large, thick-walled round tuberculated macroconidia. (From P.R. Cohen, J.C. Held, M.E. Grossman, M.J. Ross, and D.N. Silvers. *Internat. J. Dermatol.* 30(1991):104.)

## Opportunistic Mycoses

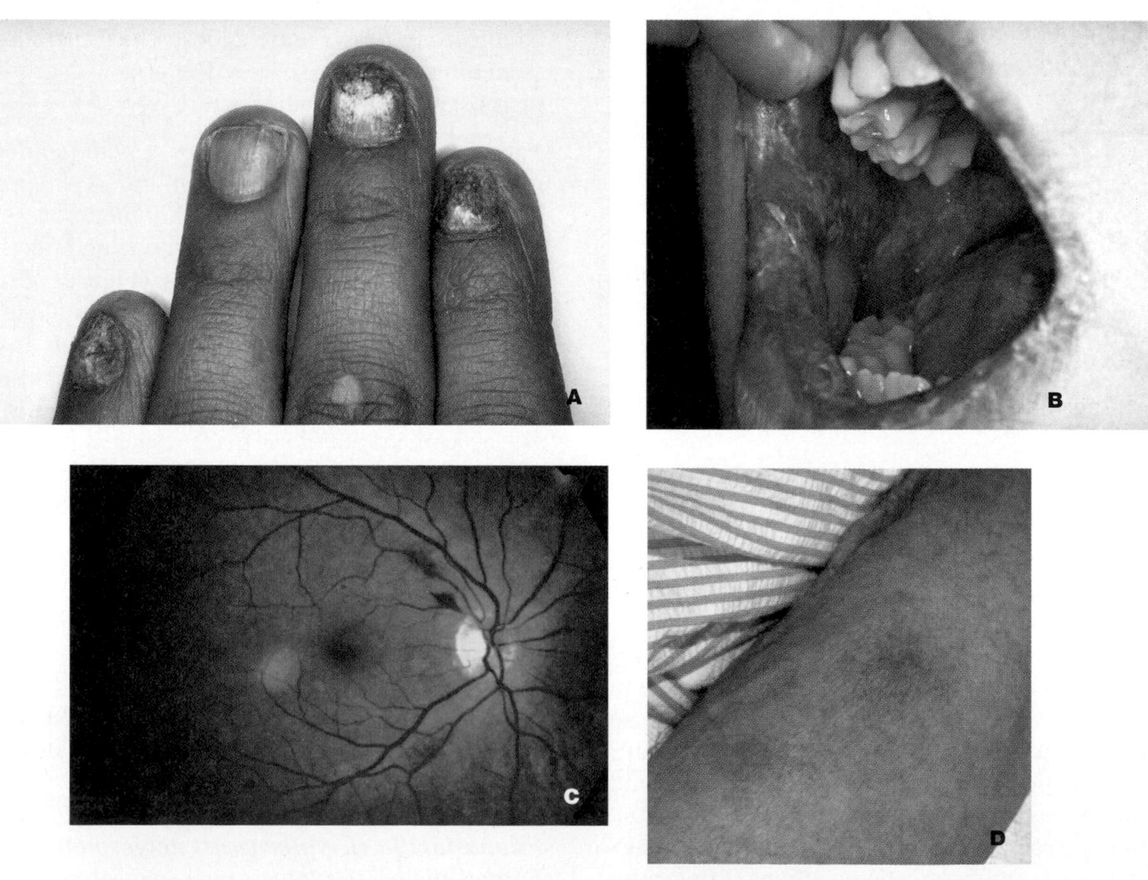

**Figure 239**

*Candida albicans* infections. (a) Candidiasis of the nails and surrounding tissues. (b) Mucocutaneous *Candida* infection of the mouth. (Courtesy of C.J. Kirkpatrick, M.D., President, Innovative Therapeutics, Inc.) (c) Retinal lesions found in a patient with candidal retinitis. (d) Skin lesions on the left forearm of a patient with disseminated candidiasis. (From D.P. McQuillen, B.S. Zingman, F. Meunier, and S.M. Levitz. *CID* 14(1992):472–478.)

## Opportunistic Mycoses *(Continued)*

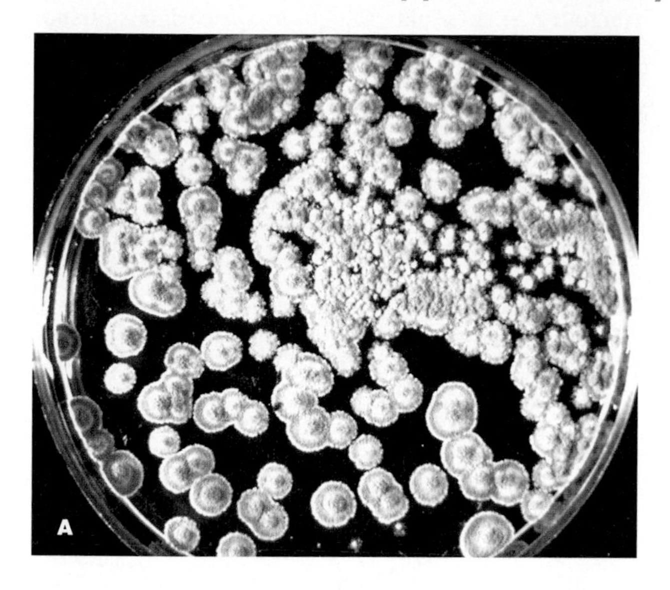


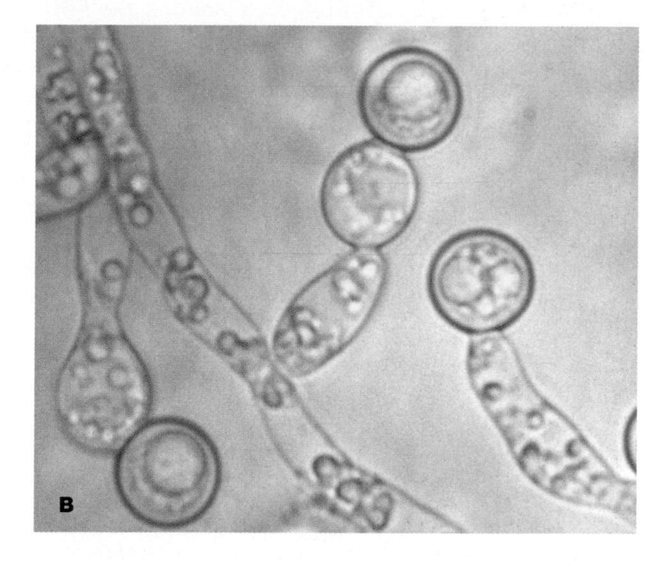


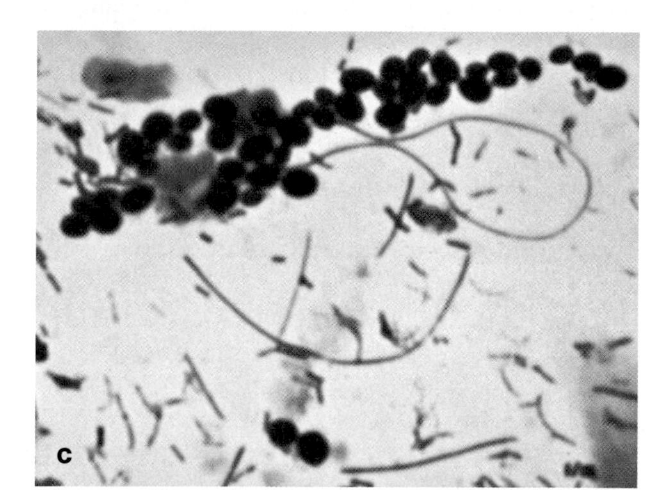


**Figure 240**
Microscopic and cultural features of *Candida albicans.* (a) *C. albicans* on Sabouraud's dextrose agar. (b) In the body and in culture *C. albicans* may develop threadlike hyphae and the spherical chlamydoconidia (chlamydospores). (c) A Gram stain of a vaginal thrush specimen showing the oval shape of the yeast form of *C. albicans*.

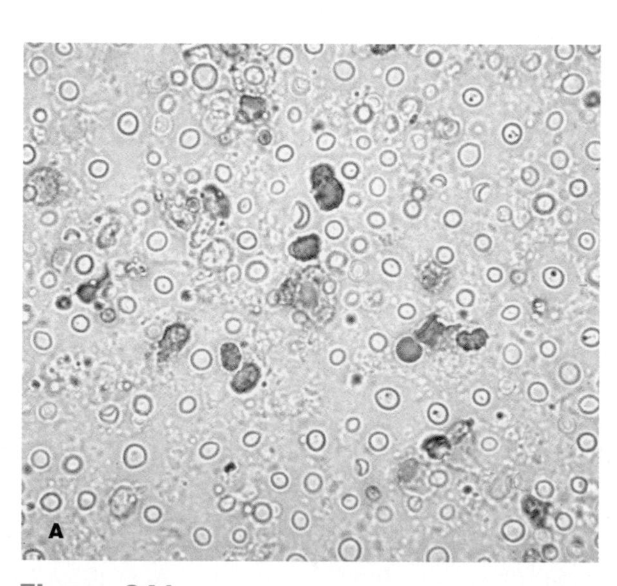


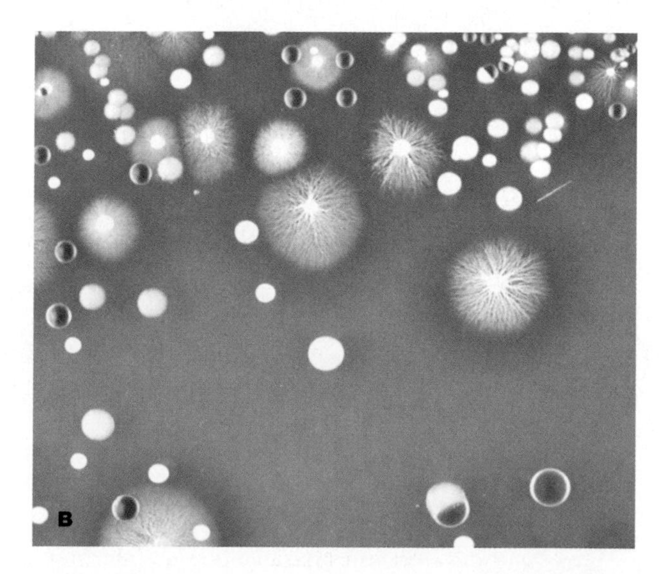


**Figure 241**
*Filobasidiella (Cryptococcus) neoformans.* (a) A microscopic view showing numerous, round, cryptococci obtained from a patient with meningitis. (b) Orange to brown colonies of *F. neoformans* on *Guizotia abyssinica* (bird seed) creatinine agar. (Courtesy of Dr. F. Staib, Former Chief, Mycology Unit, Robert Koch Institute.)

**Transmission.** The disease agent is acquired through aerosols. Association with infected pigeons and bird-related habitats is a major factor.

**Morphology.** In the tissues of an infected host, the yeast occurs as a blastoconidium with a capsule (Figure 241a). In culture, the fungus produces mucoid colonies (Figures 241b and 242a) that contain encapsulated yeast cells, usually revealed by India ink preparations (Figures 214 and 242c). Special selective and differential media also are used for isolation purposes. Bird seed agar causes the yeast to produce the pigment melanin, a distinguishing feature (Figures 241b and 242a).

**Pathology.** Six general clinical types of cryptococcosis are known. They can involve the lungs, central nervous system, gastrointestinal system, bone, skin, and mucous membranes.

**Diagnosis.** Direct microscopic examination and culture of appropriate clinical specimens (Figure 241a), serological tests to demonstrate capsular material, and nucleic acid probes are used for diagnosis.

### *Pneumocystis carinii*

*Pneumocystis* infection is caused by the unclassified microorganism, *Pneumocystis carinii.* Currently, it has not been established as to whether the organism is a fungus or a protozoan. *Pneumocystis carinii* is included here as a probable fungus because of the results obtained with ribosomal RNA analyses showing a closer relationship to fungi than to protozoa.

**Transmission.** Infection is usually acquired by inhalation of droplets containing *P. carinii.* The organism is found worldwide. It is unlikely that lower animals are significant sources of infection.

**Morphology.** *Pneumocystis carinii* develops and reproduces extracellularly, from cyst to trophozoite (active metabolizing form) to cyst. Both forms can be found in the lungs (Figure 243a).

**Pathology.** *Pneumocystis* does not produce symptomatic infection in normal, healthy individuals. The organism can exist in an inactive or latent state unless the host becomes immunocompromised, as could be the case with premature infants, AIDS patients, and individuals receiving immunosuppressive treatment for tissue transplants, cancer, and other related conditions.

Symptoms of infection include fever, dry cough, difficulty in breathing, and progressive respiratory failure.

**Diagnosis.** *Pneumocystis carinii* infection diagnosis is generally made by radiography and demonstration of the organism in sputum or lung tissue (Figure 243).

## Opportunistic Mycoses *(Continued)*

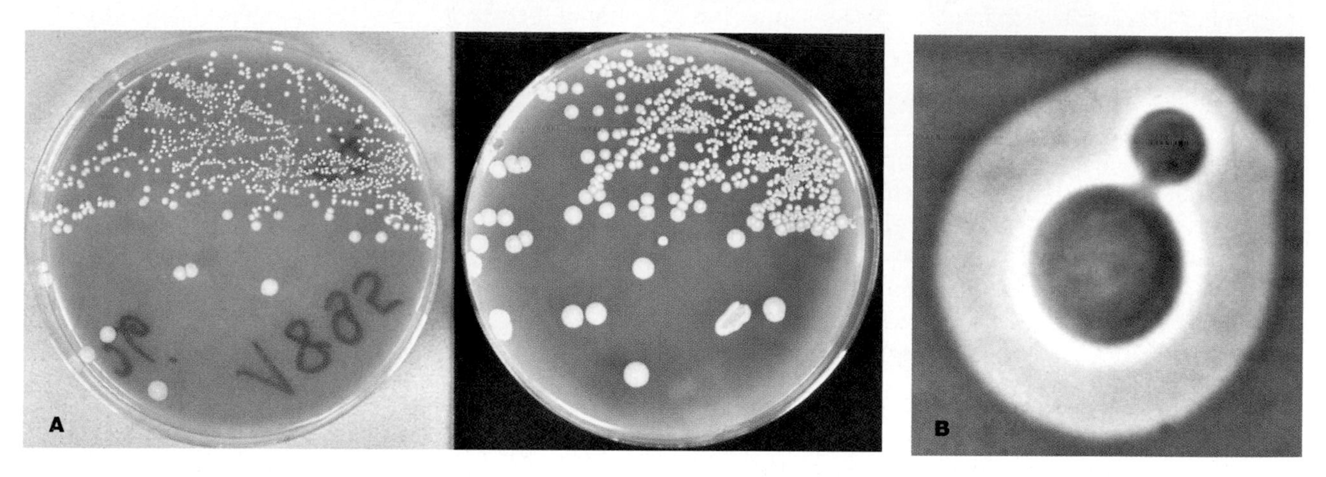


**Figure 242**
(a) *Filobasidiella neoformans* (small orange to brown) colonies on *Guizotia abyssinica* creatinine agar (left). The nonpigmented colonies are *Candida albicans.* The plate on the right primarily contains *C. albicans.* (b) A microscopic view of *F. neoformans* showing a bud and a surrounding halo-like capsule. (From F. Staib and M. Seibold. *Mycoses.* 31(1988):175–86.)

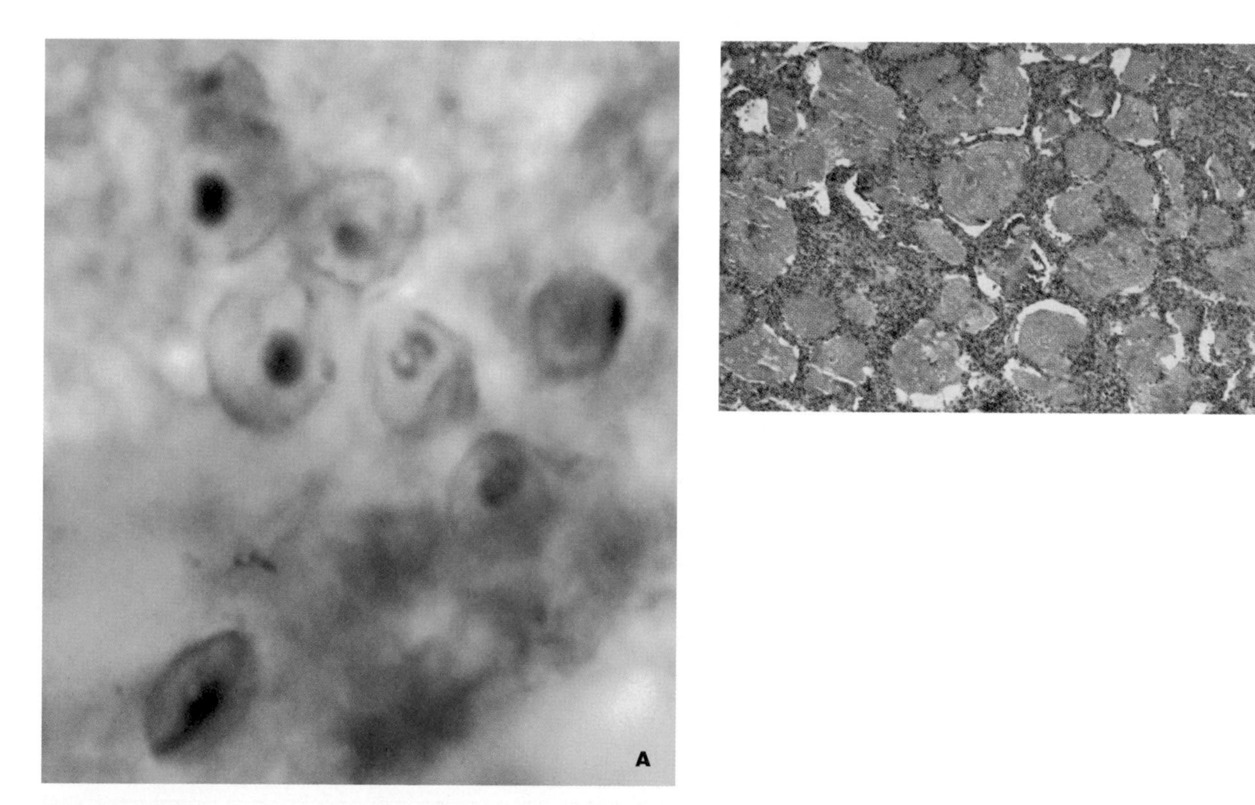


**Figure 243**
Microscopic views of *Pneumocystis carinii.* (a) A bronchial wash stained by the Gomori-methenamine-silver nitrate method. (b) The organism in tissue (Courtesy of Dr. L. Alpert, Pathology Department, The Sir Mortimer B. David Jewish General Hospital.)

SECTION 5

# Algology (Protists—The Algae)

*All we know is still infinitely less than all that still remains unknown.*

—*William Harvey, De Mota Cordis, dedication*

In the early twentieth century, most aquatic plant life was lumped together and collectively referred to as algae. By the 1920s it became clear that algae contained several related yet distinct groups of microscopic and massive forms of life (Figures 244 through 250). The organisms once grouped under aquatic plants, or algae, are classified as shown in Table 15.

The five-kingdom classification system established the kingdom of Protista. According to this system the golden algae, diatoms, euglenoids, and dinoflagellates are placed into the Protista, and all others are placed into the plant kingdom. The general features of algae are compared in Table 15.

## Properties and Distribution

Algae are photosynthetic forms of life. Most algal forms are free-floating and free-living. Many of the single-celled species are suspended in vast numbers in various bodies of water. Occasional spurts of growth known as algal blooms (Figure 244a and Figure 244b) may disrupt aquatic communities by increased accumulations of their waste products (Figure 244b).

Collectively, the photosynthetic types are phytoplankton, the food producers which form the basis of nearly all of the food webs in aquatic environments.

**Table 15** Properties of Major Groups of Algae

| Algal Group | Common Name | Morphology | Pigments | Major Habitats |
|---|---|---|---|---|
| Chlorophyta | Green algae | Unicellular to leafy | Chlorophylls *a* and *b* | Freshwater, soils, a few marine |
| Euglenophyta | Euglenoids | Unicellular, flagellated | Chlorophylls *a* and *b* | Fresh water, a few marine |
| Chrysophyta | Golden brown algae, diatoms | Unicellular | Chlorophylls *a, c,* and *e* | Fresh water, marine, soil |
| Phaeophyta | Brown algae | Filamentous to leafy, occasionally massive and plantlike | Chlorophylls *a* and *c,* xanthophylls | Marine |
| Pyrrophyta | Dinoflagellates | Unicellular, flagellated | Chlorophylls *a* and *c* | Fresh water, marine |
| Rhodophyta | Red algae | Unicellular, filamentous to leafy | Chlorophylls *a* and *d*, phycocyanin, phycoerythrin | Marine |

# Representative Algae and Their Distinctive Properties

## Chlorophyta (Green Algae)

The green algae bear the greatest structural and biochemical resemblance of all the algae to plants and may be their nearest relatives (Figure 245). This group shows more diversity than other algal groups. They include microscopic colonial forms (Figure 245d), filamentous forms (Figure 245b), desmids (Figure 245e), and the larger sea lettuce (Figure 245a). Green algae have both asexual and sexual types of reproduction. Some, such as the filamentous green alga *Spirogyra,* form conjugation tubes to allow the cellular contents from one cell to pass into another (Figure 245c).

## Euglenophyta (Euglenoids)

Euglenoids can be found in fresh water, stagnant ponds, and lakes. These microorganisms contain a large number of organelles, including chloroplasts, contractile vacuoles, and light-sensitive eyespots (Figure 246a). *Euglena* also have a type of firm but flexible outer layer, the pellicle and a flagellum. *Euglena* species also can be grown in artificial media and in the presence of a light source (Figure 246b).

## Chrysophyta (Golden Brown Algae and Diatoms)

Many of the known species of golden algae are single celled, are photosynthetic, and have scales or skeletal elements of silica. Many are important in freshwater habitats.

Diatoms are plentiful in aquatic habitats, and many are photosynthetic. The structure and organization of these algae are quite distinctive (Figure 247a). Each diatom has a shell consisting of two perforated, glasslike structures that overlap like a pillbox (Figure 247b). Substances move in and out from the plasma membrane through these cell wall perforations. Sediments at the bottoms of lakes and seas contain deposits of finely crumbled diatoms (Figure 247a), which have accumulated there for about 100 million years. These diatom remains are used as abrasives, filters, and insulating materials.

## Phaeophyta (Brown Algae)

Brown algae are found in temperate or cool marine waters. Depending on their photosynthetic pigments, they may appear as olive-green, golden, or dark brown (Figure 248). Brown algae range in size from microscopic filamentous forms to giant kelps. Some of these organisms also exhibit tissue differentiation, including anchoring structures (**holdfasts**), stemlike parts (**stipes**), and photosynthetic, leaf-shaped parts (**blades**).

## Pyrrophyta (Dinoflagellates)

The dinoflagellates are marine, or sometimes freshwater, phytoplankton. Some have flagella that fit in grooves between stiff cellulose plates at the cell surface (Figure 249). Depending on their photosynthetic pigments dinoflagellates may appear yellow-green, green, brown, blue, or red. The red forms occasionally undergo population explosions and color seas red or brown. The well-known red tide is an example.

## Rhodophyta (Red Algae)

The red algae are especially abundant in warm currents or tropical seas and at great depths (Figure 250). Some are single-celled; others are colonial. Most species are multicellular with a filamentous, often branching organization. The cell walls of red algae incorporate a mucus-type material, which gives them a flexible, slippery texture. Agar, the polysaccharide used for solidifying bacteriological and other media, is extracted from certain species of red alga.

## Lichens

Certain algal species can enter into a beneficial relationship with fungi to form a structure known as a **lichen** (Figure 251). The fungal hypha produces a tightly woven mycelium, which houses and protects its partner the alga. The fungi absorb environmental water and minerals, which the algae use for photosynthesis. Many different arrangements and colors are displayed by lichens. However, three major types are recognized: **crustose** (crustlike) lichens, found on rocks or on bark as irregular flat patches (Figure 251a,b); **foliose** (leaflike or fibrous) lichens, curled and frequently having rootlike structures for attachment; and **fruticose** (shrublike) lichens, which are highly branched and either hang from different tree parts or originate in the soil (Figure 251c). Figure 251b shows the organization of crustose lichens.

## Algology (Protists–The algae)

**Figure 244**
Distribution of algae. (a) The light green appearance of an algal bloom on a waterway near a small town in Holland (arrow). (b) The result of an algal bloom. Note the dead fish in the midst of the accumulated algae. (c) Green algae occupying (colonizing) the microscopic air space system a few millimeters below the surface of a porous desert rock. (d) The presence of green algae on the surfaces of several stones at Stonehenge, a prehistoric ruin on the plain north of Salisbury, England. (e) Certain sea anemones (green) in an endosymbiotic relationship with green algae. Nonsymbiotic anemones are white.

## Algology (Protists–The algae) *(Continued)*

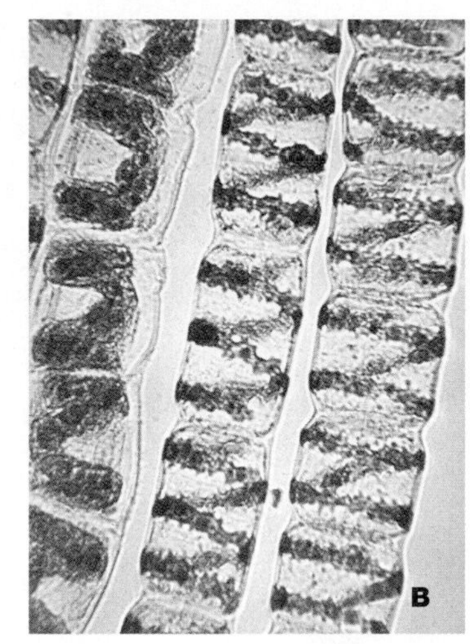


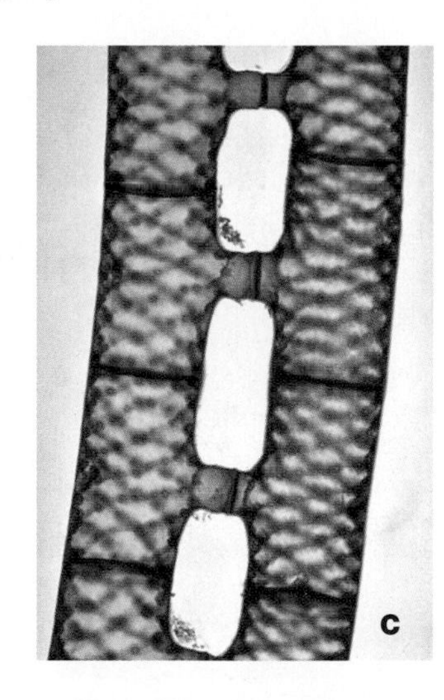


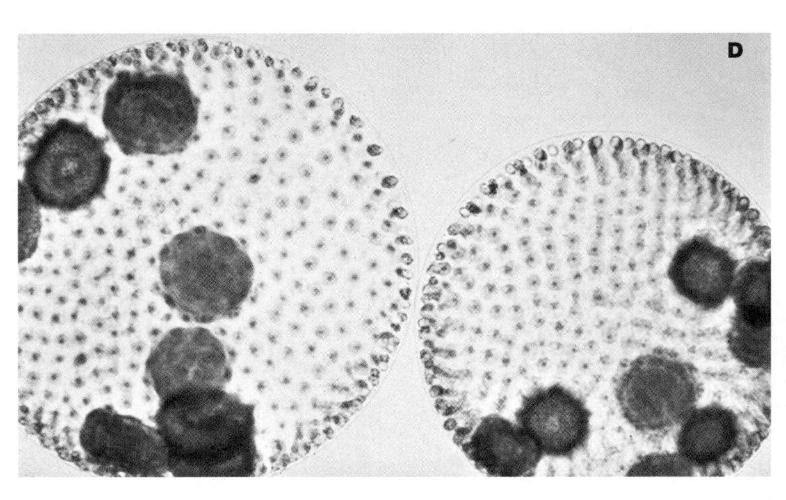

**Figure 245**
Representative green algae. (a) Sea lettuces (*Ulva lactuca*) growing in a tidal pool. (b) *Spirogyra,* an unbranched green alga, showing the typical helical (springlike) arrangements of chloroplasts. (c) Conjugation (connecting) bridges of *Spirogyra* formed during sexual reproduction. (d) A *Volvox* colony consisting of smaller interdependent cells. (e) Desmids. *Micrasterias crux-melitensis* after cell division (1,100×).

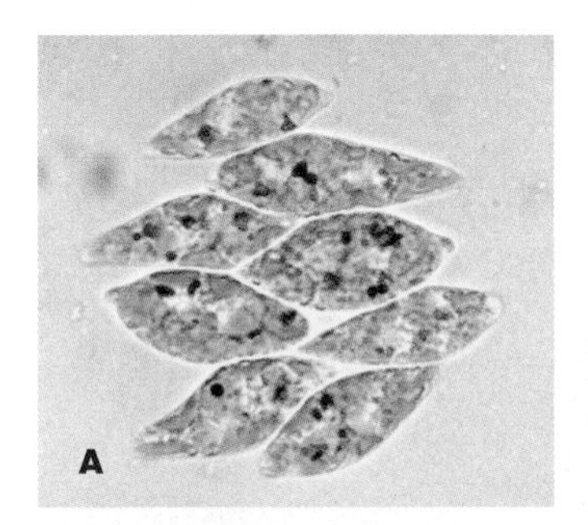

**Figure 246**
Euglenoids. (a) Several *Euglena,* showing numerous chloroplasts. These organisms have a flexible outer plasma membrane (pellicle) that allows them to easily change shape (450×). (b) The green growth of *Euglena* in *Euglena* broth. An uninocluated preparation is shown for comparison.

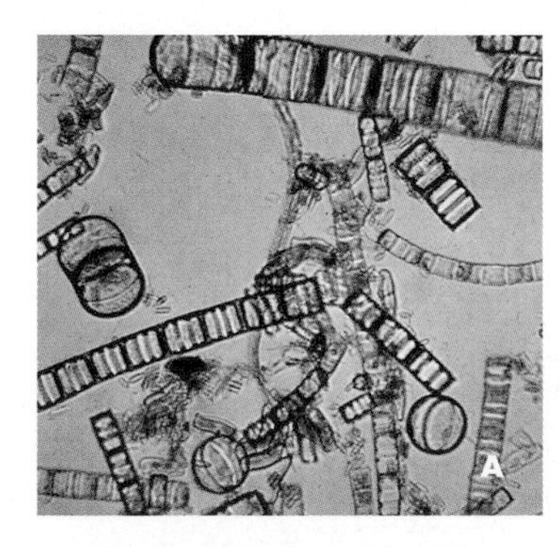


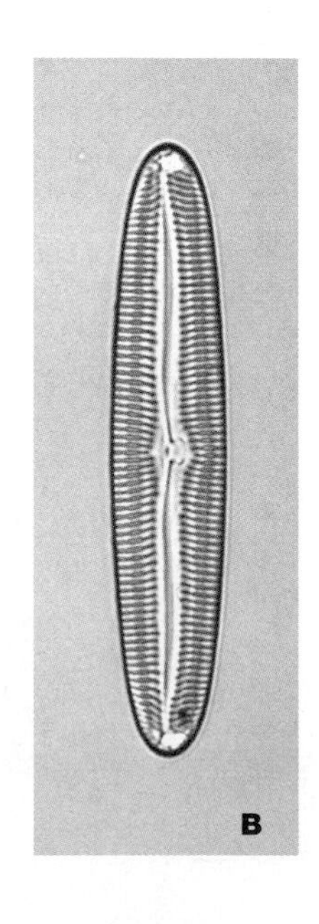


**Figure 247**
Diatoms. (a) A microscopic view of the remains of diatom cell walls. (b) The structure of a pennate diatom showing the long raphe running along the length of the diatom, the puncta (holes), and other portions of the cell wall.

## Algology (Protists–The algae) *(Continued)*

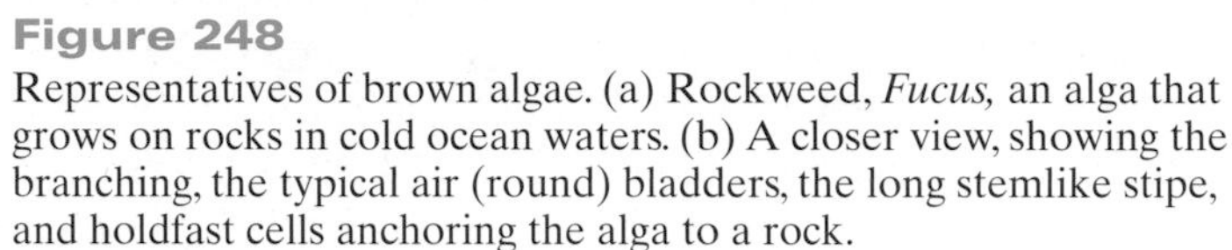

**Figure 248**
Representatives of brown algae. (a) Rockweed, *Fucus,* an alga that grows on rocks in cold ocean waters. (b) A closer view, showing the branching, the typical air (round) bladders, the long stemlike stipe, and holdfast cells anchoring the alga to a rock.

**Figure 249**
A scanning electron micrograph of the dinoflagellate *Gymnodinium catenatum.* (Courtesy of Dr. M. Ellengard, Botanical Institute, University of Copenhagen.)

**Figure 250**
Red algae are so named because of their red pigment, phycoerythrin. These organisms mainly are found in marine environments.

**Figure 251**
Lichens. (a) Several lichen species on a tree branch: the fibrous form called "old man's beard" and orange and white crustose types. (b) Crustose lichen organization. This lichen, as well as similar crustose forms, consists of four layers: (1) the upper, protective cortex formed by a thick layer of fungal hyphae; (2) a layer in which algal cells are distributed among thin-walled hyphae; (3) the medulla, a thick hyphal layer; and (4) a lower cortex containing hyphae with fine projections that attach the lichen to the rock surface (substrate). (c) A fruticose (leaflike) lichen.

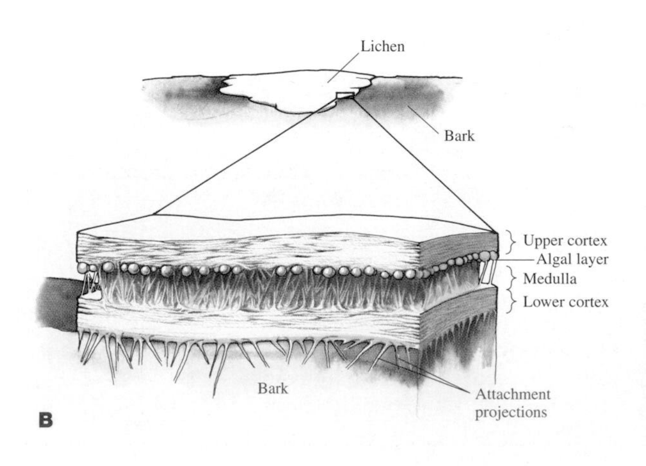

# SECTION 6 Protozoology (Protists—The Protozoa)

*Do there exist many worlds or is there but a single world? This is one of the most noble and exalted questions in the study of nature.*

*—St. Albertus Magnus*

The 25,000 or so species of protozoa were formerly placed into the animal kingdom. Today, these animal-like microorganisms are considered to be members of the kingdom Protista and are grouped into phyla that differ from one another in several respects, including their means of movement (See Table 17).

Protozoa (singular, *protozoon*) are found in many different environments. Some are present in bodies of water, where they play an important role in the food webs of natural communities. Others have mutually beneficial (**symbiotic**) relationships with higher forms of animal life or with other microorganisms.

Protozoa are also known for their harmful activities. African sleeping sickness, amebic dysentery, malaria, and toxoplasmosis are but a few of the human diseases associated with these microorganisms. Several protozoa also infect wild and domestically important animals. In severe cases, infected hosts are crippled, disfigured, and eventually die. Protozoan protists of the same species are similar to one another and none is specialized solely for feeding. Most of the individuals in a population of protists are produced by simple cell division of the parent, although sexual reproduction by the mating of two individuals also does occur.

## Structures and Functions

Protozoa are not functionally simple microorganisms. Essential functions require a division of labor of tasks involved with movement, obtaining and using nutrients, excretion and osmoregulation, reproduction, and protection. Table 16 lists and briefly describes the organelles involved in these tasks (Figures 252, 253, 257–259, 261, and 262 show several of them.)

**Table 16** Protozoan Structures

| Structure | Function(s) |
|---|---|
| Cilia | Movement |
| Contractile vacuole | Excretion and osmoregulation |
| Cytopage | Elimination of indigestible material |
| Cytosome (mouth) | Food gathering |
| Flagella | Movement, sense reception |
| Food vacuoles | Digestion |
| Macronucleus | Regulation of metabolism and development |
| Micronucleus | Overall cellular control and regulation of reproductive process |
| Pellicle (strong cellular covering) | Protection against chemicals, drying, and mechanical injury |
| Pseudopodium | Movement |
| Tentacle | Protection and trapping of food |
| Trichocyst | Defense and food capture |
| Undulating membrane (formed with plasma membrane and flagellum) | Movement |

## Trophozoites and Cysts

Several pathogenic protozoan species as well as free-living ones found in bodies of water appear in a normal, active feeding form known as the **trophozoite** (Figures 253, 255, and 258a). Some protozoa form **cysts** to overcome the effects of various chemicals, food deficiencies, temperature or pH changes, and other harsh environmental factors. This stage also serves as a means of reproduction and for spreading the protozoa (Figures 255c and 258b). Trophozoite and cysts stages also are important to the diagnosis of protozoan diseases.

## Classification

Various properties of protozoa are used in their classification. Among them are method of obtaining nutrients; method of reproduction; cellular organization, structure, and function; biochemical analyses of nucleic acids and proteins from specific cellular structures; and organelles of locomotion (Table 17).

All parasitic and either medically or agriculturally important protozoa are placed into specific phyla including the Sarcomastigophora, Ciliophora, Apicomplexa, and Microsporidia (Table 17).

**Table 17** Characteristics of the Major Groups of Protozoa

| Phylum (Subphylum) | Common Name | Means of Movement | Habitats |
|---|---|---|---|
| Sarcodina | Amoebae | Pseudopodia | Fresh water and marine, some parasitic |
| Mastigophora | Flagellates | Flagella | Fresh water, some parasitic |
| Ciliophora | Ciliates | Cilia | Fresh water and marine, some parasitic |
| Apicomplexa | Sporozoa | Generally nonmotile except for certain sex cells | Mainly animal parasites |
| Microsporidia | Microspora | Nonmotile | Intracellular parasites |

# Representative Protozoa, Their Distinctive Properties, and Selected Disease States

## Sarcomastigophora

The phylum Sarcomastigophora includes the two subphyla, **Sarcodina** and **Mastigophora,** and contains many important disease-causing species.

### *Sarcodina*

Members of this subphylum, unlike other protozoa, have no definite shape and move by forming pseudopodia (Figure 252a). Their simple cells change form as they move. The nucleus, contractile vacuole, and food-containing vacuoles shift about with the protozoon as it moves.

The many kinds of protozoa that constitute the Sarcodina are found in all bodies of water. One of the most interesting is the large subgroup called the **foraminifera** (Figure 252c). These microorganisms form shells made of lime or of substances such as sand from the surrounding waters. Approximately 1,800 species, found mainly in salt water, are known.

Other members of the Sarcodina, the parasitic amoebae, may be found in most kinds of animals. The most important forms to attack humans include *Entamoeba histolytica* (Figure 253), *Acanthamoeba* (Figure 255), and *Naegleria. Entamoeba coli* (Figure 255c), another amoeba, is nonpathogenic.

#### *Entamoeba histolytica*

*Entamoeba histolytica* is one of six parasitic amoebae of the genus *Entamoeba* known to infect humans. Infections occur worldwide but are most prevalent in the tropics. Humans are the major reservoir of infection. *Entamoeba histolytica* causes amebiasis, amebic liver abscess (Figure 254), and cutaneous amebiasis.

**Transmission and Life Cycle.** Ingestion of cyst-contaminated food and drink and fecal-oral contact are the most common means of infection. The use of human feces for fertilizer is also an important source of infection. Flies and cockroaches can also spread the pathogen.

**Morphology.** *Entamoeba histolytica* is an enteric pathogen that exists in either a trophozoite or cyst stage. The trophozoite contains a single nucleus with a small distinctive central body, known as a **karyosome**, and nuclear material distributed evenly in the nuclear envelope (Figure 253b). The resistant cyst can contain several similar nuclei and some deeply staining bundles of RNA called **chromatoidal bodies.**

**Diagnosis.** Diagnosis of infections depends primarily on examinations for *E. histolytica* cysts in stools, tissue biopsy for trophozoites, and serological tests.

### *Acanthamoeba*

*Acanthamoeba* species (Figure 255a,b) cause several diseases, including a form of amebic encephalitis, keratitis (corneal ulcers), and skin ulcers. The organism is found worldwide.

**Transmission.** Infections may be acquired by inhalation or by direct contact with contaminated soil, water, or solutions contaminated with the organism. Keratitis usually is seen in contact lens users who have used nonsterile saline solutions for cleaning purposes.

**Morphology.** *Acanthamoeba* species occur in trophozoite and cyst stages.

**Diagnosis.** *Acanthamoeba* infections are diagnosed by finding the organism in tissue biopsies (Figure 255a), in scrapings, or in freshly drawn cerebrospinal fluid.

## Mastigophora

The Mastigophora, or flagellates, are mostly unicellular and usually possess at least one flagellum at some stage of their life cycle (Figure 257b). These flagella are used for locomotion, for obtaining food, and as sense receptors.

The flagellates include more than half of the living species of protozoa and are an extremely variable group. They are believed to be the oldest of the eukaryotic organisms and the ancestors of the other major forms of life.

Free-living Mastigophora are common in both fresh and salt water. Many others inhabit the soil or the intestinal tracts of some animals. Some Mastigophora are free-living, commensal, mutualistic, or parasitic. Several flagellates are parasitic and medically important. These include *Leishmania* species, *Giardia lamblia, Trichomonas,* and trypanosomes.

### *Leishmania*

**Leishmaniasis** is a general term for diseases caused by *Leishmania* species. The clinical disease depends on the species involved and the immunologic status of the host. The spectrum of clinical conditions includes: local infections of the skin and bone (Figure 256), subcutaneous tissue, and regional lymph nodes; **mucocutaneous leishmaniasis,** or **espundia,** spreading infections of the oronasal mucosa; and **visceral leishmaniasis,** or **kala azar,** disseminated infection involving visceral organs of the mononuclear phagocyte system.

**Transmission and Life Cylce.** *Leishmania* species are transmitted by sand flies of the genus *Lutzomyia* in the New World and *Phlebotomus* in the Old World. Animal reservoirs include wild rodents, sloths, and various meat-eating mammals. In vertebrates, *Leishmania* are obligate, intracellular parasites. They invade macrophages and reproduce in membranous structures that surround the protozoon during the invasion process.

**Morphology.** The first invading form is the **promastigote** (Figure 257b), which is a long flagellated cell that develops extracellularly inside the digestive tracts of sand flies. Once inside the membranous structure, the promastigote changes into an **amastigote** (Figure 257a), which is a round, nonmotile form that multiplies rapidly. Amastigotes are released when parasitized cells burst; they then spread to neighboring host cells. All *Leishmania* species in humans are morphologically similar.

**Diagnosis.** Diagnosis involves the demonstration of amastigotes in tissues such as bone marrow or white blood cells. Cultures of blood and bone marrow also are of value. Such cultures may require 4 weeks for the isolation of organisms. Serological tests, such as immunofluorescence, and nucleic acid probes are satisfactory for indirect diagnosis.

### *Giardia*

*Giardia* is a binucleate flagellated protozoon that causes intestinal infection in mammals, birds, reptiles, and amphibians. *Giardia lamblia* (Figure 258) is the cause of giardiasis in humans.

**Transmission and Life Cycle.** Giardiasis is spread by contaminated water and food. *Giardia* has a simple life cycle, consisting of an infectious cyst (Figure 258b) and a vegetative trophozoite (Figure 258a). After a cyst is ingested, it excysts in the small intestine to form two trophozoites. Each trophozoite divides into two new cells in the small intestine, and these forms are responsible for symptoms of the infection. Some of the trophozoites transform into cysts. The cycle is completed when such cysts are passed in the feces and ingested by another host.

**Diagnosis.** Microscopic examination of stools for cysts and trophozoites is usually the first diagnostic test. Cysts are found in most patients with giardiasis. Removal of intestinal fluid and tissue biopsies also are of value.

## Protozoology (Protists–The Protozoa)
## Sarcomastigophora
## Sarcodina

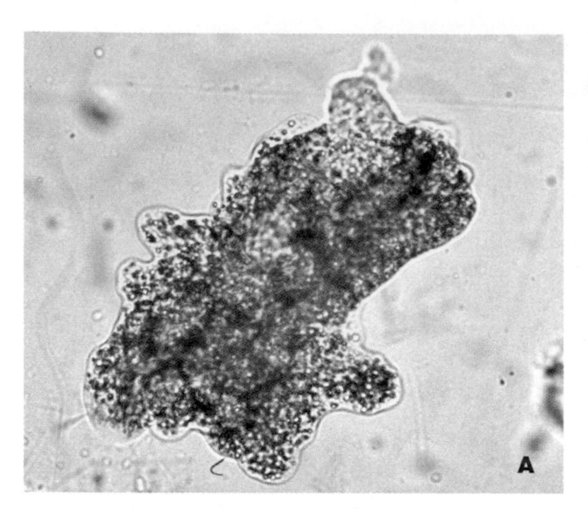
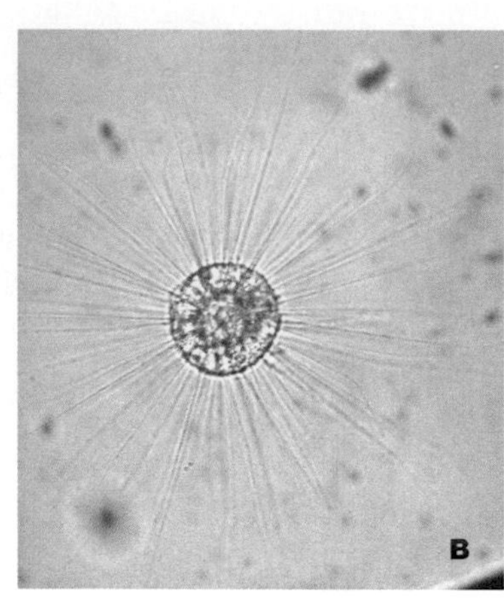

**Figure 252**
Examples of members of Sarcodina. (a) A temporary wet-mount of *Amoeba* species. Note the presence of the extended pseudopodia. (b) *Actinophrys.* (c) Examples of foraminifera.

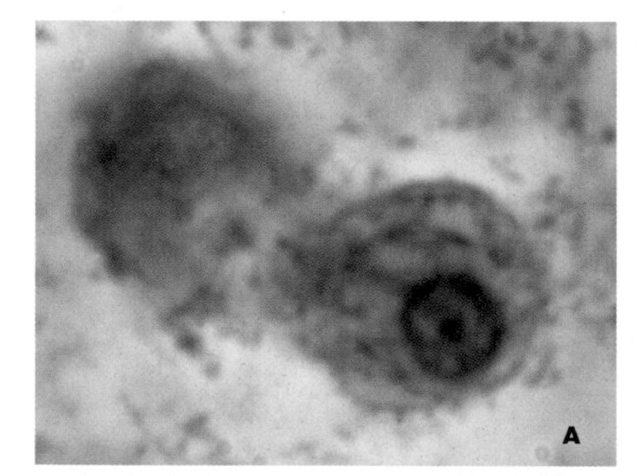
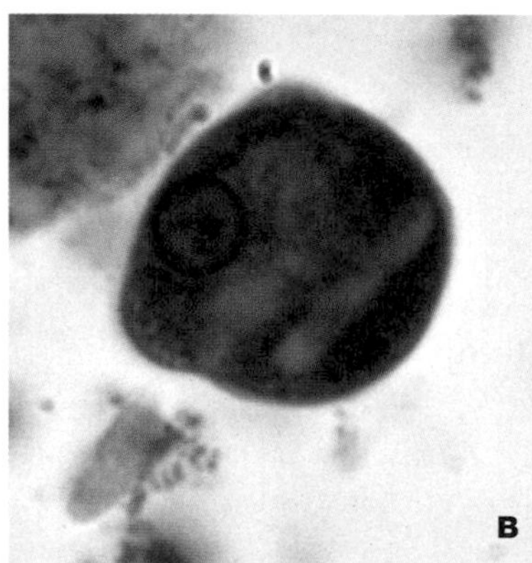

**Figure 253**
Microscopic views of *Entamoeba histolytica.* (a) A trophozoite emerging from a cyst in a stool specimen. (b) A stained preparation showing the typical wagon wheel nucleus.

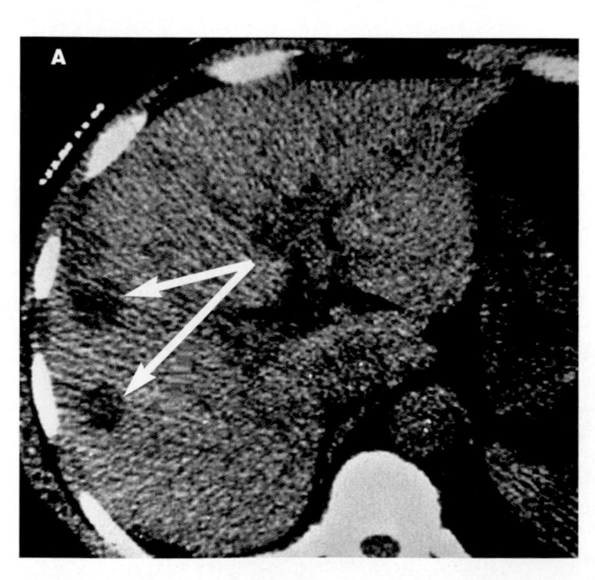
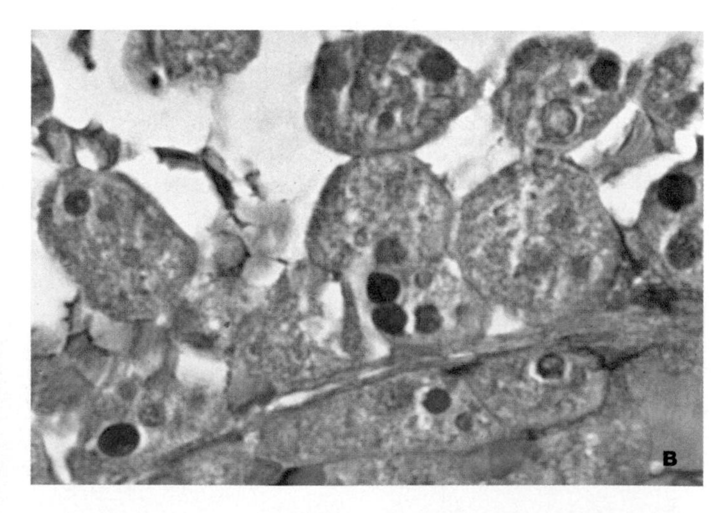

**Figure 254**
Clinical and diagnostic features of amebiasis. (a) A CAT scan showing liver destruction (arrows). (b) A stained preparation of the large intestine showing oval to round amoebae in the tissue. (Courtesy of Dr. L. Alpert, Pathology Department, The Sir Mortimer B. David Jewish General Hospital.)

## Protozoology (Protists–The Protozoa)
## Sarcomastigophora
## Sarcodina *(Continued)*

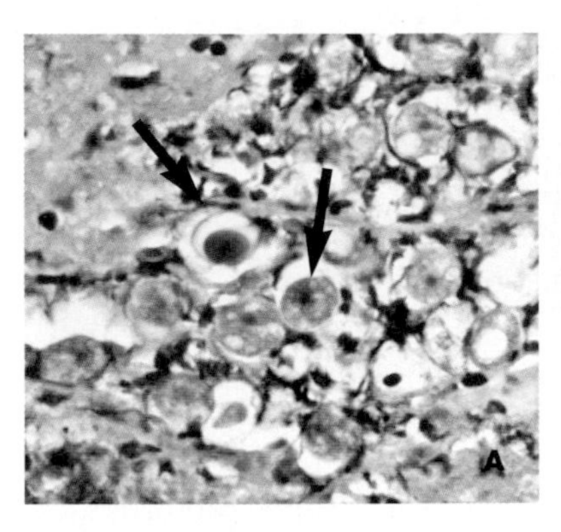
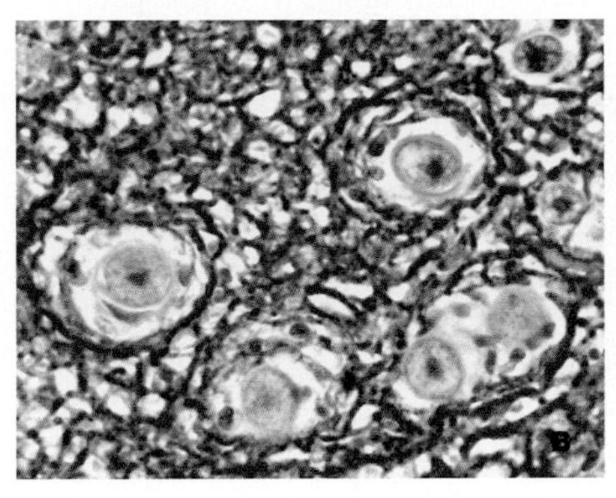
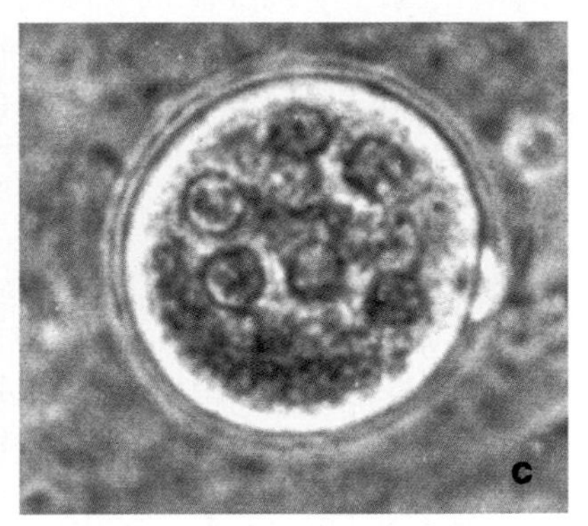

**Figure 255**
Other pathogenic and nonpathogenic amoebae. (a) Stained *Acanthamoeba* trophozoites at the base of a peptic ulcer. (b) Clusters of *Acanthamoeba* (40×). (Courtesy of Drs. K.L. Thamrasert, S. Khunamornpong, and N. Morakote.) (c) *Entamoeba coli* multinucleated cyst.

## Mastigophora

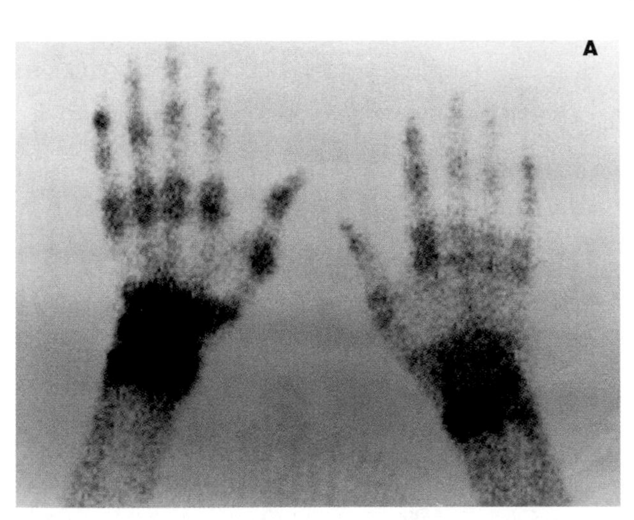
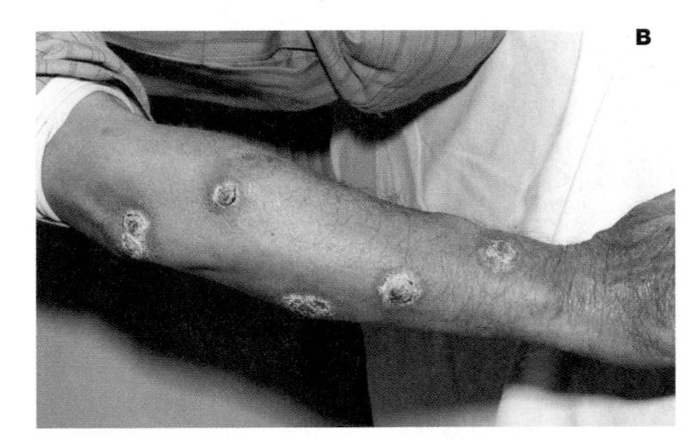

**Figure 256**
Leishmaniasis. (a) A bone scan of a patient with *Leishmania* infection. The darkened areas, which are the sites of infection, show increased radioactively labeled (radiopharmaceutical) material. (From A. Perell'o Roso. *CID* 22(1996):1113–14). (b) A case of cutaenous leishmaniasis. Six ulcers covered with white crusts are shown. (Courtesy of Dr. M. Al-Taqi, University of Kawait.)

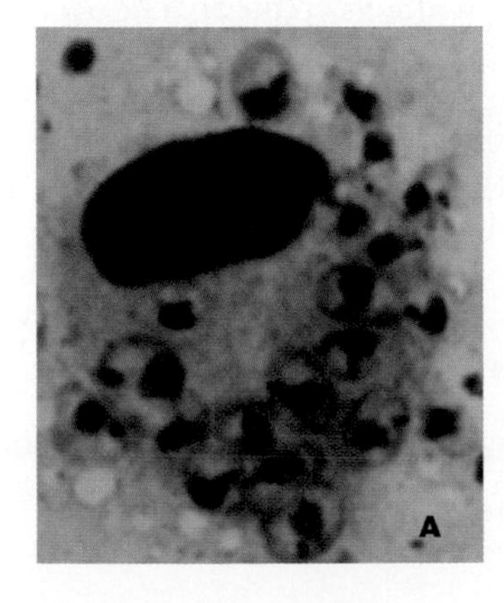
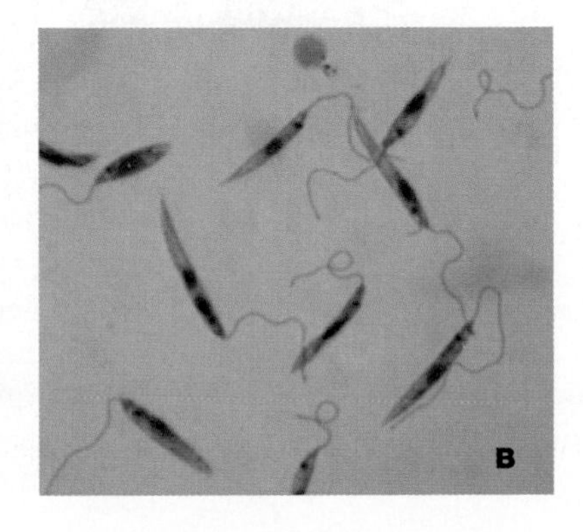

**Figure 257**
*Leishmania tropica.* (a) A specimen from a patient, showing small amastigotes in a mononuclear cell. (b) *L. tropica* promastigotes. (Courtesy of Dr. M. Al-Taqi, University of Kuwait.).

***Trichomonas vaginalis***

*Trichomonas vaginalis* is the cause of a common urogenital disease, trichomoniasis in women. Infected women usually experience a vaginitis, with a foul-smelling discharge and small, bleeding lesions.

**Transmission and Life Cycle.** Most cases of trichomoniasis are acquired by sexual intercourse. Fomites, such as shared towels, also are possible sources of the protozoon. Neonatal infections also have been reported.

**Morphology.** *Trichomonas vaginalis* is a pear-shaped organism (Figure 259). It has four anterior flagella and a fifth flagellum that forms the outer edge of an undulating membrane. The organism has only a trophozoite stage and divides in the urogenital tract by splitting in half.

**Diagnosis.** Demonstrating the organism in vaginal secretions, washings, or scrapings is generally used for diagnosis. Culture and immunofluorescence are also of value.

***Trypanosoma brucei rhodesiense***
***T. b. gambiense***

Human African trypanosomiasis (sleeping sickness) is a systemic and central nervous system infection caused by two geographically distinct forms of trypanosomes. *Trypanosoma brucei gambiense* (West Africa) and *T. brucei rhodesiense* (East Africa). Infected individuals exhibit fever, headache, and enlarged lymph glands early in the disease. Menigoencephalitis develops later.

**Transmission and Life Cycle.** The trypanosomes are spread through the bites of tsetse flies, which are members of the genus *Glossina* (Figure 260b). Both domestic and game animals are major reservoirs.

**Morphology.** The form of the protozoon found in the human is the **trypomastigote**. It is slender and spindle-shaped measuring about 15 μm in length, and has an undulating membrane extending the full length of the protozoon and ending in a flagellum (Figure 260a). The protozoon does not have a cyst stage.

**Diagnosis.** In the early stages of the disease, demonstrating trypanosomes in blood and lymph node specimens is most successful. Spinal fluid also is of value in cases of early central nervous system invasion. Culture, laboratory animal inoculations, and serological tests such as immunofluorescence are usually successful for diagnosis.

***Trypanosoma cruzi***

American trypanosomiasis, also known as Chagas' disease, is caused by *T. cruzi* (Figure 261a), a protozoon with a large animal reservoir in South and Central America.

Clinical conditions resulting from infection include local inflammation at the site of inoculation, severe inflammation of the heart tissue, encephalitis, and multi-organ involvement.

**Transmission.** *Trypanosoma cruzi* is usually transmitted to humans by blood-sucking triatomid (reduviid) bugs (Figure 261b) that deposit infectious feces at the site of a bite or on the mucosal surface. Transmission may also occur with the infection of contaminated human blood products or transplacentally from infected mother to fetus.

**Morphology.** In the blood *Trypanosoma cruzi* appears as short, stumpy cells with undulating membranes and free flagella. These forms are able to infect a variety of cells in which they become small, round protozoa lacking flagella.

**Diagnosis.** *Trypanosoma cruzi* can be easily demonstrated microscopically in stained blood smears in the early stages of the disease. Xenodiagnosis, a procedure in which laboratory-raised reduviid bugs feed on a suspected patient, also is used. The gastrointestinal contents of these arthropods are examined for organisms 1 to 2 months later. Polymerase chain reactions (PCR) and nucleic acid probes are expected to replace xenodiagnosis.

## Ciliophora

Members of the phylum Ciliophora are characterized by the presence of numerous cilia on their surfaces (Figure 262). These cilia function both in moving the organism and in obtaining food. The beating strokes of the cilia, as in the case of the *Paramecium,* cause the cell to revolve as it swims. Ciliates also have a definite shape due to the presence of a sturdy but flexible outer covering, the pellicle (Figure 262c).

Of all the protozoa, the ciliates are the most specialized because they have organelles that carry out particular vital processes. These organelles include trichocysts, the macronucleus, the micronucleus, and contractile vacuoles. (see Table 16).

Ciliates are found in both fresh and salt water. Some are free-living; others, such as *Balantidium coli,* are parasitic.

### *Balantidium coli*

*Balantidium coli* causes balantidiasis. It is the only ciliate known to cause human infections. This organism produces ulcerations in the large intestine similar to those found in amebiasis.

**Transmission and Life Cycle.** Infection is acquired by the ingestion of cyst-contaminated food or water. Human-to-human transmission also is common.

Although the infection is found in a variety of mammals, pigs are the major reservoir. Many infected persons have a history of contact with pigs.

## Protozoology (Protists–The Protozoa)
## Mastigophora *(Continued)*

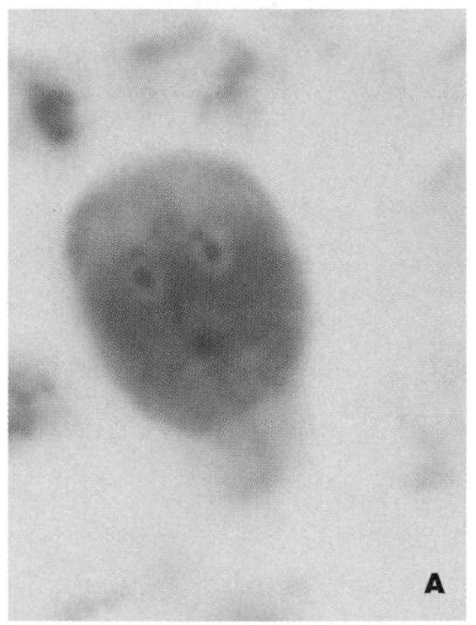


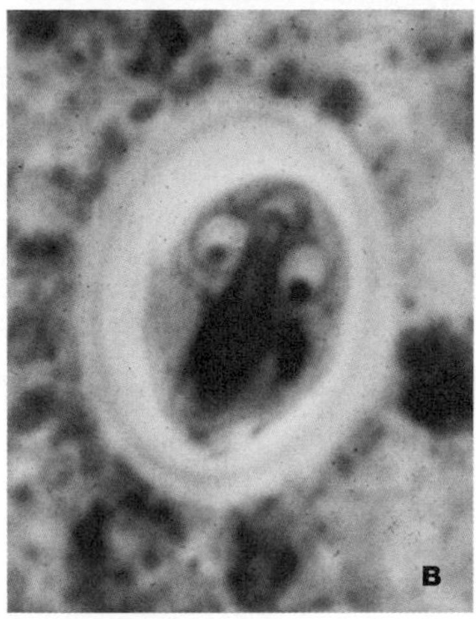


**Figure 258**
Stained preparations of *Giardia lamblia* in fecal smears. (a) A trophozoite. (b) A cyst.

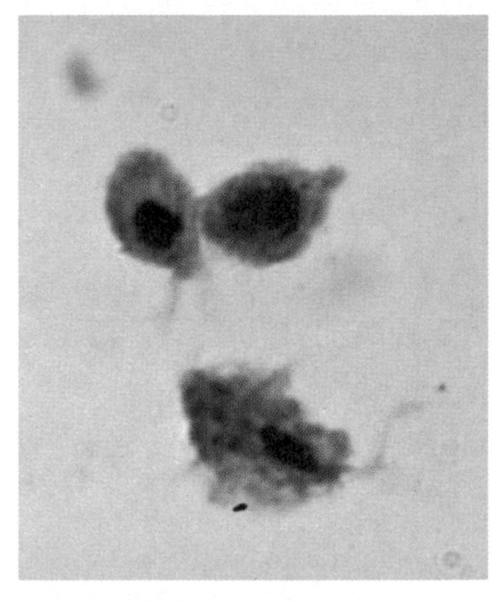

**Figure 259**
*Trichomonas* in a stained smear.

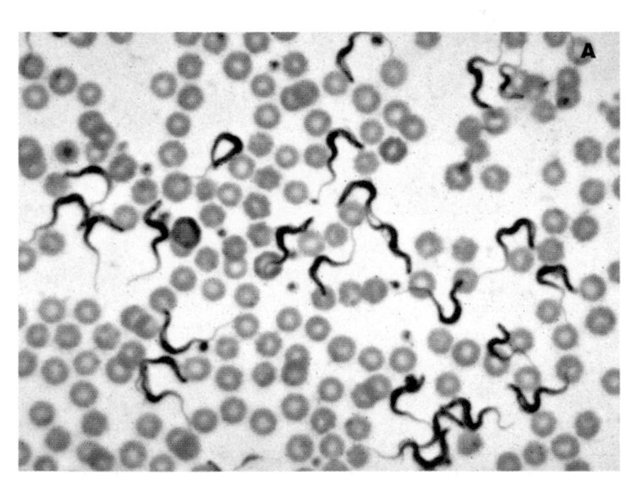


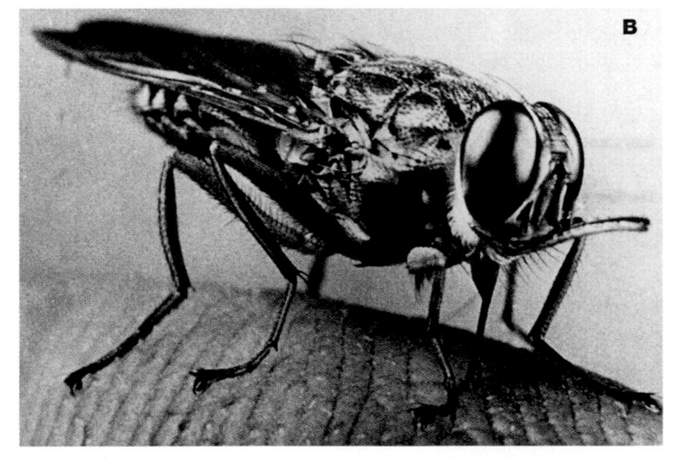


**Figure 260**
African sleeping sickness. (a) The causative agent, *Trypanosoma brucei gambiense* in a blood smear. (b) The tsetse fly (*Glossina* species) on the tip of a finger. (Courtesy of the World Health Organization.)

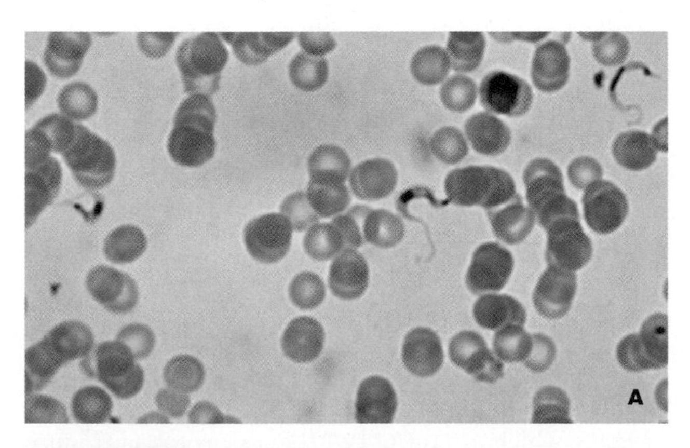

**Figure 261**
Chagas' disease. (a) The small causative agent, *Trypanosoma cruzi,* surrounded by numerous red blood cells. (b) Cone-nosed bugs of the family Reduviidae.

**Morphology.** *Balantidium coli* has both a trophozoite and a cyst stage.

**Diagnosis.** Finding trophozoites (Figure 263) or cysts in stool specimens is diagnostic.

## Apicomplexa

Members of the phylum Apicomplexa, also called sporozoa, are intracellular parasites that contain several organelles organized into an **apical complex.** This structure helps the parasite penetrate into host cells.

All sporozoa are parasitic, absorbing nutrients from their hosts. Some are intracellular; others are found in body fluids or in various body organs. Adult sporozoa have no organelles for movement.

Both asexual and sexual reproduction occur among sporozoa. A number of sporozoa undergo sporulation, producing numerous small, infective spores called **oocysts**. Infected spores reach a susceptible host by way of food, water, or arthropod bites. Spores typically contain one or more smaller individual organisms called **sporozoites**. Many sporozoa have complicated life cycles; certain stages take place in one host, and other stages in a different host. An example of such a cycle involves mosquitoes and the different species of *Plasmodium,* which cause malaria not only in humans but also in several other animals (Figure 264).

Several other pathogens in the Apicomplexa include *Toxoplasma gondii* (Figure 267), *Babesia* species (Figure 268a), *Theileria* species (Figure 268b), *Isospora* (Figure 269), and *Cryptosporidium* (Figure 270).

### Plasmodium

Human malaria is caused by four *Plasmodium* species: *P. falciparum* (Figure 265), *P. vivax* (Figure 266), *P. ovale*, and *P. malariae. Plasmodium falciparum* accounts for the majority of deaths.

Malaria is endemic throughout the major tropical areas of the world.

**Transmission and Life Cycle.** Malaria is transmitted through the bites of female *Anopheles* mosquitoes. Blood transfusion and sharing of contaminated hypodermic syringes and needles are other modes of transmission.

Infective forms of *Plasmodium* (sporozoites) are introduced into the host by the mosquito while it is biting. These sporozoites are motile, and enter the host's circulatory system. In about one hour, after leaving the circulation the sporozoites enter liver cells, where they initiate the first phase of the asexual reproductive cycle, *exoerythrocytic schizogony.* Schizogony refers to the formation of a *schizont*, a stage resulting from the repeated division of the nucleus. The parasite enlarges and eventually fragments into a large number of daughter cells called *merozoites*. Released merozoites from the liver then begin the second phase of the asexual reproductive cycle, *erythrocytic schizogony*, by invading red blood cells (RBCs) in the circulatory system and continuing to reproduce asexually and forming schizonts. Upon maturing the erythrocytic schizonts rupture and release merozoites. These forms then, in turn invade other uninfected cells and continue the reproductive cycle. Reproductive cycles occur with typical regularity for each of the *Plasmodium* species and result in schizont stages with characteristic numbers of merozoites. The cycles occurring in both the mosquito and the human are shown in Figure 264. The various (RBC) stages of two species, *P. falciparum* and *P. vivax,* are shown in Figures 265 and 266, respectively.

**Diagnosis.** The diagnosis of malaria is generally made by identifying the parasites in blood smears.

### Toxoplasma gondii

*Toxoplasmosis* is a worldwide zoonosis caused by *Toxoplasma gondii* (Figure 267). The cat is the definite host, with a broad range of mammals, including humans, serving as secondary hosts.

**Transmission and Life Cycle.** *Toxoplasma gondii* can be acquired directly by ingestion of fecal oocysts shed by infected cats or indirectly by ingestion of poorly cooked cyst-containing meats. Transmission of the protozoon also can occur by congenital passage of *T. gondii,* organ transplantation, transfusion with contaminated blood, and laboratory accidents.

Cats acquire *T. gondii* by eating infected rodents and birds that have tissue cysts or by ingesting oocysts from feces of other cats. The parasite reproduces sexually, resulting in the production of infectious oocysts that are excreted in the cat's feces. When humans or other animals ingest such oocycts, organisms escape and develop into trophozoites called **tachyzoites**. These tachyzoites are elongated, crescent-shaped cells that actively penetrate body organs and tissues (Figure 267c). Eventually, some tachyzoites form cysts (**bradyzoites**) in brain, skeletal muscle, and heart tissues.

**Diagnosis.** Serological tests are the principal means of diagnosis.

### Babesia

**Babesiosis** is a malaria-like illness caused by *Babesia* species, which are intraerythrocytic protozoa (Figure 268a). *Babesia,* together with members of the genus *Theileria* (Figure 268b), are referred to as **piroplasmas**

## Protozoology (Protists–The Protozoa)
## Ciliophora

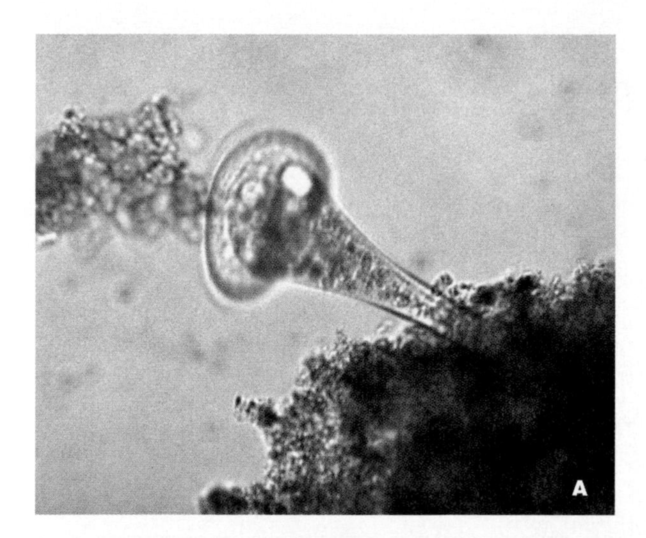


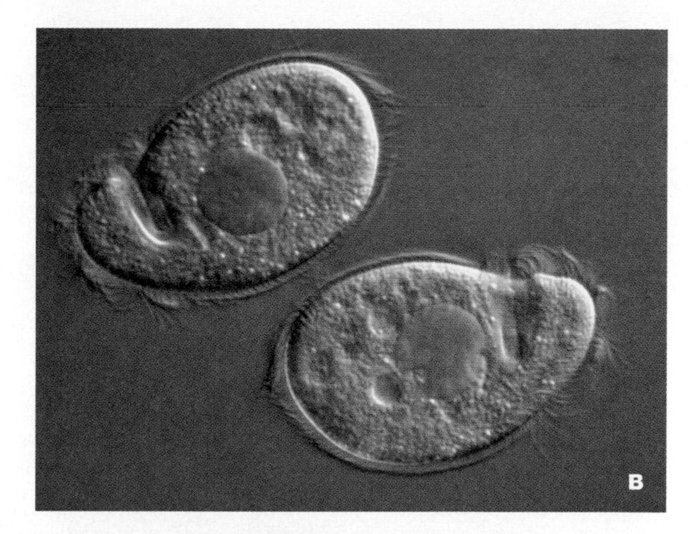


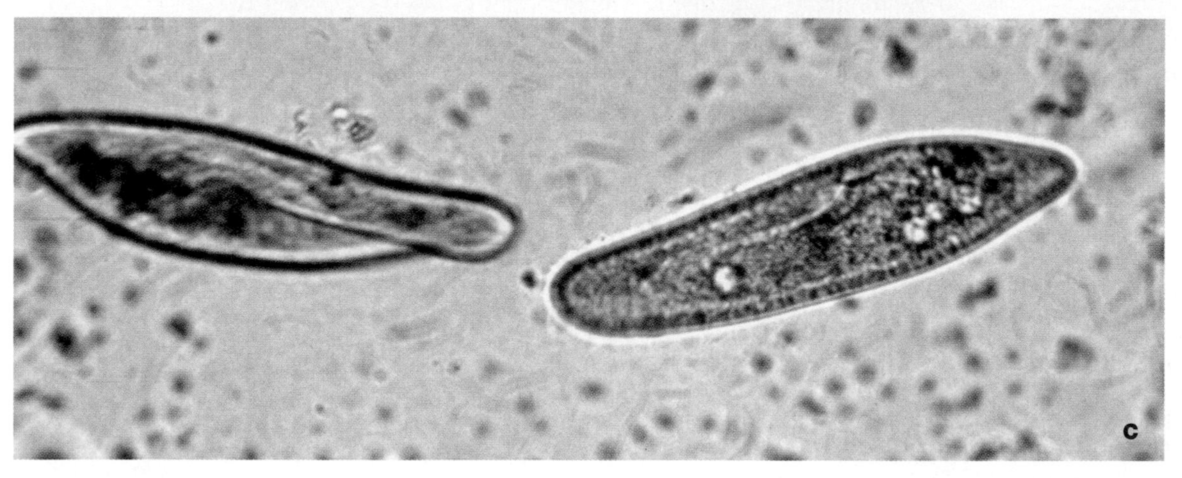


**Figure 262**
Ciliates. (a) *Vorticella.* (b) The marine anaerobe *Plagiopyla frontata,* with an obvious circular to oval nucleus. (c) *Paramecium.*

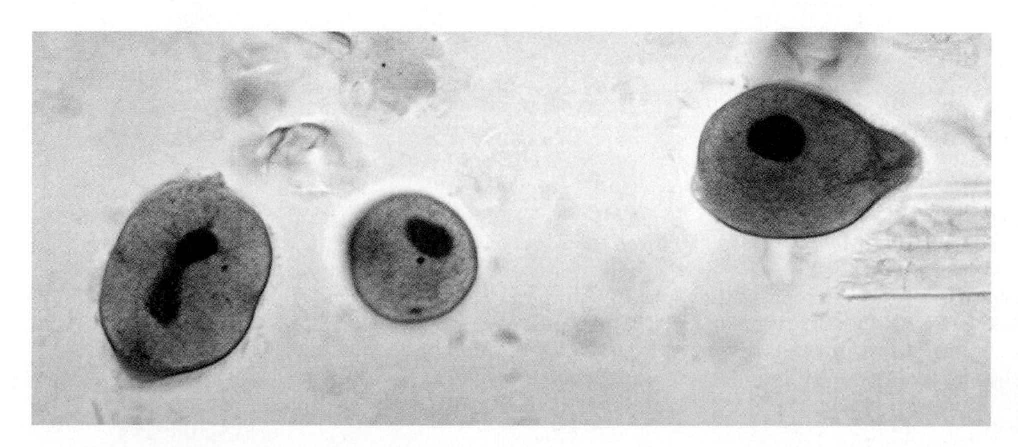

**Figure 263**
The pathogenic *Balantidium coli* (trophozoite).

## Protozoology (Protists–The Protozoa)
## Apicomplexa

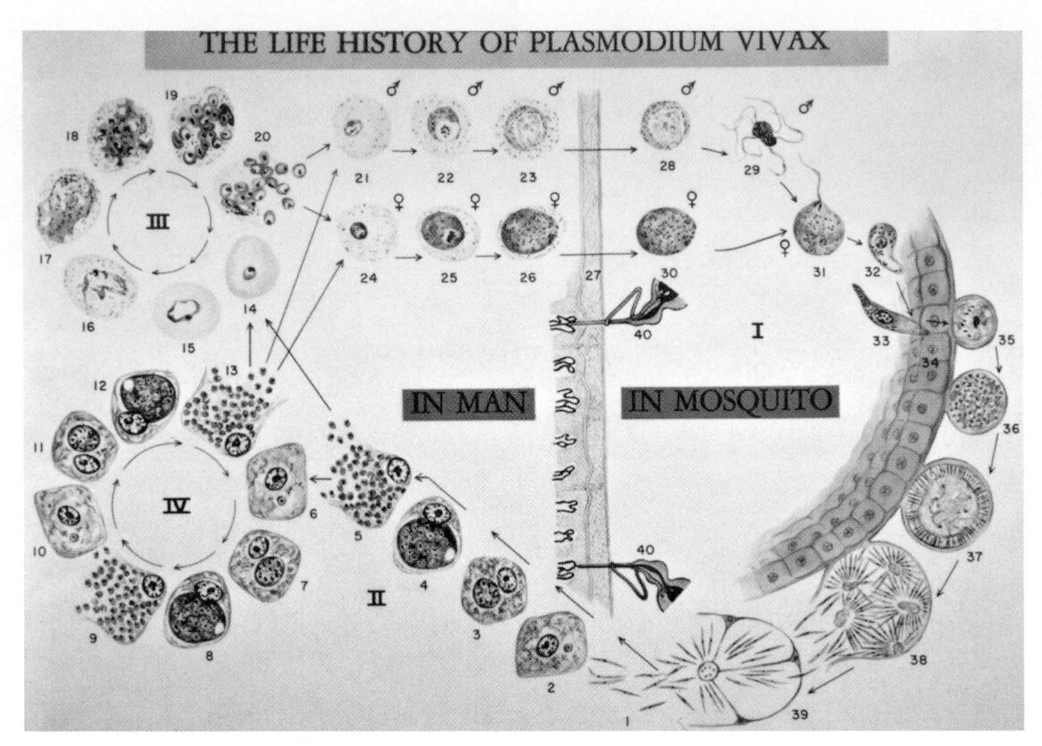


**Figure 264**
The life cycle of *Plasmodium* in a mosquito (right) and in the human (left). The human is infected by small elongated cells (no. 1), the sporozoites, which are produced in the *Anopheles* mosquito and invade the salivary gland of the insect (nos. in part I). The female mosquito introduces sporozoites into the human bloodstream when she takes a blood meal. The newly introduced parasites are removed from the blood by organs such as the liver and spleen (nos. 2–13, in II and IV). Here they multiply and produce other infective forms that are released into the bloodstream to attack and carry out the asexual cycle in red blood cells (nos. 14–26, in III). The outcome of this cycle is the formation of sex cells (gametocytes) and other infective units for red blood cells (nos. 24–26). If these sex cells are ingested by another female mosquito, they mature and participate in the sexual reproductive phase by forming a zygote (nos. 28–31). This fertilized cell undergoes development within the mosquito's stomach lining (I), where it enlarges and forms a large number of sporozoites. The cycle then continues.

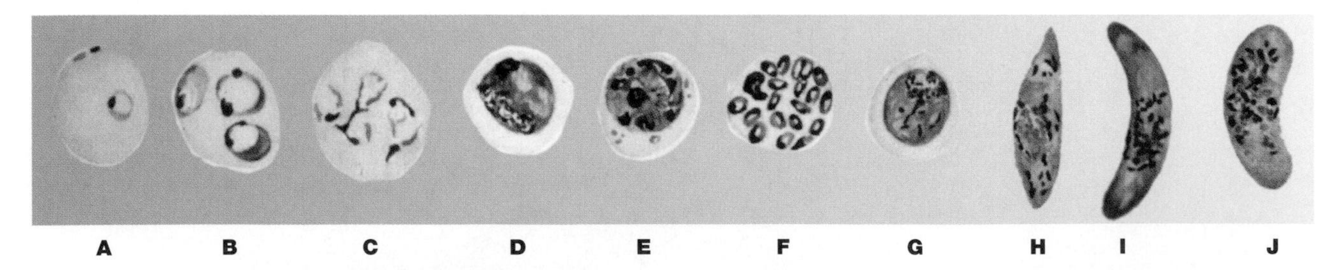


**Figure 265**
(The descriptions given here are adapted from the 1960 U.S. Department of Health, Education and Welfare publication *Manual for the Microscopical Diagnosis of Malaria in Man.*) *Plasmodium falciparum.* (a) A single erythrocyte showing a double infection with young trophozoites. The parasite close to the center of the red cell is a "signet ring" form, and the organism located at the periphery is referred to as a "marginal form." (b) One red blood cell with three somewhat more developed trophozoites. (c) The parasites shown are called estivo-automnal forms. (d) The parasite is undergoing initial chomatin (red area) division. (e,f) Mature schizonts with merozoites. Note the number of small merozoites. (g,h). These stages are representative of the successive events that take place in gametocyte (sex cell) development. Such forms generally are not found in the peripheral circulation. (i) A mature macrogametocyte (female sex cell). (j) A mature microgametocyte (male sex cell).

## Protozoology (Protists–The Protozoa)
## Apicomplexa *(Continued)*

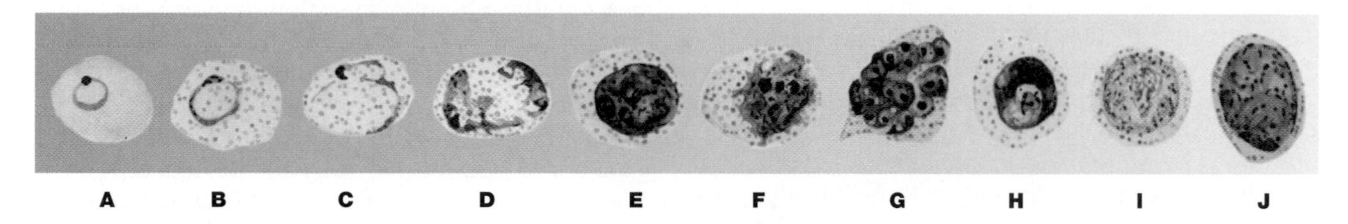


**Figure 266**
*Plasmodium vivax*. (a) A typical signet ring shape. (b) An enlarged erythrocyte with a ring form of trophozoite. The cell also contains Schüffner's stippling. Such stippling may not always be present in infected red blood cells. (c) Another trophozoite form. (d) The presence of two amoeba-shaped trophozoites. (e,f) Erythrocytes showing progressive stages in schizont division. (g) A mature schizont. Note the number of merozoites. (h) A developing gametocyte. (i,j) Mature microgametocyte, macrogametocyte, male and female sex cells, respectively.

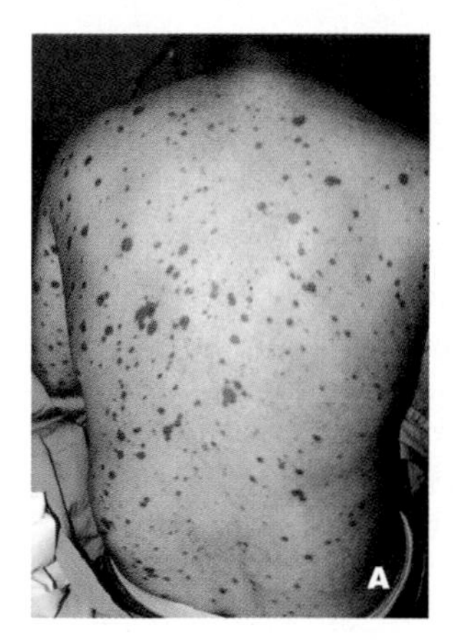


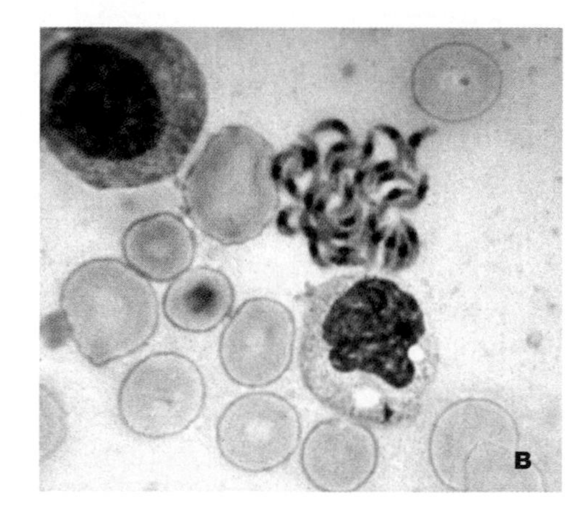


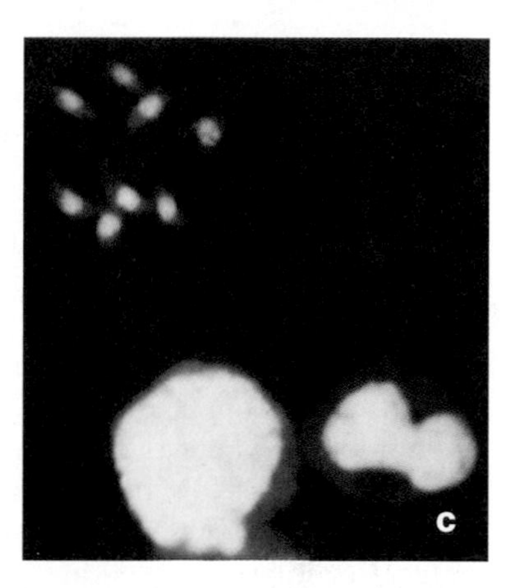


**Figure 267**
Toxoplasmosis. (a) The appearance of skin involvement with toxoplasmosis. (From P.J. Spagnuolo et al. *CID* 14(1992):1084–88.) (b) The differently staining *Toxoplasma gondii* in a bone marrow smear. White and red blood cells are also shown. (c) Extracellular *T. gondii* in a blood specimen shown by immunofluorescence. The protozoon can develop intracellularly as well as extracellularly; note the crescent shape of this microorganism.

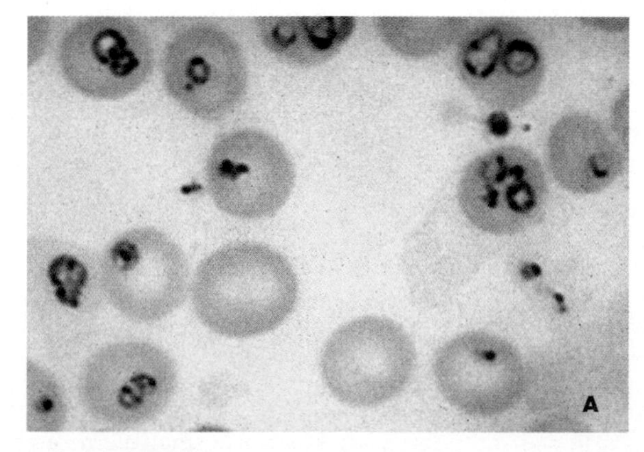


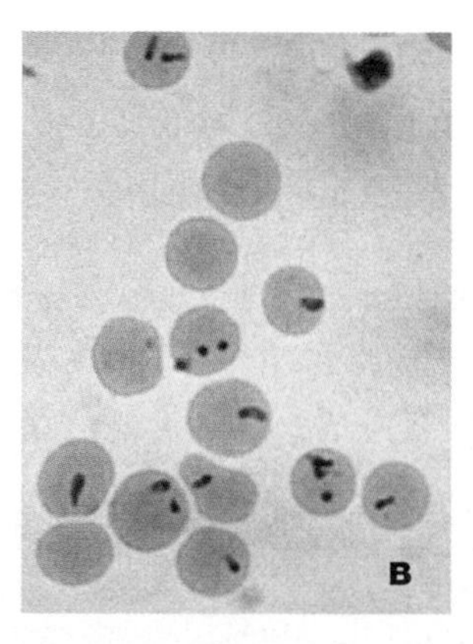


**Figure 268**
Piroplasmas in blood smears. (a) *Babesia microti*. (b) *Theileria parva*. Note these organisms are located within several of the circular red blood cells.

because of their pear-shaped red blood cell stages. Both genera infect a wide variety of wild and domestic animals. *Babesia microti* (in the United States) and *B. divergens* (in Europe) cause infection in humans.

**Transmission.** Young forms (nymphal) of *Ixodes dammini* (ticks) transmit the *Babesia* species. Infections also can be acquired by transfusions with infected blood. Many human infections caused by *B. microti* are either asymptomatic or subclinical. Severe and even life-threatening infections with high fevers and hemolytic anemia occur in immunocompromised and elderly individuals.

**Diagnosis.** Diagnosis is made by identifying the parasites in stained blood smears. The presence of four pear-shaped infective units, a tetrad, called the *Maltese cross* is diagnostic (Figure 268).

### *Isospora belli*

**Isosporiasis** is caused by *Isospora belli.* The disease, which takes the form of a long-lasting diarrhea, is uncommon in the United States, but it occurs frequently in South America and Southeast Asia.

**Transmission and Life Cycle.** Infection is acquired by ingestion of oocysts (Figure 269). Oocysts are excreted in stools and contain sporocysts consisting of four sporozoites. After they have been ingested, the sporozoites escape from the oocyst and penetrate the intestinal lining, where they reproduce and eventually cause the chronic, disabling diarrhea. AIDS patients and others with some type of immunosuppression are quite susceptible to infection.

**Diagnosis.** Diagnosis is dependent on finding oocysts in stool specimens. Modified acid-fast stains of materials are of value.

### *Cryptosporidium parvum*

*Cryptosporidium parvum* causes **cryptosporidiosis,** an infection of the epithelial cells lining the human digestive tract resulting in diarrhea. This organism is recognized as a major cause of diarrhea worldwide.

**Transmission and Life Cycle.** *Cryptosporidium parvum* is acquired by ingestion of infectious forms, oocysts, in contaminated water; by contact with infected lower animals; or by person-to-person transmission.

*Cryptosporidium parvum* has a life cycle similar to that of *Isospora belli.* In the immunocompetent host, illness is self-limited. However, in individuals with immunosuppression, infection may result in prolonged life-threatening cholera-like diarrhea.

**Diagnosis.** Modified acid-fast and other staining techniques are used with stool specimens for diagnosis (Figure 270). Immunofluorescent techniques also are of value.

## Microsporidia

Microsporidia are obligate intracellular protozoa that parasitize numerous species of vertebrates and invertebrates and may coexist as commensals with their animal hosts. The epidemiology and pathogenesis of microsporidial disease in humans are unclear. Since 1985, increasing numbers of cases of human microsporidiosis, mostly in HIV-infected or other immunocompromised patients, have been reported. Most infections reported in AIDS patients are caused by Enterocytozoon species (Figure 271).

**Transmission.** Intestinal and systemic infections are probably acquired through the ingestion of spores.

**Diagnosis.** Diagnosis usually is made by demonstrating the organisms in biopsies or in staining fecal preparations. Currently, electron microscopy is required to identify microsporidia species.

## Protozoology (Protists–The Protozoa)
## Apicomplexa *(Continued)*

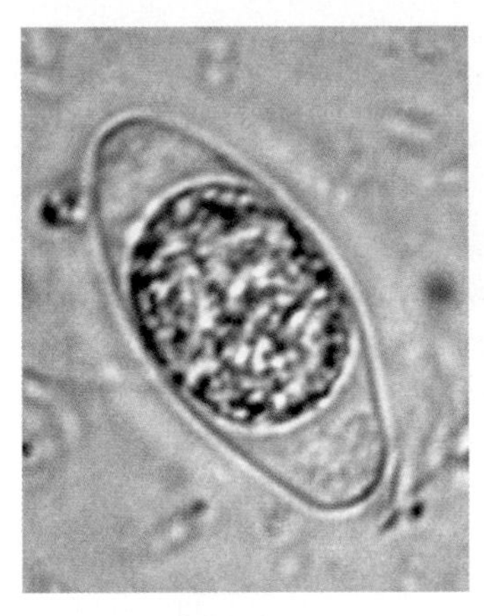

**Figure 269**
*Isospora* sporulated oocyst.

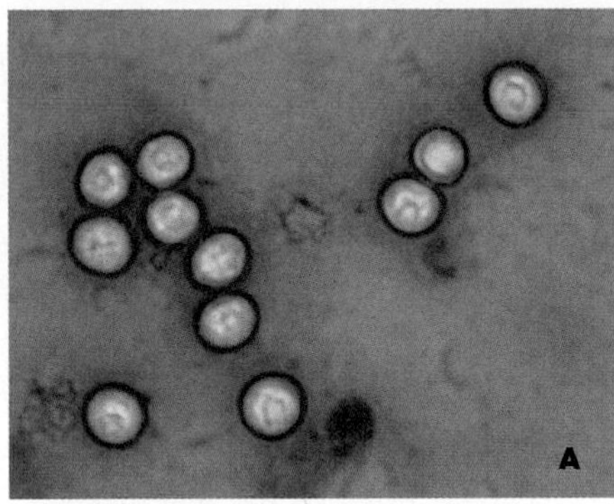


**Figure 270**
*Cryptosporidium parvum* oocysts in human stool. (a) The results of a rapid negative stain procedure with methyl green. The methyl green acts as a background for the oocysts, which do not stain. (b) Red oocysts stained by a modified acid-fast technique (1,000×). (Courtesy of Drs. M. Scaglia and G. Chichino, Laboratory of Clinical Parasitology. Institute of Infectious Diseases, University-IRCCS, San Matteo, Pavia, Italy.)

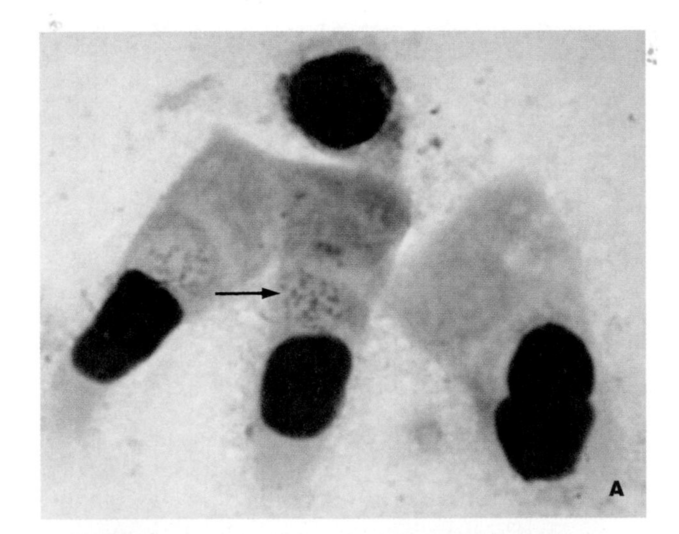


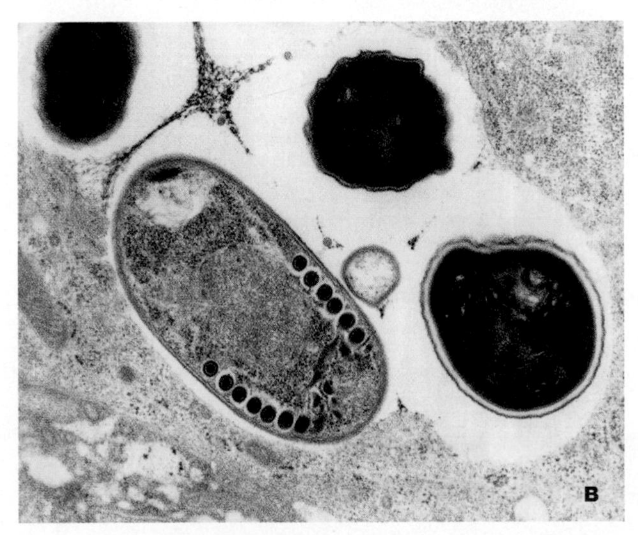


**Figure 271**
(a) A Giemsa-stained intestinal biopsy specimen showing darker microsporidia, *Enterocytozoon bieneusi* surrounding the nuclear area (1,000×). (From J.-M. Molina, et al., *J. Inf. Dis.* 167(1993):217–21.) (b) A transmission electron micrograph of *E. bieneusi.* (Courtesy of Dr. J.M. Orenstein, Department of Pathology, George Washington University.)

# SECTION 7 Virology

*They can strike anywhere, anytime. On a cruise ship, in the corner restaurant, in the grass just outside the back door. And anyone can be a carrier.*

—*Michael D. Lemonick, Time Magazine*

## Viruses and Their Effects

Viruses are very different from the microbial groups mentioned thus far. They are so small that most can be seen only with an electron microscope, and they are acellular (not cellular). Structurally, a virus particle (**virion**) contains a core made of only one type of nucleic acid, either DNA or RNA. This core may be surrounded by a protein coat. Sometimes the coat is encased by an additional layer, a lipid-containing membrane called an **envelope**. All living cells have RNA and DNA, can carry out chemical reactions, and can reproduce as self-sufficient units. Viruses can reproduce only by using the cellular machinery of other organisms. Thus, all viruses are parasites of other forms of life.

Virtually every kind of life can be infected by viruses—vertebrate and invertebrate animals, plants, prokaryotes, and eukaryotic microorganisms such as fungi, protozoa, and certain algae (Figure 272). There are even some "satellite" viruses, which are considered in a sense parasites of other viruses. Until about 1972, it was generally believed that the smallest infectious disease agents were viruses. This view changed in the 1980s with the discovery of smaller and less complex agents called **viroids**. A number of plant diseases are known to be caused by viroids, which consist of small uncovered (naked) molecules of ribonucleic acids. The discovery of subviral agents of diseases apparently did not end with viroids. Another smaller and quite different disease agent, the **prion** (Figure 273) has been shown to be the cause of scrapie, a neurological disease of sheep and goats. Other prion-caused diseases include mad cow disease and Creutzfeld-Jakob disease. Viruses are unlike any other form of microorganism. This difference is obvious not only from their submicroscopic size but also from other differences related to the way they function (Table 18).

## Basic Structure of Viruses

Each mature virus particle or virion has a characteristic morphology—basic structure and organization. This feature of viruses is best studied by electron mi-

**Table 18** Properties of Viruses and Other Microorganisms

| Microbial Group | Cell Wall | Microbial Components | | Growth Requirements | |
|---|---|---|---|---|---|
| | | *Internal membrane parts* | *Nucleic acid content* | *Cultivation in or on artificial media* | *Require living cells* |
| Algae | Present | Present | DNA, RNA | Yes | No |
| Bacteria | Present | Absent | RNA, RNA | Yes | Some |
| Fungi | Present | Present | DNA, RNA | Yes | No |
| Protozoa | Absent | Present | DNA, RNA | Yes | Some |
| Viruses | Absent | Absent | DNA or RNA[a] | No | Yes |
| Viroids | Absent | Absent | RNA | No | Yes |
| Prions | Absent | Absent | None | No | ? |

[a] Individual virus particles contain either DNA or RNA, never both.

# VIROLOGY
## Viruses and Their Effects

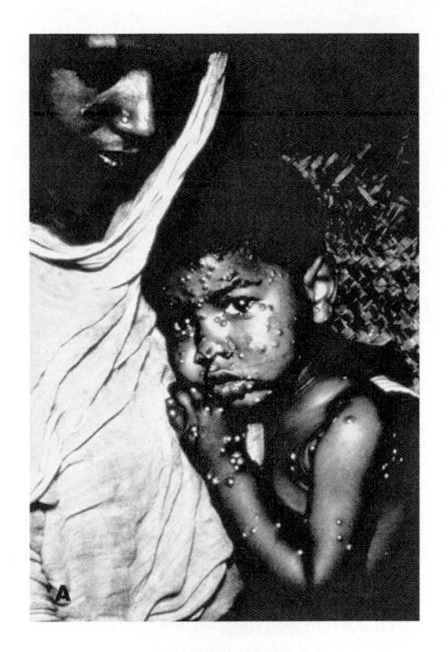


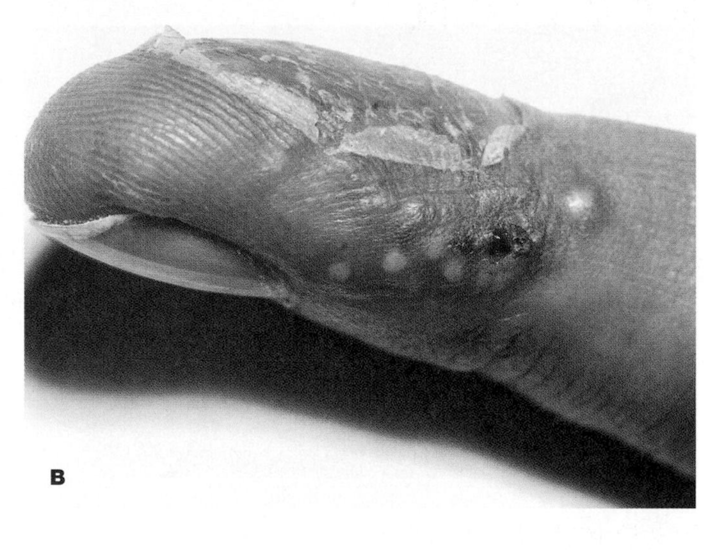


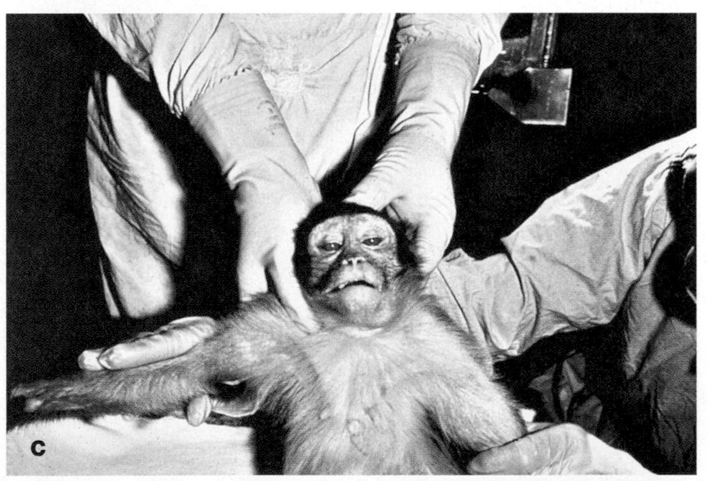

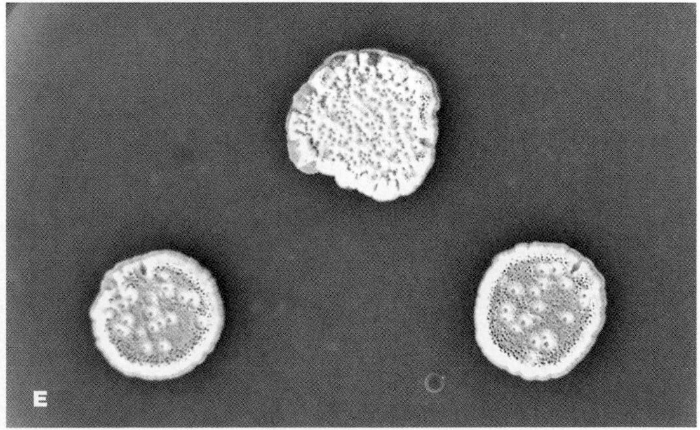


**Figure 272**
The effects of viruses. (a) Smallpox. A victim of a human disease eradicated by immunization. (Courtesy of the World Health Organization). (b) Herpetic whitlow, a herpes simplex virus type 1 infection of the finger. (c) A simian version of chickenpox. (d) A virus-infected plant. The yellow areas contain virus. (e) Colonies of the bacterium *Streptomyces azureus* with plaques (surface holes) caused by bacterial viruses.

# VIROLOGY
## Basic Structure of Viruses

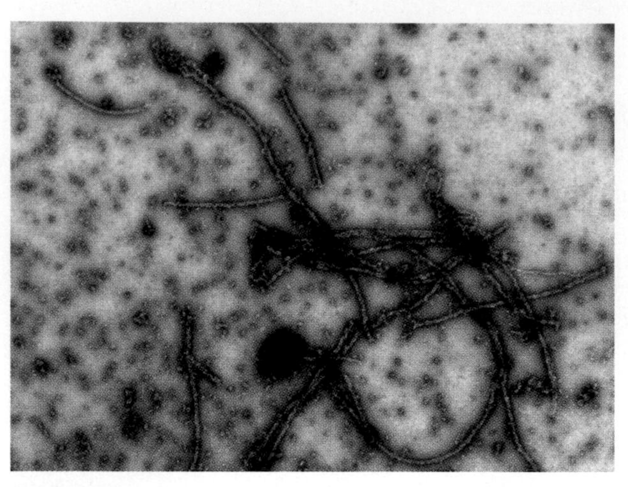

**Figure 273**
A transmission electron micrograph showing negatively stained long particles from a scrapie-infected animal. (Courtesy of Dr. H.K. Narang, The London Hospital Medical College.)

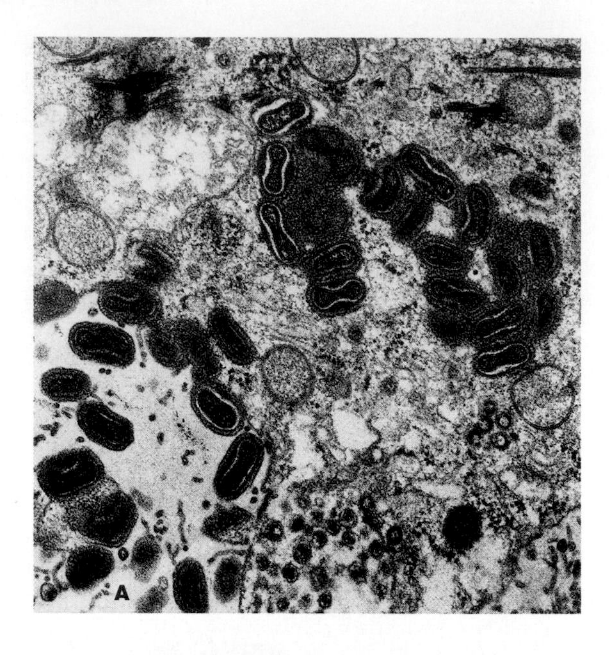


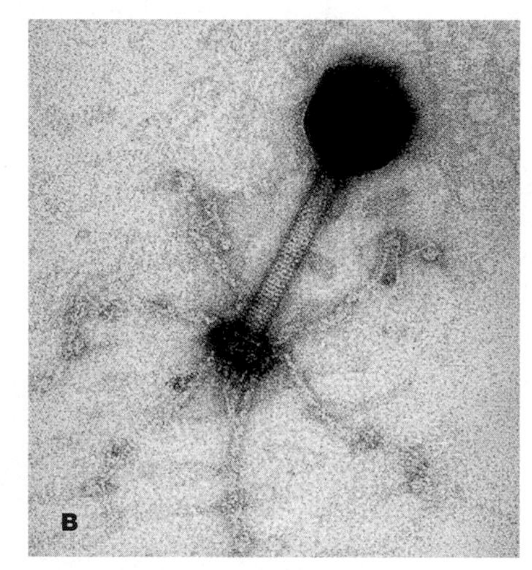


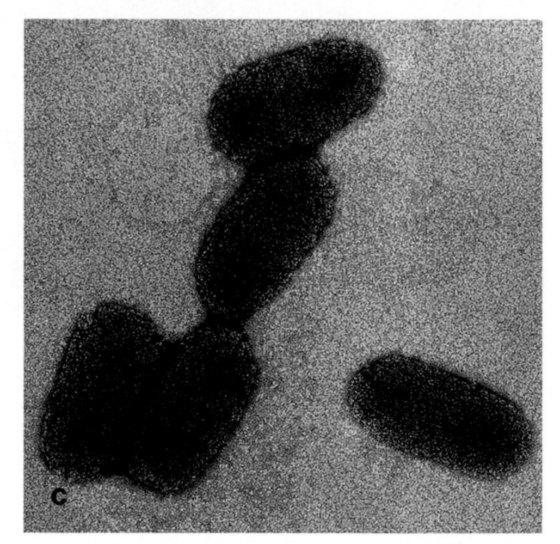


**Figure 274**
Electron microscopic views of negatively stained viruses. (a) A poxvirus, showing several viral components. (b) A bacteriophage (bacterial virus). (c) Sowthistle yellow vein virus of plants, showing surface proteins.

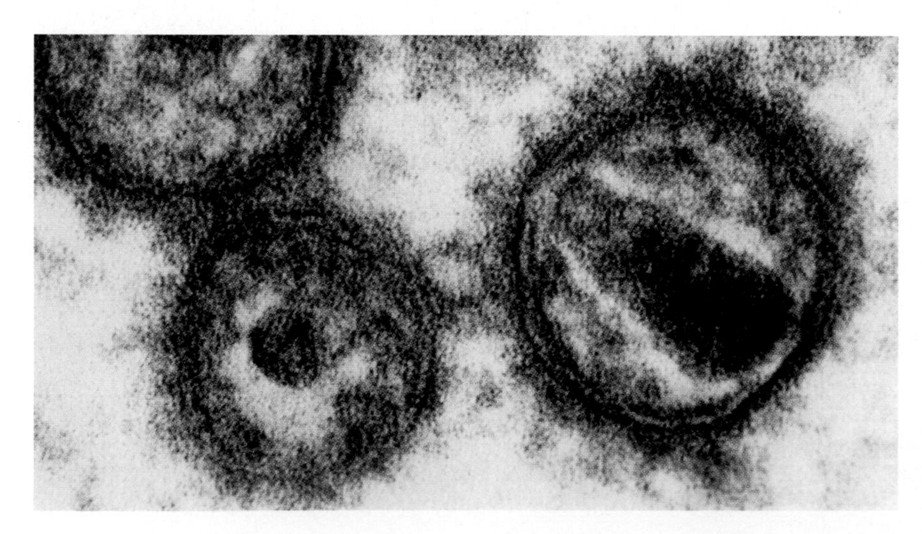

**Figure 275**
An ultrathin section (slice) of human immunodeficiency virus (HIV).

croscopy and techniques such as **negative staining** (Figure 274) and **thin sectioning** of infected tissues (Figure 275).

The viral DNA or RNA forms a central core of a virion, which, depending on the virus, may or may not be encased in a protein shell known as thc **capsid**. Each capsid is constructed from a definite number of subunits called **capsomeres,** which are grouped together to form a characteristic shape, or morphology (Figure 274b,c). Some virions may in turn be wrapped in a membranous envelope derived from the membranes of host-infected cells (Figure 274a). Viral envelopes may have additional structures associated with them.

Several basic viral morphological types are recognized: (1) icosahedral (20 triangular faces) capsid, (2) icosahedral capsid and an outer envelope, (3) helical (springlike) capsid, (4) helical capsid and outer envelope, (5) binal, a combination of icosahedral and helical capsids (Figure 274b), and (6) nucleic acid core surrounded by an envelope (Figure 275). Many virus particles lacking a basic shape or containing accessory structures are called **complex**.

## Cultivation

The replication (multiplication) of viruses differs significantly from the reproduction mechanisms of other microorganisms. Because viruses do not have the cellular structures of eukaryotic and prokaryotic cells, they are incapable of independent activity outside of a host cell. Viral replication can occur only within a host cell. Replication of animal viruses occurs within animal cells (Figure 277), plant viruses within plant cells, and bacterial viruses (bacteriophages) within bacteria (Figure 279). Viruses exhibit a high degree of specificity for host cells.

Viral replication begins with the **attachment** of the virus to a susceptible and compatible host cell. This step is followed by a series of events including (1) the entry of the viral **genome** (genetic material) into the host cell, (2) the synthesis of viral proteins and nucleic acid molecules under the direction of the viral genome, (3) the assembly of new viral particles within specific cellular sites (Figure 278a), and (4) the release of mature viruses from the host cell.

Certain viruses acquire a covering of host cell membranes (envelope) as they emerge (Figures 274a and 278b). Virus release usually causes the death of host cells.

### Animal Virus Cultivation

A number of laboratory animals can be used for animal virus cultivation. These include rats, guinea pigs, hamsters, and mice. Chicken, duck, and turkey embryonated eggs also are of value and provide an excellent experimental system for isolating viruses, for obtaining large quantities of viruses for vaccines and diagnostic reagents, and for studying viral replication mechanisms and the effects of drugs on such processes. The chicken embryo (Figure 276a) provides several tissues and cells that readily support viral replication. These include the yolk sac, amnion, and the chorioallantoic membrane (Figure 276b). Viruses replicating in chicken embryos may cause a variety of visible destructive effects. These include death, embryonic growth defects, and localized areas of membrane damage resulting in bleeding or the formation of discrete opaque areas known as **pocks** (Figure 276c). A pock contains large concentrations of viruses.

### Cell (Tissue) Culture

Although the technique of growing tissues outside of an animal or plant (*in vitro* cell or tissue culture) is almost as old as the study of viruses, early virologists could not use this technique because of problems of contamination by bacteria and fungi. It was only after three advances in virology that tissue culture gained routine acceptance for the cultivation of viruses. These were the introduction of antibiotics, which greatly reduced the contamination problem; development of an excellent, defined growth medium for cells; and the introduction of the enzyme trypsin to free cells from fragments of tissue so that they could be grown in single-cell layers. Cells are introduced into growth containers (tubes, Petri plates, and so on) so that they can form a monolayer, or a single sheet of cells (Figure 277a). Monolayers can be examined macroscopically and microscopically for signs of viral infection. Only the cultivation of animal tissues will be considered.

Two types of cell cultures are in common use: primary and continuous. **Primary cultures** generally consist of cells taken from normal embryonic, fetal, and adult animal tissues. Primary cultures have several characteristics of the original tissue from which they were obtained but will be limited in the number of times they can be subcultured. Eventually, primary cultures die or mutate into different cell strains. **Continuous cultures** are derived from mutant cell lines or cancerous tissues. Such cell lines can undergo an unlimited number of divisions.

Virus-infected cells often develop abnormally and show visible changes in appearance. These changes are known as **cytopathic effects** (CPEs) and include (1) abnormal cellular rounding and detachment (Figure 277b); (2) cell destruction (Figure 277b); (3) syncytium, a joining of several cells or giant cell formation (Figure 277c), and (4) inclusion body formation consisting of masses of viruses or damaged host cell organelles (Figure 277e). Viruses also can be detected

## Cultivation

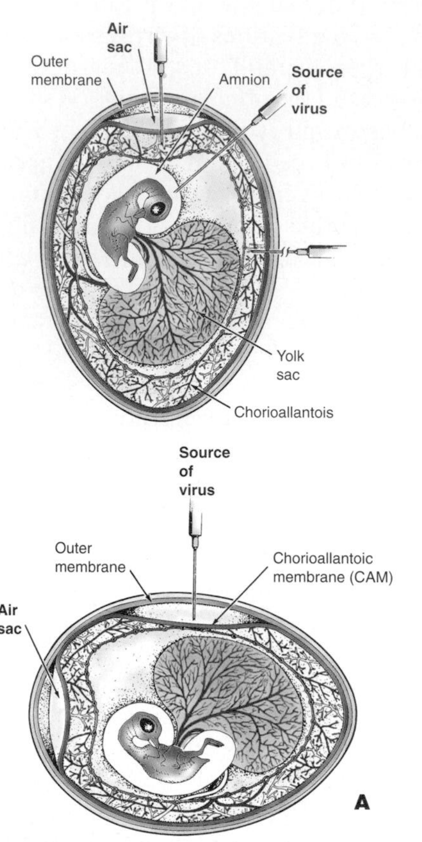


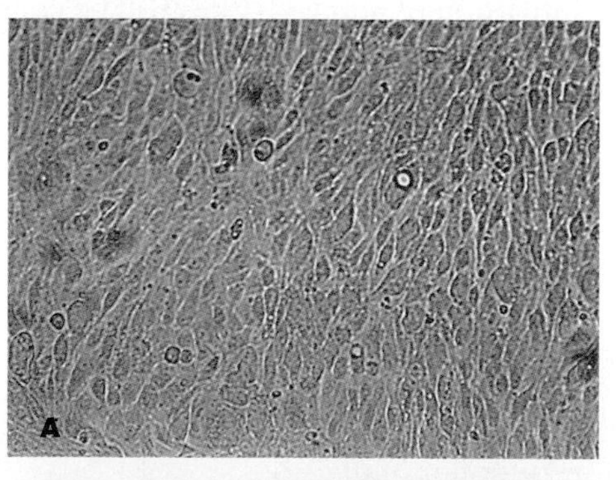


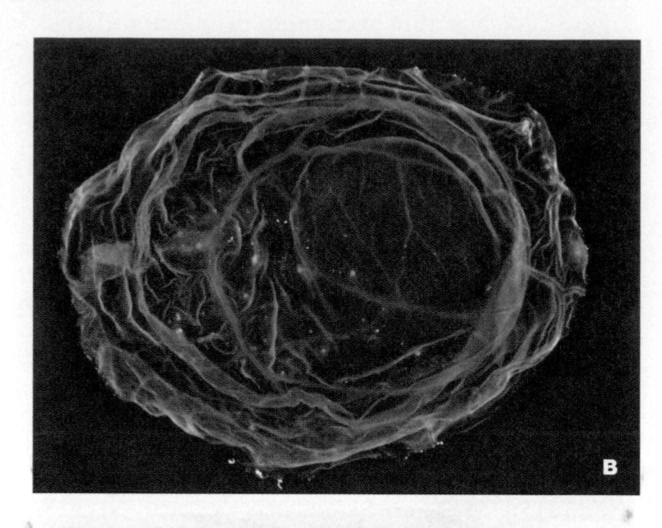


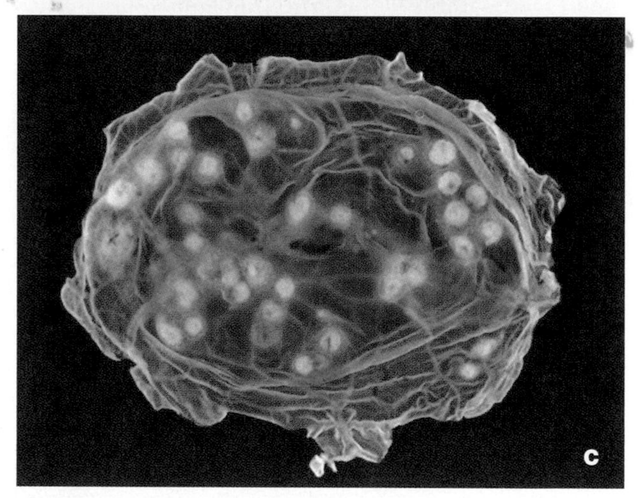


**Figure 276**
Embryonated chicken egg inoculation. (a) Cutaway diagram of the different parts of the animal and the chorioallantoic membrane (CAM) site of inoculation are shown. (b) The appearance of a normal CAM. (c) An infected CAM showing pocks, localized dense areas, and clouding.

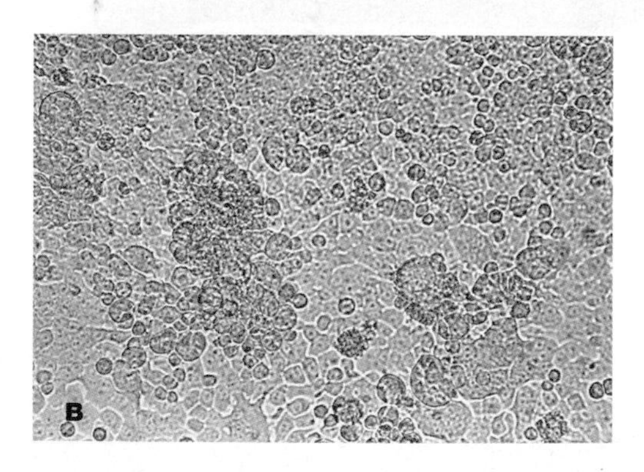


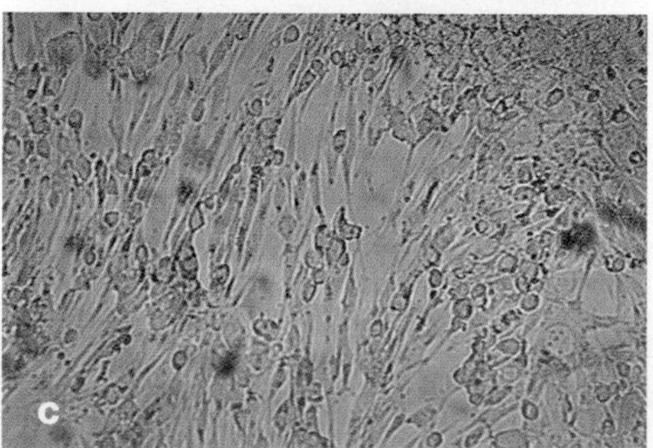


**Figure 277**
Light microscopy of cytopathic effects. (a) Unstained normal rabbit corneal cells growing in a continuous layer. (b) Destruction of cell layer by adenoviruses. (c) Giant cell (multinuclear) formation caused by rubella virus.

## Cultivation *(Continued)*

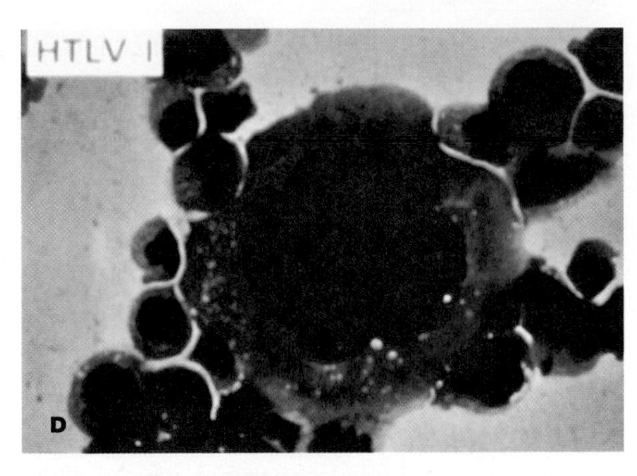


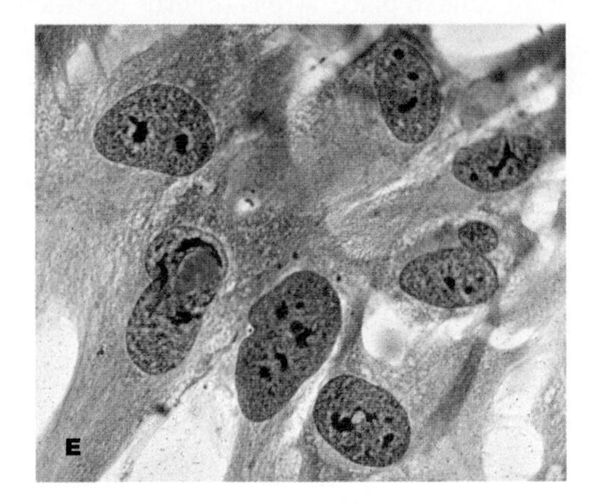


**Figure 277** ***(Continued)***
(d) Stained giant cell formation caused by human T- lymphotrophic virus, type 1. (From R.C. Gallo. *J. Inf. Dis.* 164(1991):235–43) (e) Eosinophilic nuclear inclusion body formation (arrow).

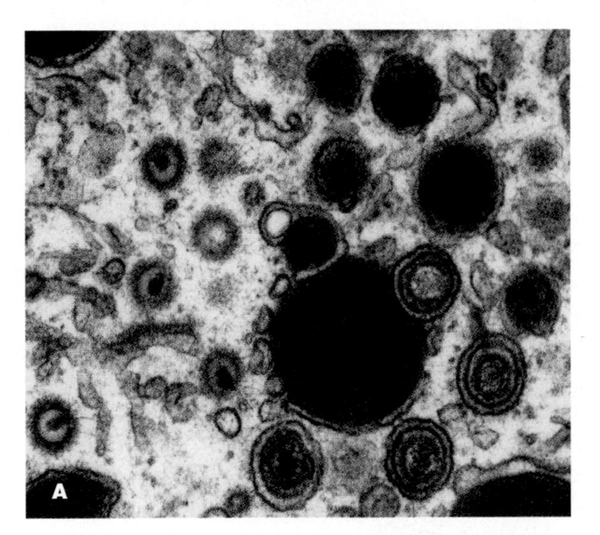


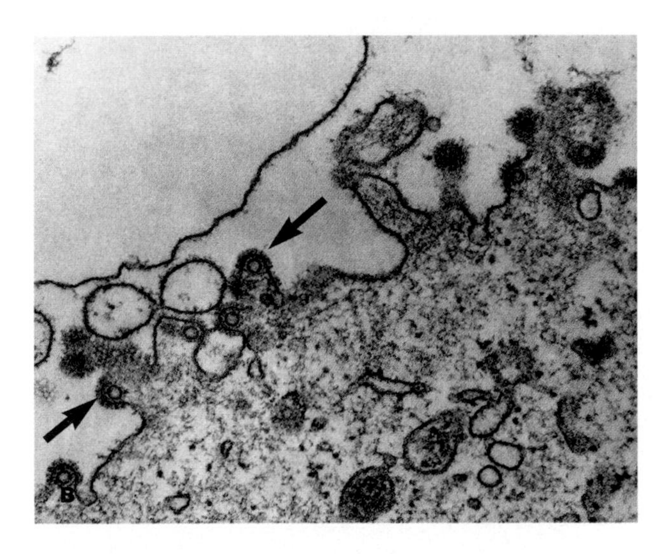


**Figure 278**
Electron microscopic views of viral replication. (a) Numerous cytoplasmic virus particles in different stages of development. Complete virions are shown at the bottom of the figure. (From A. Grefte et al. *J. Inf. Dis.* 168(1993):116–18.) (b) Viruses budding from cell surfaces (arrows).

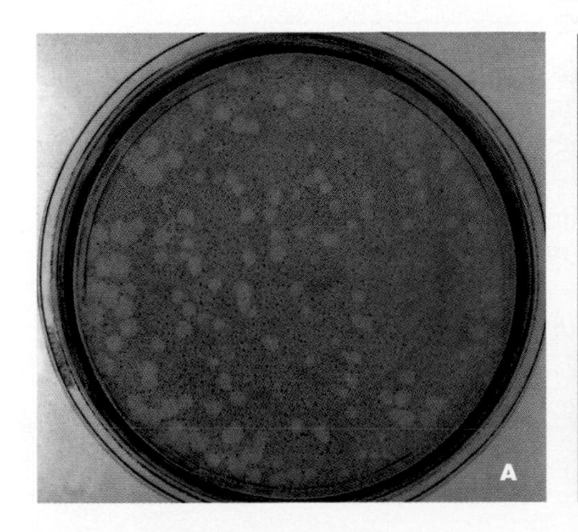


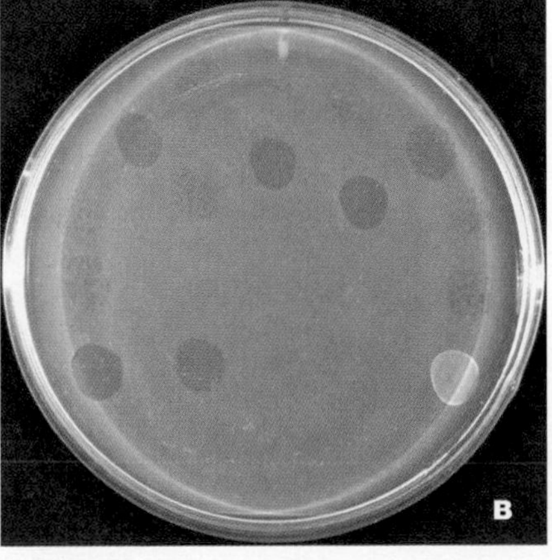


**Figure 279**
Bacteriophage application. (a) Result of virus number determination by the plaque assay procedure. (b) Plaque formation. The effects of bacteriophages on a lawn of susceptible bacteria (clear areas) lead to their identification.

by overlaying infected cells with a soft agar preparation to hold cells in place. Areas of virus destruction appear as clear, well-defined holes in the monolayer. The clear patches are called **plaques**. Similar techniques can be used to detect or determine the numbers of bacteriophages in specimens (Figure 279a).

Another procedure known as **bacteriophage (phage) typing** can be used to identify or to distinguish among bacterial strains. A set of specific bacteriophages is tested against cultures of different strains of a bacterial species. Because different strains of bacterial species are generally killed by different bacteriophages in the set used, it is often possible to distinguish one strain from another. The identification is based on the appearance of plaques caused by a specific phage (Figure 279b).

## Classification

The taxonomy of viruses undergoes changes as more is learned about their properties. Only the taxonomic principles and selected features of animal viruses are briefly considered here.

The most widely used taxonomic criteria are based on four physical viral characteristics: (1) the nature of the nucleic acid in the virion, (2) virus morphology, (3) the presence or absence of a viral envelope, and (4) virion size. Other characteristics, such as the effects on host cells and immunologic properties, provide a basis for placing viruses into taxonomic families, the names of which end in "viridae." There are 12 families of RNA-containing viruses and six families of DNA-containing viruses. Selected features of representatives from these two groups are briefly considered here.

## Diagnosis

Proving the specific viral cause of an infection is generally expensive, time-consuming, and difficult to do in most clinical settings. Therefore, laboratory virological tests are usually reserved and used for situations in which the results will significantly contribute to the management of infections.

### Laboratory Diagnosis

A wide variety of laboratory tests are available for viral disease diagnosis and can be grouped into four categories: (1) morphology, (2) serology, (3) virus isolation, and (4) nucleic acid technology.

Morphological tests involve direct examination of appropriate clinical specimens for the presence of viruses or their parts. They include electron microscopy and immunofluorescence techniques.

Serological tests include procedures used to show the presence of viral antigens, increases in antibody concentration during recovery, or the appearance of specific antibody (IgM) during the early (acute) phase of the disease.

Virus isolation techniques involve inoculation of appropriate clinical specimens into laboratory animals or cell cultures and the subsequent demonstration of specific virus replication (Figures 276 through 278).

Nucleic acid techniques are used to detect the presence of virus genetic material (genome) in cells or secretions. Nucleic acid probes are quite effective (Figure 280). A probe is defined as a fragment of DNA or RNA, typically labeled with radioactive material, that can be applied to show the presence of a complementary sequence of DNA or RNA in cells or secretions. This reaction is generally referred to as **nucleic acid hydridization.**

# Representative Viruses, Their Distinctive Properties, and Selected Diseases

## RNA Viruses

### *Rotaviruses*

Rotaviruses (Figure 281) are the only members of the Reoviridae that produce significant human disease in the United States. Rotaviruses cause diarrhea among infants and young children.

**Morphology and Genome Properties.** The virion consists of double capsid layers that are icosahedral. There is no envelope. The capsomeres of the virus particle give it a striking wheel-like appearance (Figure 281). The rotavirus genome is a double-stranded RNA molecule.

**Transmission.** Rotaviruses are transmitted by the fecal-oral route.

### *Rabiesvirus*

The rabiesvirus is the most important member of the Rhabdoviridae. The disease rabies is an acute encephalitis.

**Morphology and Genome Properties.** All rhabdoviruses are enveloped and have a helical capsid (Figure 282). These viruses exhibit a bullet-shaped morphology. The genome consists of a single-stranded RNA molecule.

**Transmission.** Rabiesviruses are transmitted by the bite of infected animals.

### *Influenzaviruses*

Three species of Orthomyxoviridae that cause the human respiratory disease influenza are influenza viruses A, B, and C.

**Morphology and Genome Properties.** Influenzaviruses have an enveloped virion with a helical nucleocapsid (Figure 283). The units of the envelope contain hemagglutinin and neuraminidase molecules. The viral genome consists of a segmented single stranded RNA molecule.

**Transmission.** Influenzaviruses are spread by aerosols and contaminated objects. Human-to-human transmission is common.

### *Hantavirus*

Hantaviruses take their name from the Hantaan River in Korea. These members of the Bungaviridae cause hemorrhagic fever often complicated by varying degrees of acute kidney failure. A newly isolated virus is known to cause a respiratory illness that can be difficult to distinguish from influenza.

**Morphology and Genome Properties.** Hantaviruses are enveloped RNA viruses (Figure 284). The genome consists of a segmented single-stranded RNA molecule.

**Transmission.** Hantaviruses are believed to be acquired by humans most often by inhalation of infected rodent excreta (urine, feces, and saliva), by contamination of the eye, or by direct contact with open cuts. Deer mice are believed to be the main reservoirs for these viruses.

### *Hepatitis A Virus*

Hepatitis A (Figure 285), a member of the Picornaviridae and also known as *Hepatovirus*, is the cause of viral hepatitis.

**Morphology and Genome Properties.** Picornaviruses are nonenveloped and have an icosahedral capsid. The genome consists of a nonsegmented single-stranded RNA molecule.

**Transmission.** The major means of transmission for hepatitis virus A is the fecal-oral route.

### *Retroviruses*

Retroviruses represent a unique family of viruses. Members of the Retroviridae use DNA in their replication. Genetic information flows from the viral RNA genome to DNA and then back to viral RNA. This reverse flow of genetic information depends on the specific viral enzyme reverse transcriptase and is the basis for the family designation. The viruses in this group include several cancer-causing agents and the human immunodeficiency viruses (HIVs). HIV is the major focus here.

Infection with HIV type 1 and 2 disable the human immune system and thus predispose individuals to a wide range of opportunistic infections (Figure 286a) and certain types of cancers. Acquired immunodeficiency syndrome (AIDS) results from HIV infection and the spectrum of opportunistic infections.

**Morphology and Genome Properties.** HIV, a lentivirus, is an enveloped virus with a nucleocapsid of varying morphology (Figure 286b). This virus has a centrally located "nucleoid" (nuclear acid core). The genome consists of two identical single-stranded RNA molecules.

**Transmission.** HIV infection generally can be acquired sexually, by transmission from an infected mother to her fetus, through breast feeding, and by exposure to contaminated blood or body fluids.

## DNA Viruses

### *Papillomaviruses*

Papillomaviruses, members of the Papovaviradie, are mainly of interest because of their cancer-producing capabilities. Certain human papillomavirus types have been associated with cervical and anogenital cancers. The main targets in papillomavirus infections are the skin and mucosal epithelial linings. Genital warts are caused by members of this viral group. Infections do not result in cell death but in cell reproduction leading to tumor (**wart**) formation (Figure 287a).

**Morphology and Genome Properties.** Papillomaviruses are nonenveloped virions with icosahedral nucleocapsids (Figure 287b). The genome consists of a single double-stranded DNA molecule.

**Transmission.** Genital warts are transmitted through sexual activities and direct contact with infected individuals. The source of the virus is the wart.

## RNA Viruses

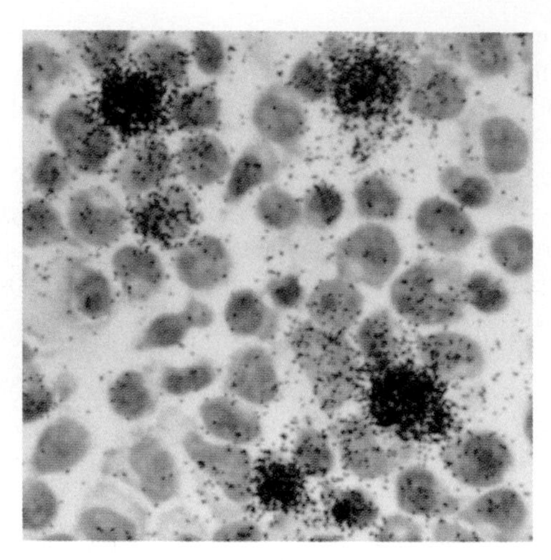

**Figure 280**
The results of applying a nucleic acid probe. The clusters of black dots indicate the location of virus-infected cells. A DNA probe was used to identify viral messenger RNA within the cells.

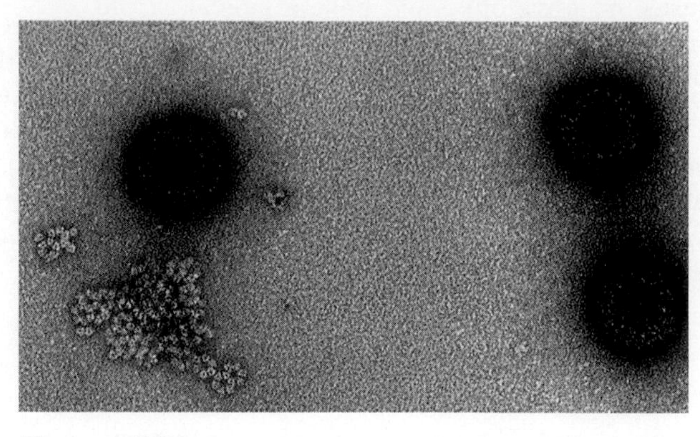

**Figure 281**
An electron micrograph of negatively stained rotavirus capsids (160,000×). (From Y. Hosaka et al. *J. Electron Microsc.* 40(1991):407–10.)

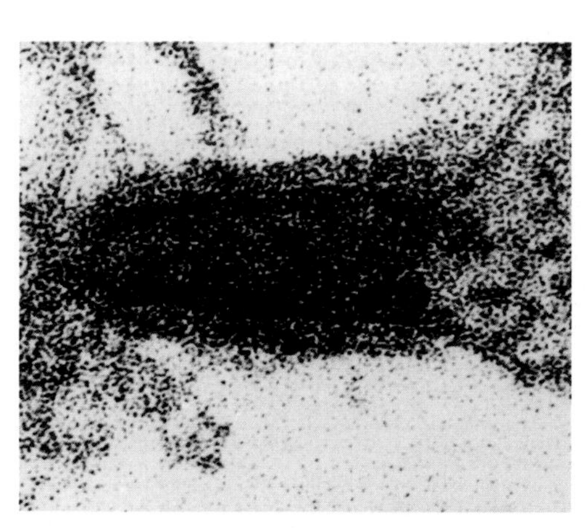

**Figure 282**
A tissue section showing a rabies virus (*Lyssavirus*) budding from a plasma membrane of an infected cell (150,000×).

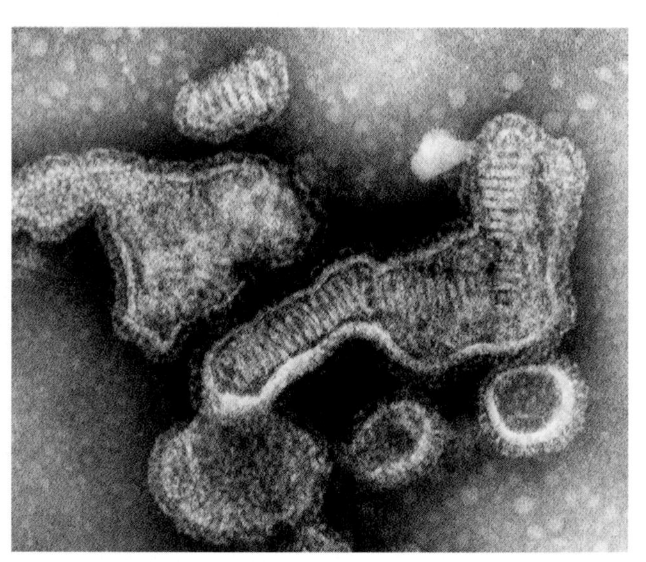

**Figure 283**
A negative stain of influenza virus (260,000×).

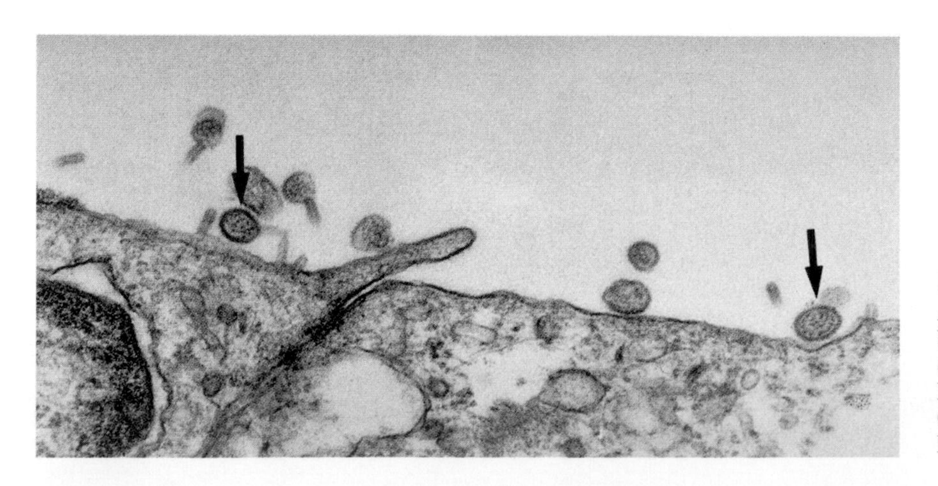

**Figure 284**
A section of cultured cells showing budding *Hantavirus* particles. (Courtesy of Drs. J.M. Hughes, C.J. Peters, B.W. Cohen, and J. Mahy, Centers for Disease Control and Prevention.)

## RNA Viruses (Continued)

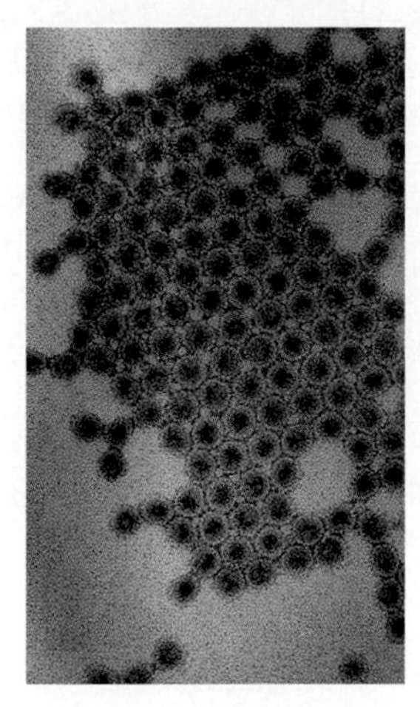

**Figure 285**
A transmission electron micrograph of Hepatitis A virus, *Hepatovirus*. (100 nm)

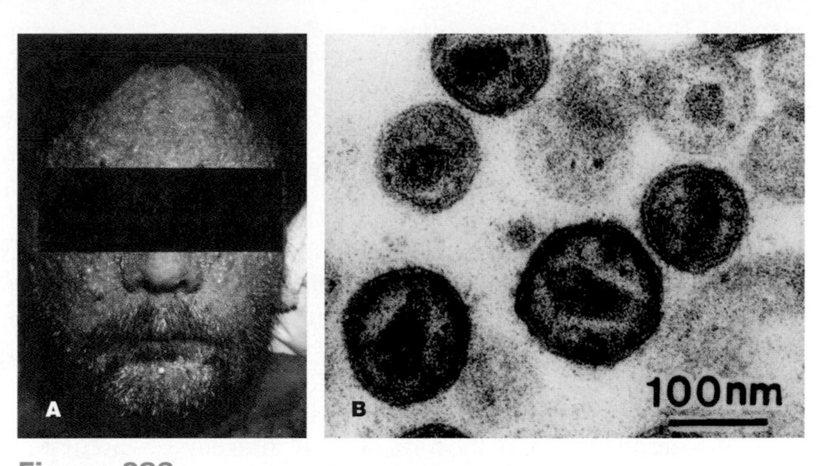


**Figure 286**
Retroviruses. (a) A case of cystic acne with increased severity in an HIV-infected patient. (From A.G. Martin et al. *Brit. J. Dermatol.* 126(1992):617–20.) (b) A tissue section of human immunodeficiency virus particles.

## DNA Viruses

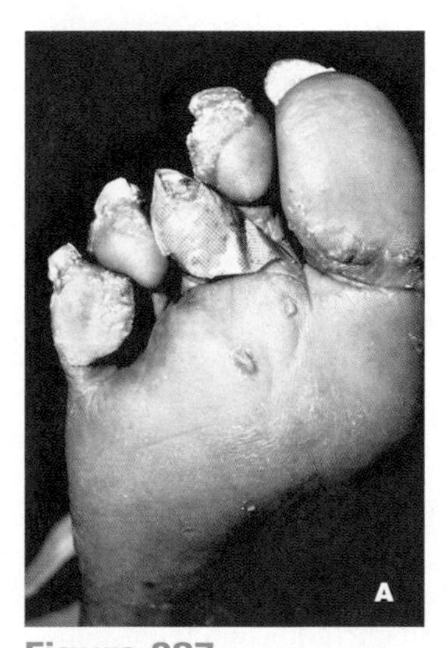


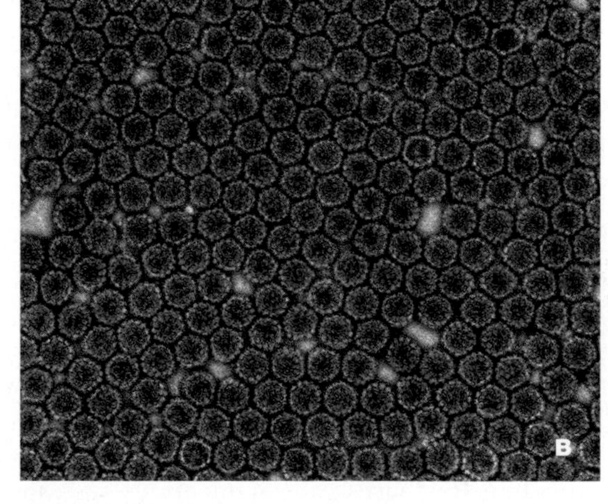


**Figure 287**
Papillomaviruses. (a) A case of plantar warts. (b) The nonenveloped virions of human papillomavirus (100,000×). (From T. Iwasaki et al. *J. Path.* 168(1992):293–300.)

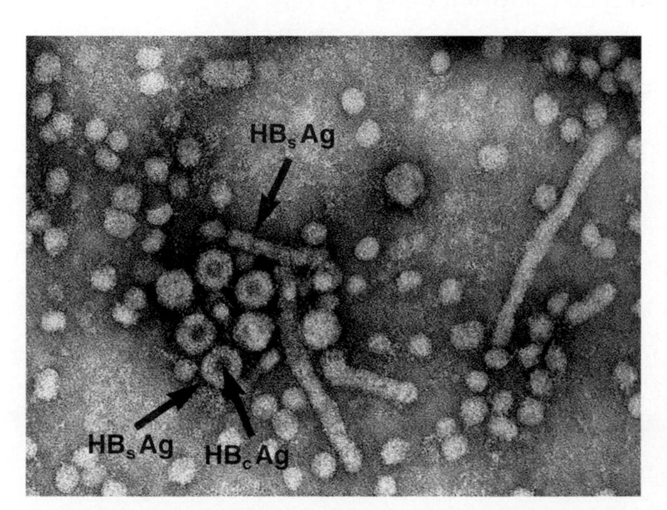


**Figure 288**
Hepatitis B virus, *Hepadnavirus*. Virions showing circular and filamentous forms.

### *Hepatitis B Virus*

Hepatitis B virus (HBV) infections are common and worldwide in distribution. This virus, a member of the Hepadnaviridae, is one of several hepatitis viruses pathogenic for humans. The virus causes severe liver damage. Some infections may become chronic. The virus also is strongly associated with liver cancer.

**Morphology and Genome Properties.** Hepatitis B virus exhibits several morphologies, including spherical and filamentous forms (Figure 288). It is an enveloped virion with an icosahedral capsid. The genome consists of a partially double-stranded DNA molecule.

A virion can contain outer-shell surface antigens ($HB_sAg$), capsid proteins in the form of a core antigen ($HB_cAg$), and a derivative of the core antigen ($HB_gAg$). Figure 288 shows the relationship of certain parts.

**Transmission.** Hepatitis B virus is transmitted through contact with contaminated blood. Transfusions with contaminated blood or blood products, sexual activity with infected persons, and transplacental infection (from infected mother to fetus) are the major means of transmission.

### *Herpes Viruses*

Eight members of the family Herpesviridae are known to infect humans: herpes simplex virus types 1 and 2 (HSV-1 and HSV-2), varicella-zoster virus, cytomegalovirus, Epstein-Barr virus, and human herpesviruses 6, 7 (Figures 289 through 292) and 8 (Figure 295). All of these viruses persist for life after the primary infection and are reactivated at intervals during the infected individual's lifetime.

Herpesvirus-6 has been associated with disorders involving white blood cell increases, febrile illness, hepatitis, and non-Epstein-Barr virus-caused infectious mononucleosis. This virus also has been shown to be the cause of the skin infection exanthum subitum or roseloa infantum.

Herpesvirus-7 has been isolated from patients with chronic fatigue syndrome, roseola infantum, enlarged livers and spleens, and decreased blood platelets

Herpesvirus-8 mainly has been isolated from patients with Kaposi's sarcoma and certain other forms of cancer.

**Morphology and Genome Properties.** Herpesviruses share a similar morphology. The virus particles consist of a central core containing linear double stranded DNA; an icosahedral capsid of 162 capsomeres surrounding the core with an envelope that, on the outside, displays glycoprotein spikes; and material called the **tegument** filling the space between the capsid and the envelope (Figure 289d,f). The envelope is derived from the host cell's nuclear membrane.

### *Herpes Simplex Viruses*

Herpes simplex viruses (HSV) types 1 and 2 are the most frequent causes of disease. Infection involves the body surface, where, in addition to replicating and causing cellular destruction (Figure 289a,b,f), the virus enters sensory nerve endings. Common sites for herpetic lesions are the skin and mucous membranes around the mouth (Figure 289a) and genital openings, the oral cavity itself, the cornea, the cervix, the anus, and the urethra.

Important secondary targets include the brain, the meninges, and the cornea. Herpetic whitlow (see Figure 272b) is a special type of herpetic skin infection. It involves the finger, may be quite painful, and is commonly found among health care workers. Recovery from the first infection results in latency.

**Transmission.** Herpes simplex virus infection is acquired through direct contact with active herpetic lesions. Neonatal infections also can occur through contact with HSV in the birth canal of infected mothers.

### *Varicella-zoster Virus (Varicellovirus)*

Chickenpox and herpes zoster (shingles) are different clinical effects of infection by the same virus, varicella-zoster virus (VZV; Figure 290f). Chickenpox is generally regarded as a childhood disease. Infections, however, can be severe and even fatal in high-risk groups such as immunocompromised persons and newborns. Recovery from primary infections leads to latency. Herpes zoster results from reactivation of the virus in sensory nerve ganglia (Figures 290a,b). Its incidence and severity increase with age, and it seldom occurs in the young. Once activated, the virus moves down the sensory nerves until it reaches the skin and produces lesions similar to those of chickenpox. The lesions are confined to the areas affected in chickenpox, including thoracic, lumbar, and facial regions (Figure 290a–d). Infected cells may show the presence of prominent red (eosinophilic) intranuclear inclusions called Cowdry A bodies (Figure 290e).

**Transmission.** Infection with VZV is acquired by aerosol.

### *Cytomegalovirus*

Cytomegalovirus (CMV) is considered to be the most important cause of congenital infections in the United States and a major opportunistic disease agent in organ transplant recipients and in immunosuppressed individuals (Figure 291). The most frequent manifestation of CMV in patients with AIDS is a sight-impairing condition, known as chorioretinitis, that can result in blindness. In addition, CMV is an occasional cause of infectious monocucleosis in apparently normal individuals.

Cytomegalovirus causes a systemic primary infection and becomes widely distributed in the body. Most

## DNA Viruses *(Continued)*

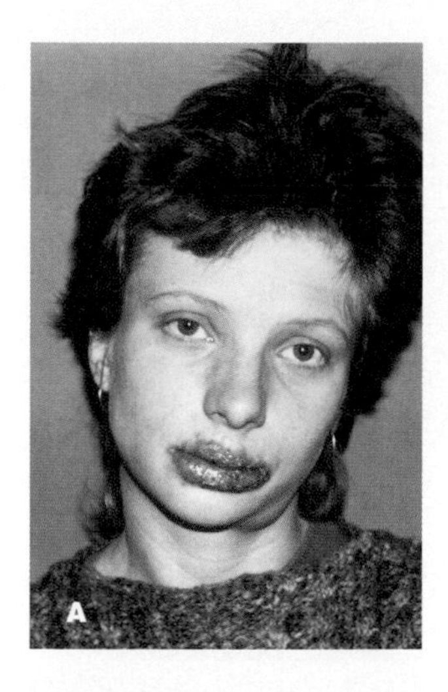

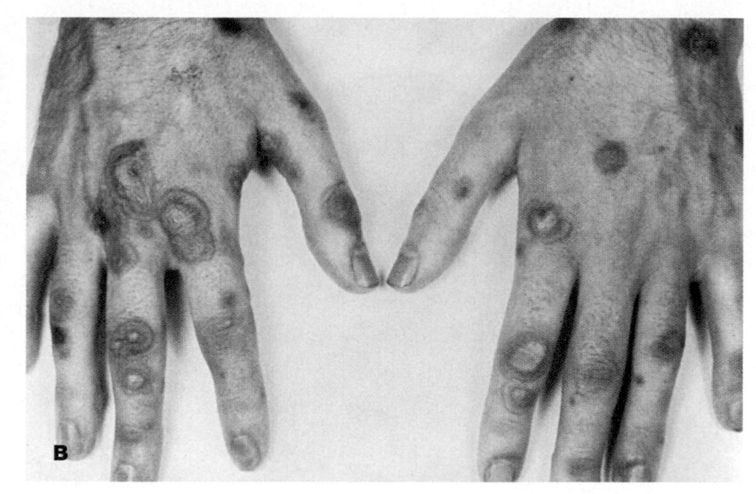

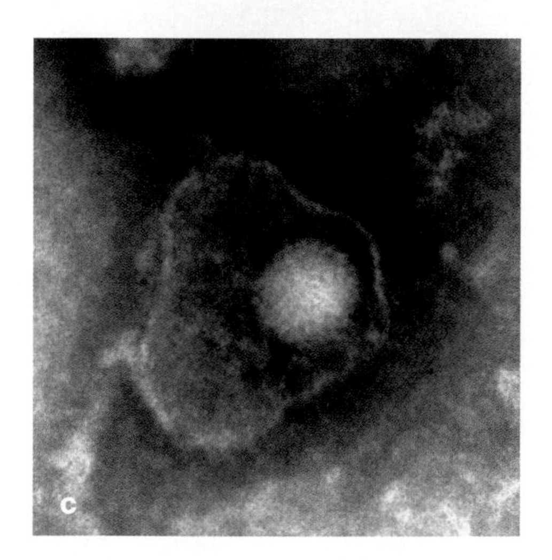

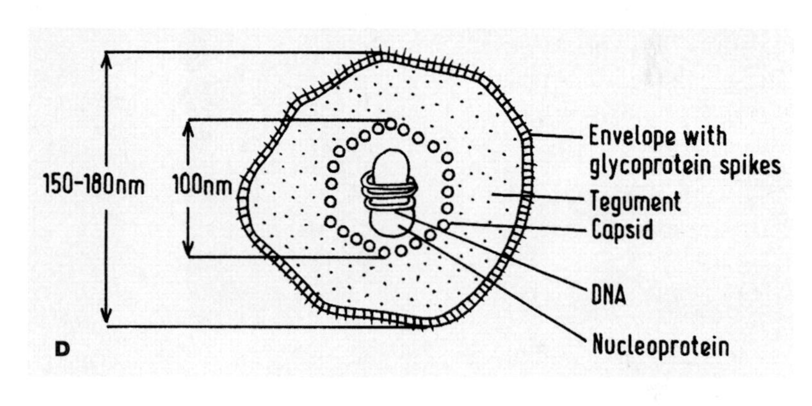


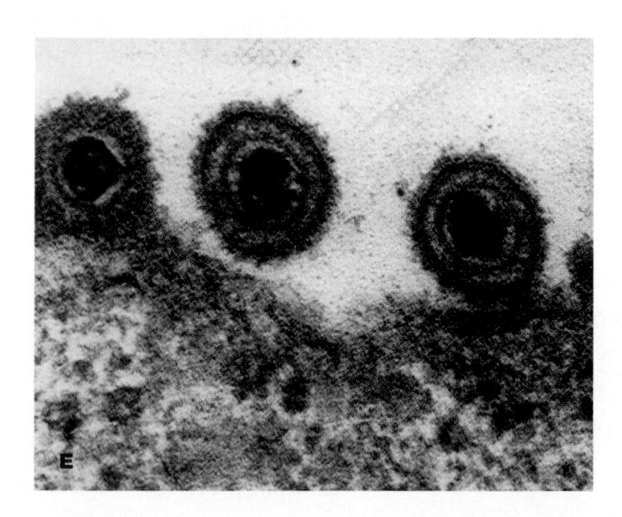

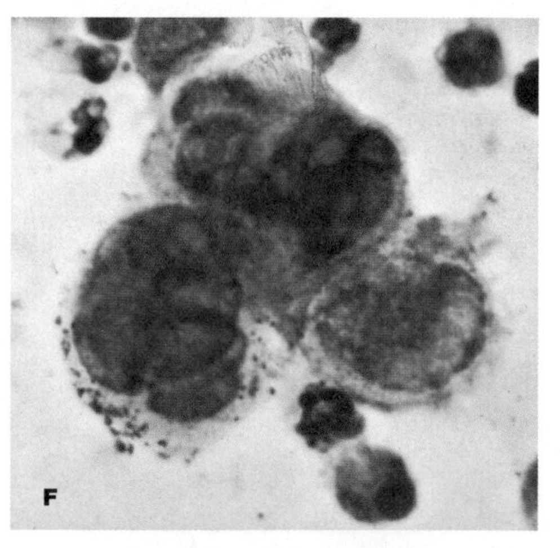

**Figure 289**

Herpes simplex virus (HSV) type 1 (*Simplexvirus*). (a) A severe crop of blisters (vesicles) in a primary herpes infection. (b) Typical target lesions associated with a form of recurrent HSV infection. (c) A negatively stained HSV particle shown by transmission electron microscopy (120,000×). (d) A diagrammatic view of herpes simplex type 1 virus. (From J.P. Vestey and M. Norval. *Clin. & Exp. Dermatol.* 17(1992):221–37.) (e) An electron micrograph showing virus particles. (f) The results of the Tzanck test with a tissue smear showing multinucleated giant cells typical of herpes infections.

## DNA Viruses *(Continued)*

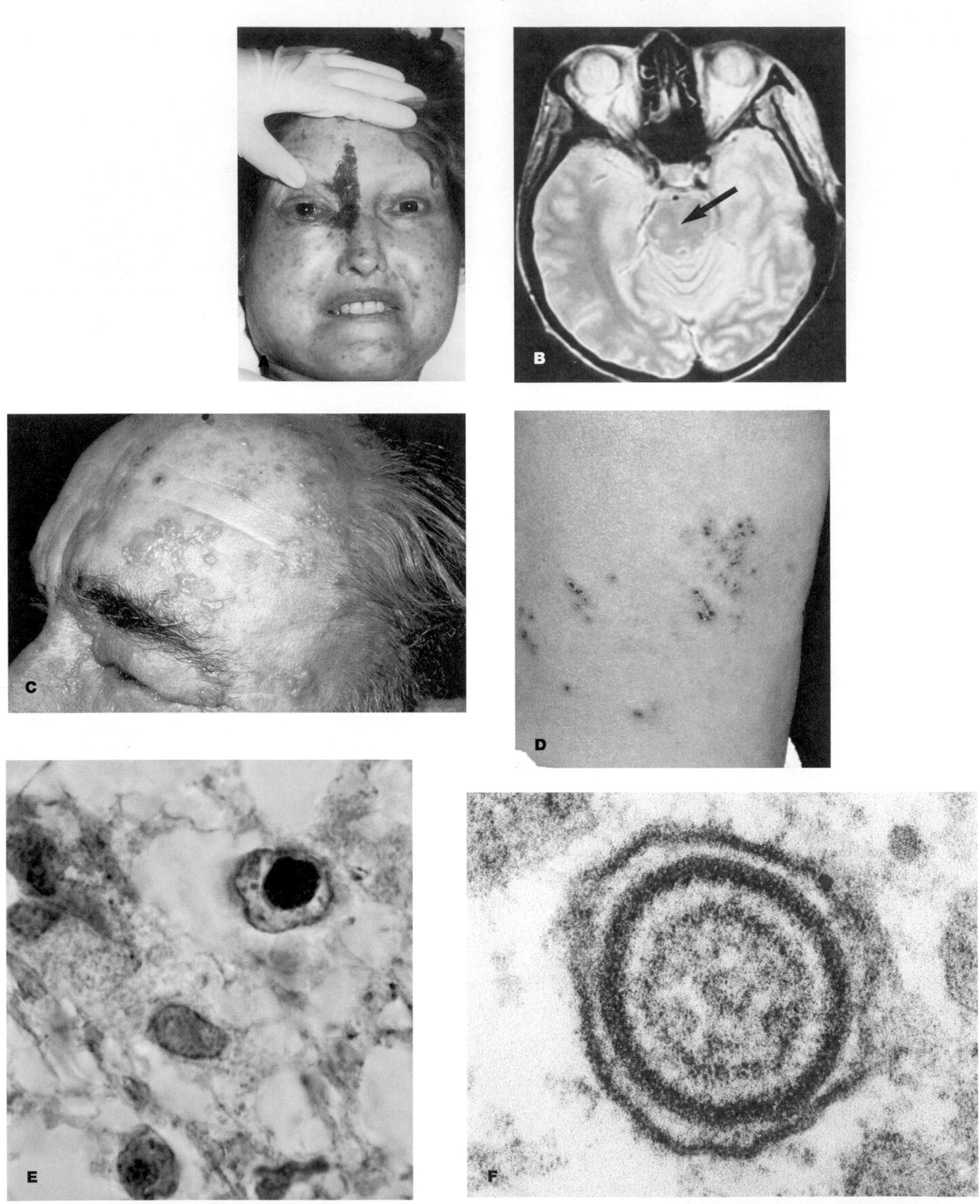


**Figure 290**
Varicella-zoster virus (varicellovirus). (a) A case of herpes zoster ophthalmicus and optic nerve dysfunction. (b) A magnetic resonance image showing central nervous system involvement (arrow). (From F.J. Lexa et al. *AJNR* 14(1993):185–90.) (c,d) Appearance of shingles on different parts of the body. (e) Cowdry type A intranuclear eosinophilic inclusion within a neuron (400×). (From A. Moulignier et al. *CID* 20(1995):1378–80.) (f) Electron micrograph of a *Varicellovirus* virion, showing the almost centrally located nucleic acid core and surrounding envelope.

of such infections are without signs and symptoms. Severe fetal infections known as **Cytomegalo-inclusion disease** occur mainly in pregnant women experiencing a primary infection.

The development of immunity results in latency, but the virus can be reactivated by the use of immunosuppressive agents during pregnancy.

**Transmission.** Cytomegalovirus is spread by direct contact with infectious body fluids such as saliva, urine, semen, and cervical secretions.

### *Epstein-Barr Virus*

Epstein-Barr (EBV) is the major cause of infectious mononucleosis. It is also associated with other diseases, including hairy leukoplakia (Figure 292a), nasopharyngeal cancer, African Burkitt's lymphoma, Hodgkin's disease, and salivary gland cancer. The primary targets of the virus are the salivary glands, where susceptible B lymphocytes become infected. These infected cells spread the virus (Figure 292b) by way of the lymphatics and blood to distant lymphoid and related organs.

Infectious mononucleosis is generally a self-limiting disease.

**Transmission.** Epstein-Barr virus infection is acquired through direct contact with body fluids containing the virus and fomites.

### *Poxviruses*

Several members of the Poxviridae are of medical importance. These include smallpox, vaccinia, molluscum contagiosum (Figure 294), and catpox viruses (Figure 293b).

**Morphology and Genome Properties.** The poxviruses are large, brick-shaped or ovoid envelope viruses (Figure 293a). The genome consists of double-stranded DNA.

**Transmission.** Smallpox has been eradicated through a worldwide immunization campaign by the World Health Organization.

### *Vaccinia Virus*

Vaccinia virus is serologically related to the smallpox virus, although its exact origin is unknown. Vaccinia virus was used for smallpox immunization. It is currently used for immunization against monkeypox. The virus can cause potentially serious and lethal complications (Figure 293a).

### *Molluscum Contagiosum Virus*

Molluscum contagiosum virus (MCV) causes a benign form of skin tumors (Figure 294a,b). Although the disease is found worldwide, it is most frequently found as an easily treated disease of childhood.

**Transmission.** Transmission of MCV can occur by direct contact with contaminated articles, such as towels, and by sexual contact.

### *Kaposi's Sarcoma*

Four distinctive forms of Kaposi's sarcoma (KS) are recognized. One of these, known as epidemic KS, is the most important opportunistic tumorous growth (Figure 295) occurring in persons infected with HIV-type 1. A newly isolated herpesvirus (Herpesvirus-8) is believed to be the causative agent.

## DNA viruses *(Continued)*

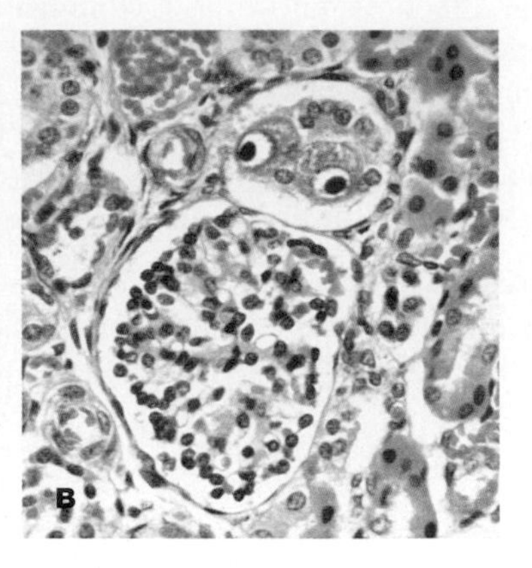


**Figure 291**
Cytomegalovirus (CMV) infection. (a) Infection of the esophagus, showing numerous ulcers and tissue damage. (From F.S. Buckner and C. Pomeroy, *CID* 17(1993):644–56.) (b) A kidney tissue section showing large intranuclear CMV inclusion bodies. The inclusions are referred to as owl's-eye cells. (From R. Herriot and E.S. Gray. *NEJM* 331(1994):649.)

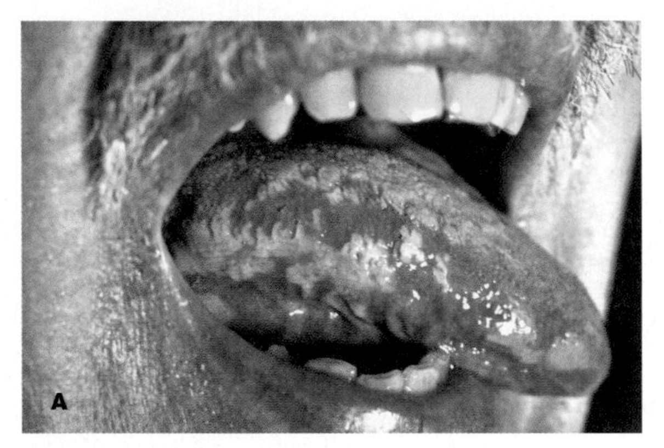


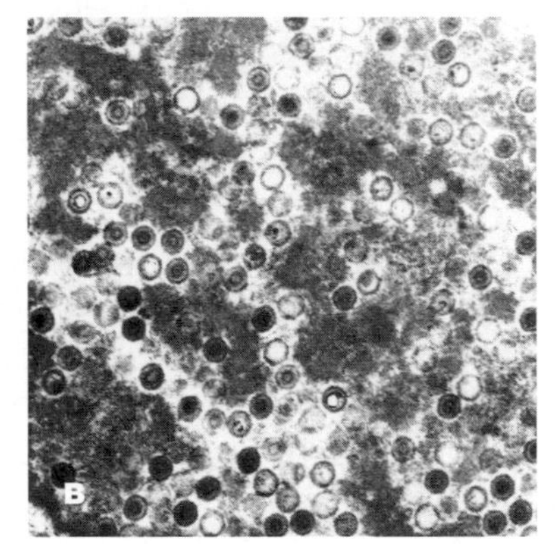


**Figure 292**
Epstein-Barr virus (*lymphocryptovirus*). (a) A case of oral hairy leukoplakia (OHL). (Courtesy of Dr. P. Itin, Kantonsspital, Basel.) (b) Electron micrograph showing large numbers of Epstein-Barr virions (20,000×). (From J. Kantakis et al. *Brit. J. Dermatol.* 124(1991):483–86.)

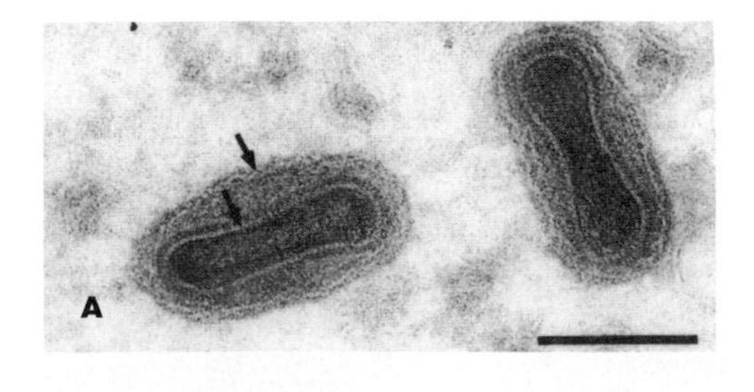


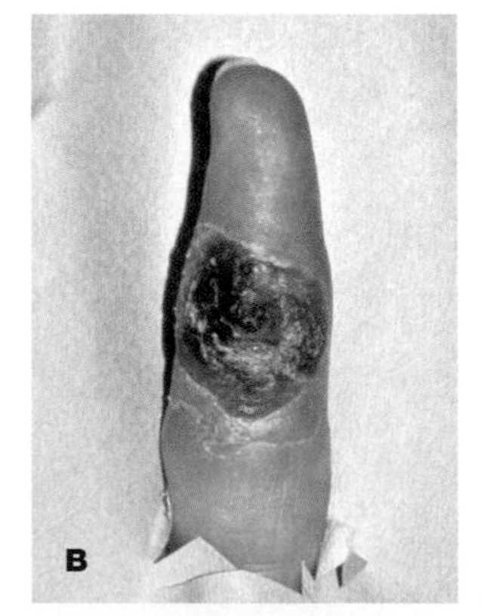


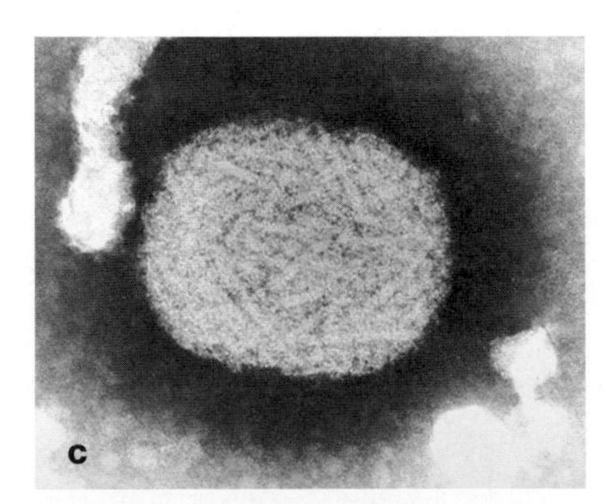


**Figure 293**
Poxviruses. (a) A transmission electron micrograph of intracellular vaccinia viruses (Orthovirus). (From B. Sodeik et al. *J. Cell Bio.* 121(1993):521–41.) (b) A case of the catpox/cowpox virion infection of the finger caused by the bite of an infected cat. (From J.B. Vestey et al. *Brit J. Dermatol.* 124(1991):74–78.) (c) A negative stain of the catpox/cowpox virion. (Courtesy of Dr. J.P. Vestey, University of Edinburgh.)

## DNA viruses *(Continued)*

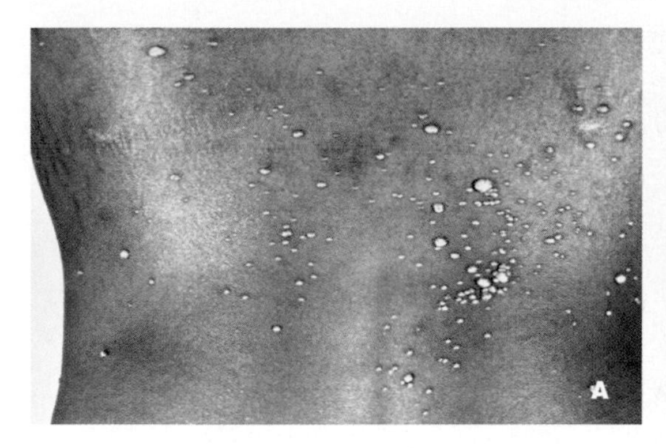


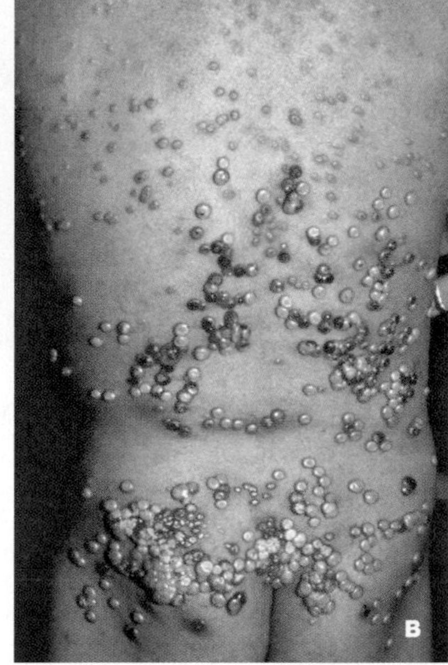


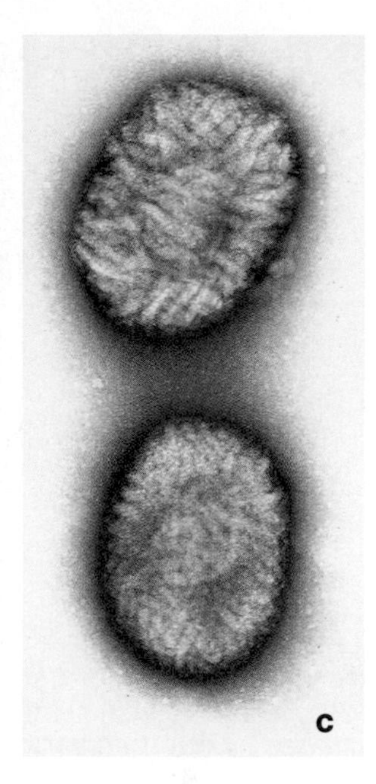


**Figure 294**
Molluscum contagiosum virus (*Molluscipoxvirus*). (a,b) Two examples of molluscum contagiosum infection. (c) The virion, shown by negative staining.

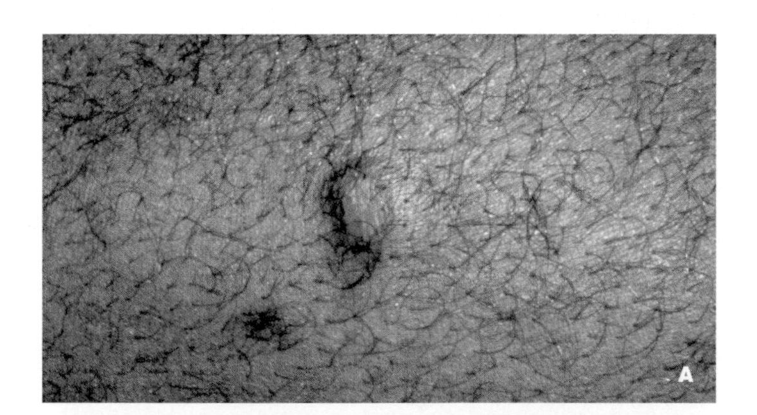


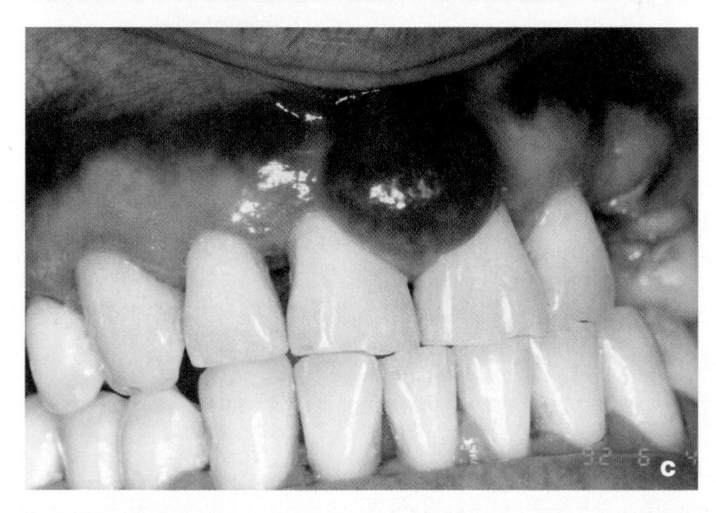


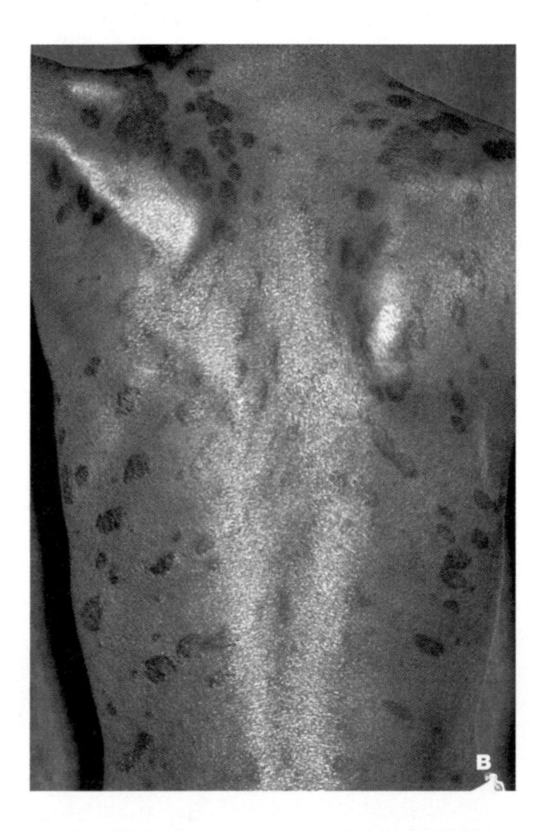


**Figure 295**
Kaposi's sarcoma (KS) a newly associated herpesvirus-8 infection. (a) A close-up of a lesion. (b) Kaposi's sarcoma in an HIV-infected person. (c) Intraoral KS. (From B.C. Muzyka and M. Glick. *NEJM* 328(1993):210.)

# SECTION 8 Helminthology

*Lines on the antiquity of parasites: "Adam…had 'em."*

*—Gilian Strickland*

Animals and plants have competed for millions of years for food and space. Over this time parasites have invaded almost every type of living host, exhibiting various degrees of dependence. The major groups of animal parasites are found among the helminths (Figure 296) and the arthropods.

Parasitic worms live in varying environments and must adapt to conditions in order to survive a host body's cellular and chemical defenses (Figure 297). A parasitic helminth's (worm) existence and survival are in large part determined by the development of certain structural and metabolic modifications. These include an especially hard outer body covering; hooks (Figure 296b), spines, cutting plates, suckers, and other structures for purposes of penetration or attachment to body tissues; various enzymes, and elaborate reproductive systems and strategies.

## Classification

The common human helminthic parasites can be placed into one of three classes on the basis of body and digestive system properties, general body organization, nature of the reproductive system, and the need for more than one host species for the completion of the worm's life cycle. These classes are the **cestodes** (tapeworms); **nematodes** (roundworms), and **trematodes** (flukes). Table 19 lists the main features of these parasites.

The general laboratory diagnosis of helminthic diseases is based on the finding and identification of ova, or eggs (Figure 298a), larvae (young or developing forms), or adult worms (Figure 298b). Skin tests and various serological tests also are of value.

**Table 19** Comparison of Cestode, Nematode, and Trematode General Features

| Feature | Cestodes (tapeworms) | Nematodes (roundworms) | Trematodes (flukes) |
|---|---|---|---|
| Shape | Tapelike, in segments | Cylindrical, unsegmented | Flat, leaflike; unsegmented |
| Suckers | Present | Absent | Present |
| Hooklets | Often present | Absent | Absent |
| Digestive canal | Absent | Present | Present, although not complete |
| Sex organs | Present in same worm | Separate | Generally present in same worm |

## Helminthology

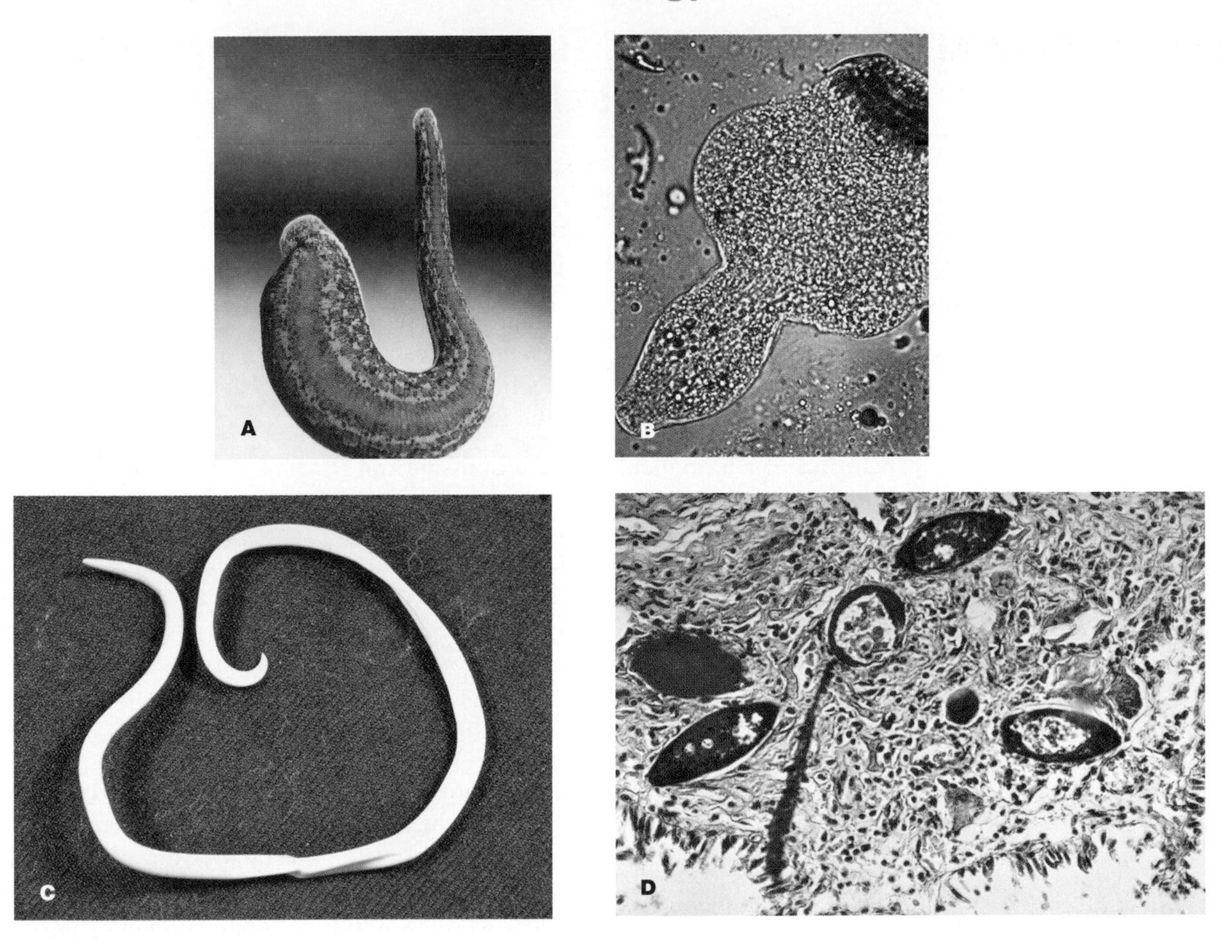


**Figure 296**
Examples of helminths. (a) Leeches such as this one, *Hirudo medicinalis,* because of their ability to repeatedly suck blood from a variety of hosts, may serve as a source of microbial pathogens (2.5×). (Courtesy of Dr. H. Mehlhorn.) (b) A microscopic view of the sheep tapeworm. Note the hooklets (dark top) on the scolex, or head. (c) An adult *Ascaris,* a roundworm, that is visible to the eye. (d) Flukes (trematodes) in a tissue specimen.

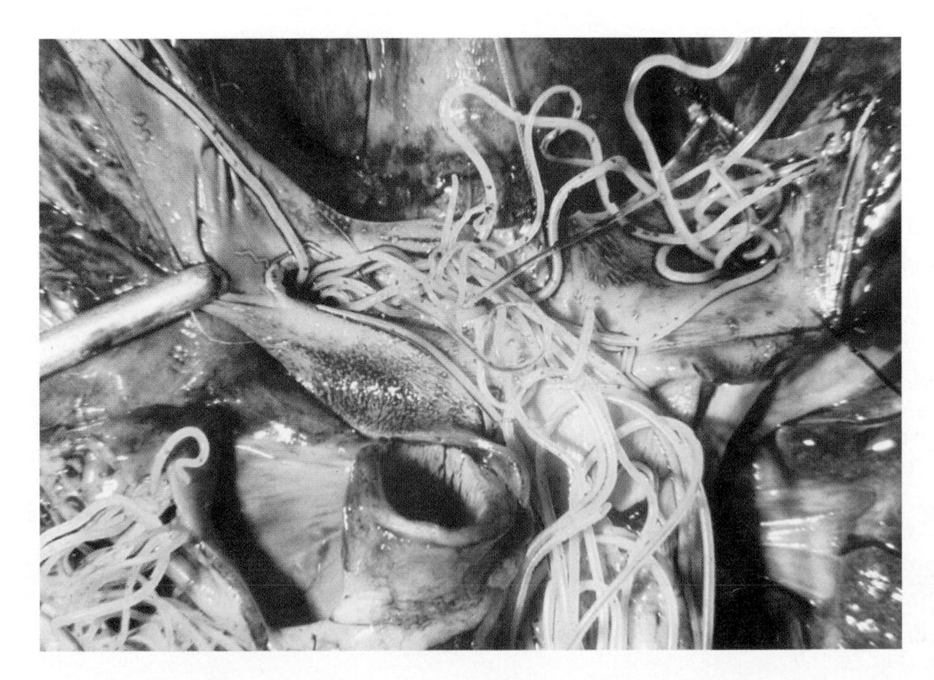

**Figure 297**
An opened heart, showing typical worms in the right portion of the organ and in the pulmonary artery (upper center and lower right side).

# Representative Helminths, Their Distinctive Properties, and Selected Diseases

## Cestodes

Cestodes, or tapeworms, have flat, ribbon-shaped bodies. They have a **scolex,** or head, at the anterior, or front, end. For purposes of attachment, scolexes have **suckers** (Figure 299) and at times, depending on the worm, **hooklets** (Figure 296b). Immediately behind the scolex is a neck region that produces reproductive body segments known as **proglottids**. Each segment contains both male and female sex organs, making the worm **hermaphroditic**. The proglottids located farther from the head are sexually mature and, by cross-fertilization of proglottids of either the same worm or another one, produce ova. Several examples of cestodes and their distinctive microscopic diagnostic properties will be described.

### *Diphyllobothrium latum*

Diphyllobothriasis is caused by the fish tapeworm, *Diphyllobothrium latum.* Most infected persons are asymptomatic. Some, however, experience a variety of gastrointestinal effects. Infection with *D. latum* also can interfere with vitamin $B_{12}$ absorption.

**Transmission and Diagnosis.** Infection is acquired through the ingestion of raw, uncooked, or pickled freshwater fish containing young forms (larvae) of the worm. Diagnosis includes finding ova (Figure 300) or typical proglottids in stool specimens.

### *Diphylidium caninum*

*Diphylidium caninum* (Figure 301) is the cause of dog tapeworm infection. Cats and humans also can be infected. Children are usually the ones infected, especially if they have been around infected dogs or cats. Diarrhea and general discomfort are typical signs and symptoms.

**Transmission and Diagnosis.** Human infection is acquired by the ingestion of infected adult fleas. Such fleas contain larval forms of the worm. Once inside the gastrointestinal tract of the host, the larval form develops into a mature tapeworm (Figure 301).

Diagnosis of *D. caninum* infection is based on finding adult worm proglottids in stools or released around the anus. The scolex of this worm contains a fleshy protrusion or swelling with one or more rows of hooks (Figure 301).

### *Echinococcus granulosis*

Echinococcosis, or hydatid disease, is caused by the sheep tapeworm, *Echinococcus granulosis.* A hydatid is the larval stage of the adult worm. The adult is very small and has only three proglottids (Figure 299). Dogs, sheep, cattle, and humans can be infected. Infection can result in blockage and interference with the functioning of organs such as the liver.

**Transmission and Diagnosis.** Humans acquire hydatid disease as a consequence of ingesting ova, usually found in infected dog feces. Sheep ingest eggs on contaminated grass. The eggs hatch in the intestine and develop into embryos, which eventually make their way to the liver. Some embryos may be carried by the bloodstream to other body organs. Embryos develop into hydatid cysts (Figure 302c). Special structures called **brood capsules** bud from the cysts and give rise to several potential adult worms known **protoscolexes** (Figure 302a). Ingestion of such brood capsules can result in the development of adult worms.

Diagnosis is best made by showing cysts through the techniques of ultrasound, magnetic resonance imaging (MRI), or computed tomography (CT scans). Surgically removing cysts and finding potential tapeworms and hooklets (Figure 302b) are diagnostic.

### *Taenia*

*Taenia solium* is the pork tapeworm (Figure 303a) and *T. saginata* is the beef tapeworm. Infected individuals generally complain of slight pain, but most are asymptomatic.

**Transmission and Diagnosis.** *Taenia solium* infection results from the ingestion of raw or inadequately cooked pork containing larval forms (cysticerci) of the worm. Similarly, *T. saginata* infection occurs as a consequence of eating raw or undercooked beef containing larvae of the worm.

Diagnosis is based on finding the characteristic proglottid associated with each *Taenia* species. The ova of the two species are identical (Figure 303b).

## Nematodes

Nematodes, or roundworms, have a cylindrical tapered body (Figure 305a). Their tubular digestive tract extends from the mouth at the anterior end to the anus at the posterior end. The sexes are separate, with males

## Helminthology *(Continued)*

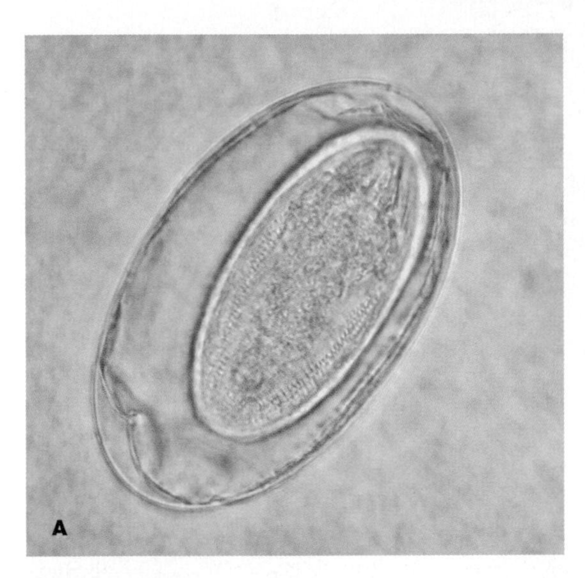


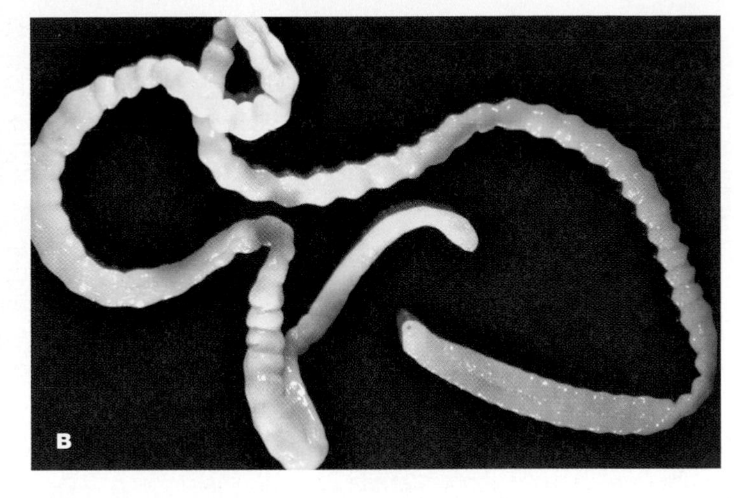


**Figure 298**
Diagnostic properties of helminths. (a) An unstained *Moniliformis moniliformis* ovum found in a stool specimen. (b) A complete adult female worm measuring 133 mm in length. (From R.C. Naafie and A.M. Marty *Clin. Microbiol. Revs.* 6(1993):34–56.)

## Cestodes

**Figure 299**
Adult sheep tapeworms, *Echinococcus granulosus.*

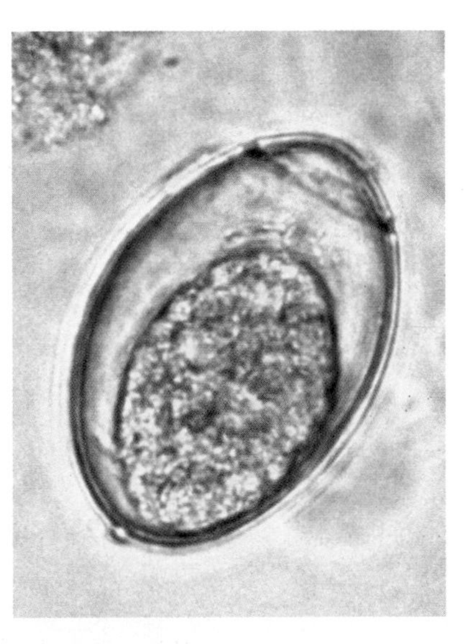

**Figure 300**
*Diphyllobothrium latum* (fish tapeworm) ovum.

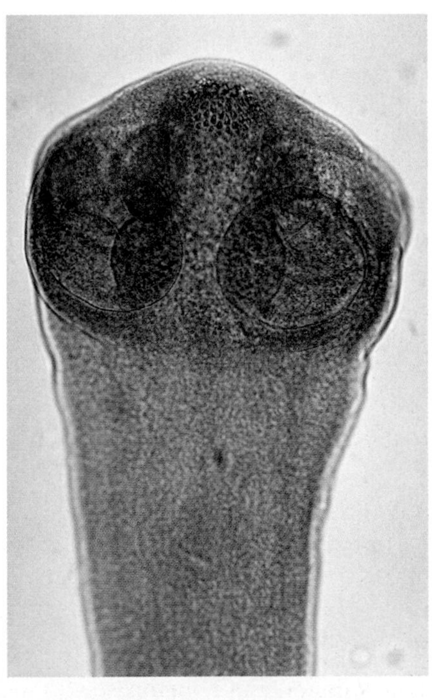

**Figure 301**
*Diphylidium caninum* adult worm. Note the round suckers and rostellum (a fleshy swelling or protrusion) with its rows of hooks near the top of scolex.

## Cestodes *(Continued)*

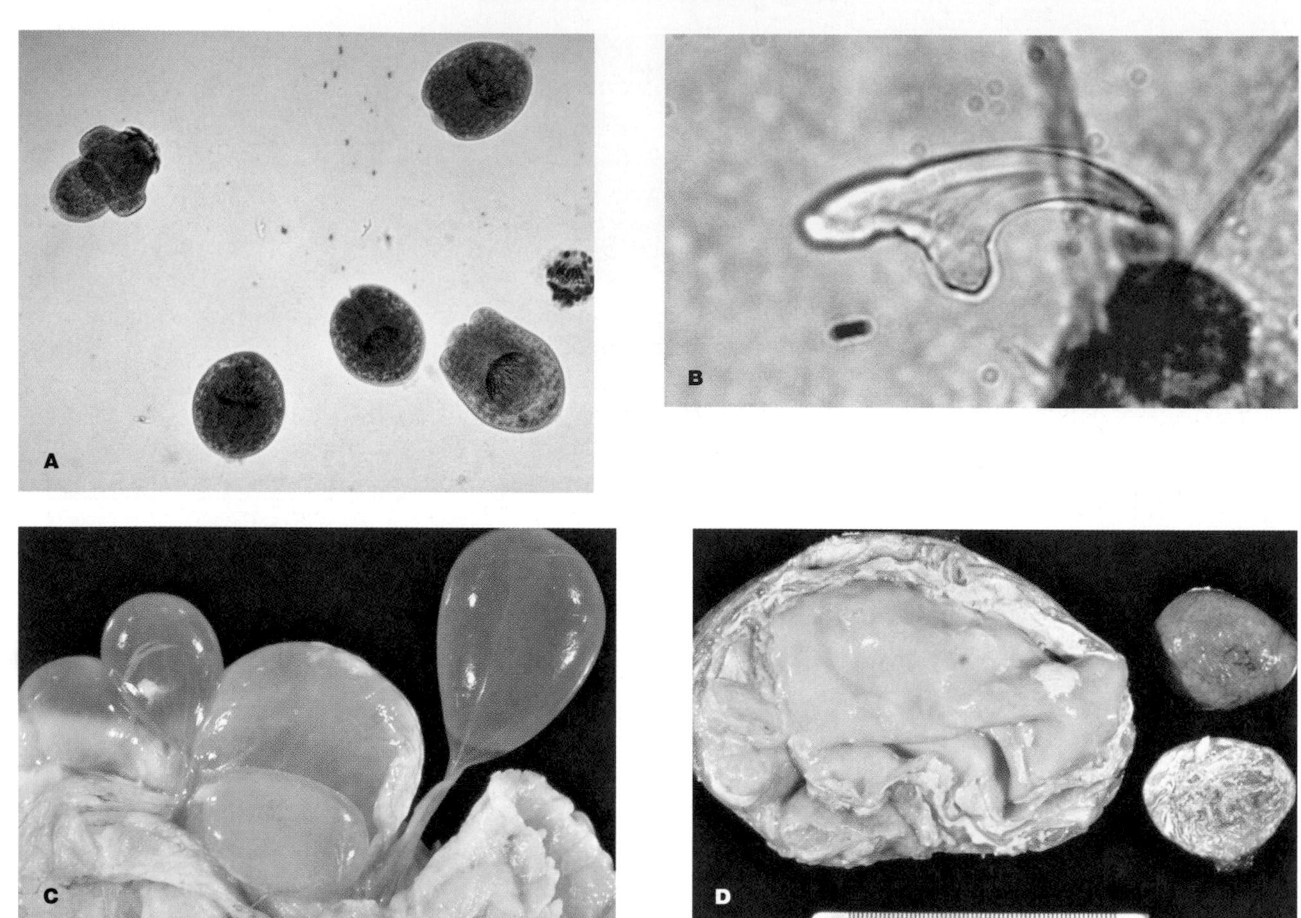


**Figure 302**
*Echinococcus granulosis* (sheep tapeworm). (a) Developing worms called protoscolexes. Note the hooklets on the scolexes. (b) A close-up of a hooklet. (c) Hydatid cysts *in situ*. (d) Two echinococcal cysts removed from the pelvis of a woman. (Courtesy of Dr. L. Alpert, Patology Department, The Sir Mortimer B. David Jewish General Hospital.)

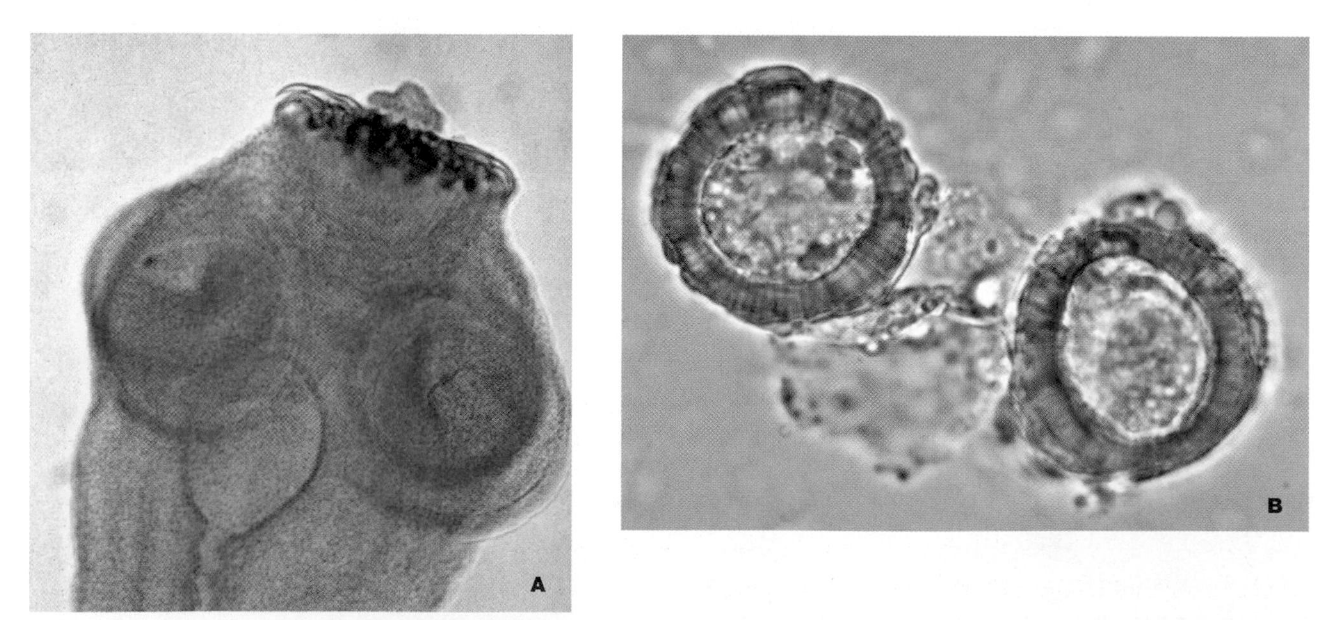


**Figure 303**
*Taenia solium* (pork tapeworm). (a) Adult worm. (b) Ova.

typically being smaller than females. Nematodes can be divided into those that typically inhabit the gastrointestinal tract of hosts and those that are found in the blood and other tissues of hosts.

### *Ancylostoma duodenale and Necator americanus*

*Ancylostoma duodenale* and *Necator americanus* are the causes of hookworm infection (Figure 304a–d). The signs and symptoms are generally proportional to the number of worms (**worm burden**) in the host. Abdominal pain, diarrhea, increased numbers of eosinophils (**eosinophilia**), and anemia can occur. Skin lesions also may be found in areas where hookworm larvae (Figure 304e) enter the host. The hookworm larvae of dogs and cats can also penetrate human skin and cause **cutaneous larva migrans** (creeping eruption). These larvae wander about in the skin, producing linear skin lesions (Figure 304e).

**Transmission and Diagnosis.** Infection is acquired by direct contact with soil that contains infective larvae. Larvae hatch from ova in the soil and often pass through two distinct stages. The first, known as *rhabditiform larvae* go on either to develop into free-living adults or to become infective *filariform larvae*. Worms having rhabditiform larvae include hookworms and *Strongyloides stercoralis* (Figure 309a). Filariform larvae develop no further unless they make contact with exposed human skin and gain entrance into the body by penetration. The most common area of such invasion is the feet. After penetration, larvae are transported via the circulatory system to the lungs. Subsequently, the larvae escape from the lungs, are swallowed, and enter the intestinal tract, where they develop into mature male and female worms. Mating results in the formation and depositing of ova (Figure 304a). Identification of the ova of the two hookworm species is based on a comparison of the mouthparts; *A. duodenale* has pairs of teeth, and *N. americanus* has cutting plates.

Diagnosis depends on finding ova or larvae or both in stool specimens.

### *Ascaris lumbricoides*

*Ascaris lumbricoides* is the cause of ascariasis (Figure 305a). Infected individuals may experience abdominal pain, diarrhea, and other gastrointestinal problems. Heavy intestinal infection may produce obstruction and liver, gallbladder, and pancreas involvement.

**Transmission and Diagnosis.** Infections are acquired by the ingestion of ova (Figure 305b) in contaminated food or water. Fomites and fecal-contaminated fingers also are possible sources of infection.

Diagnosis is based on finding ova in stool specimens (Figure 305b).

### *Enterobius vermicularis*

*Enterobius vermicularis* is the cause of enterobiasis, or pinworm. Most cases are asymptomatic. However, depending on the migration activity of pinworms, infected persons may develop severe itching around the anus or the vagina (Figure 306a).

**Transmission and Diagnosis.** Pinworm infection is acquired by the ingestion of ova (Figure 306b). Contaminated fingers, toys, and other objects may be sources of ova. Ova also may be inhaled.

Diagnosis can be made by applying cellulose double sticky surface tape to the perianal area in the morning and before the infected individual bathes or has a bowel movement. Microscopic examination of the tape should show the presence of the ova (Figure 306b).

### *Loa loa*

*Loa loa* is the cause of the filarial infection loiasis. Filaria are slender nematodes that have complex life cycles involving various insects as intermediate hosts. Adult worms migrate in subcutaneous tissues and at times move across the eye (Figure 307). Localized swellings are commonly found in infected persons. The disease is found in various parts of Africa.

**Tranmissions and Diagnosis.** *Loa loa* is transmitted by mango flies belonging to the genus *Chrysops*.

Diagnosis is based on finding young (embryonic) forms of the nematode, known as **microfilariae.**

### *Onchocerca volvulus*

*Onchocerca volvulus* is the cause of another filarial infection, onchocerciasis (river blindness). Infected individuals develop painless subcutaneous swellings over bony body parts (Figure 308). Associated skin areas also itch and thicken, and lymph nodes enlarge. Microfilariae may also involve the eye and cause visual problems. The disease is found in tropical areas including Central and West Africa, Mexico, and various South American countries.

**Transmission and Diagnosis.** *Onchocerca volvulus* is transmitted by black flies from the genus *Simulium*. Laboratory diagnosis is based on finding microfilariae in skin specimens.

### *Strongyloides stercoralis*

*Strongyloides stercoralis* causes strongyloidiasis. The cycle of this nematode is similar to that described for hookworms such as *Necator*. Infected individuals may experience respiratory problems such as coughing and difficulty in swallowing because of large numbers of larvae migrating through the lungs. Disseminated strongyloidiasis results in the involvement of other body organs. The infection is found worldwide.

**Tranmission and Diagnosis.** *Strongyloides stercoralis* infection is acquired by direct contact with soil that contains infective larvae. Such larvae can penetrate unbroken skin.

Diagnosis involves finding larvae in stool or other specimens (Figure 309b).

### *Trichinella spiralis*

*Trichinella spiralis* causes trichinosis. Early signs and symptoms of infection associated with the gastrointestinal tract and include nausea, vomiting, diarrhea, and abdominal pains. Later effects due to muscle invasion by larvae include fever, muscle pain, and tissue swelling of involved areas. Cardiovascular and central nervous system involvement also may occur.

**Transmission and Diagnosis.** *Trichinella spiralis* infection is acquired by the ingestion of undercooked pork or pork products and the tissues of wild animals containing the larvae of this nematode.

Diagnosis is usually based on clinical findings. Muscle biopsies are performed to confirm the infection (Figure 310).

### *Trichuris trichiura*

*Trichuris trichiura* the whipworm, causes trichuriasis. Severe infections, especially those that occur in children, may result in bloody diarrhea and protrusion of the anus (**prolapse**). Light worm infections are generally asymptomatic.

**Transmission and Diagnosis.** *Trichuris* infection is acquired by ingestion of whipworm ova (Figure 311). Fomites and unsanitary conditions contribute to transmission.

Diagnosis is based on finding the typical ova in stool specimens.

### *Wuchereria bancrofti*

Several nematodes cause lymphatic filariasis, also known as elephantiasis (Figure 312a). These include *Wuchereria bancrofti* (Figure 312b), *Brugia malayi,* and *B. timori.* Early signs and symptoms of infection include inflammation of lymph nodes and channels fever and general discomfort. Males experience inflammation of the epididymis and the testes. Chronic infection leads to permanent enlargement of the legs (Figure 312a). Filariasis is widely distributed throughout tropical areas including parts of Africa, Asia, Southeast Asia, Central and South America, Pacific islands, and some Caribbean islands.

**Transmission and Diagnosis.** *Wuchereria bancrofti* can be transmitted by a number of blood-sucking mosquitoes including *Culex tarsalis*.

Diagnosis is based on finding microfilariae in blood smears or other types of preparations (Figure 312b).

## Trematodes

Most trematodes, or flukes, are bilaterally symmetrical, flat, and leaf-shaped, and they generally have oral and ventral suckers for attachment (Figure 313). All species except schistosomes contain both sex organs in the same worm (hermaphroditic).

Flukes vary with respect to their life cycles. Many begin with the development of ciliated larval forms called **miracidia** (Figure 314b) within ova. These larvae escape through hatching and penetrate the tissues of snails (first intermediate host). Further development results in the formation of other larvae, which escape from snails. Depending on the trematode species, other larval forms develop in other hosts These include **sporocysts** and **rediae** (Figure 314c,d).

**Cercariae**, tail-bearing larvae (Figure 317h), represent a later stage in the life cycle of trematodes which are released from snail hosts and ready to attach to and attack susceptible hosts. With hermaphroditic flukes, cercariae round up and form **metacercariae** on aquatic plants or animals.

### *Clonorchis (Opisthorchis) sinensis*

*Clonorchis sinensis* (Figure 313), also known as *Opisthorchis,* is the Chinese liver fluke and is the cause of clonorchiasis. Most infections are asymptomatic. However, heavy fluke infections can result in bile duct blockage and gallbladder inflammation.

Signs and symptoms can include fever, diarrhea and abdominal pain.

**Transmission and Diagnosis.** Clonorchiasis is acquired by ingestion of raw, pickled, and dried freshwater fish containing metacercariae. Diagnosis is based on finding ova in stool specimens (Figure 313b).

### *Fasciola hepatica*

*Fasciola hepatica,* the sheep liver fluke, causes fascioliasis. The infection is found in sheep-raising areas of the world. Adults flukes cause damage to and blockage of bile ducts and the gallbladder.

**Transmission and Diagnosis.** Infection is acquired by the ingestion of unwashed, raw aquatic vegetation containing *F. hepatica* metacercariae.

Diagnosis generally is based on finding ova in stool specimens (Figure 314a).

## Nematodes

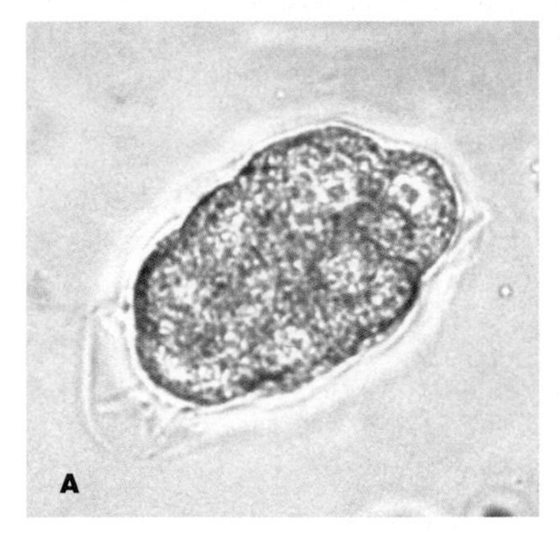


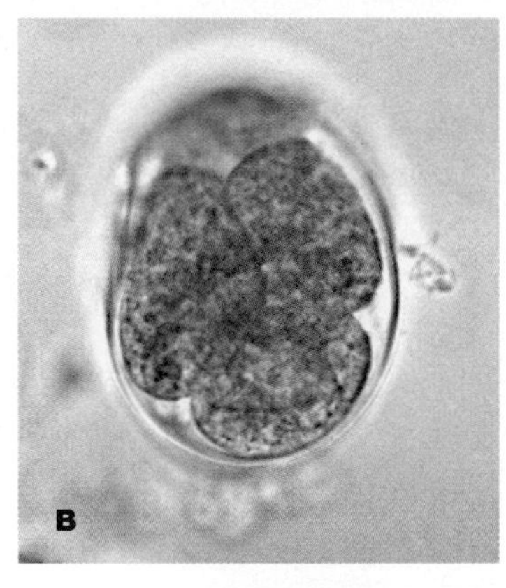


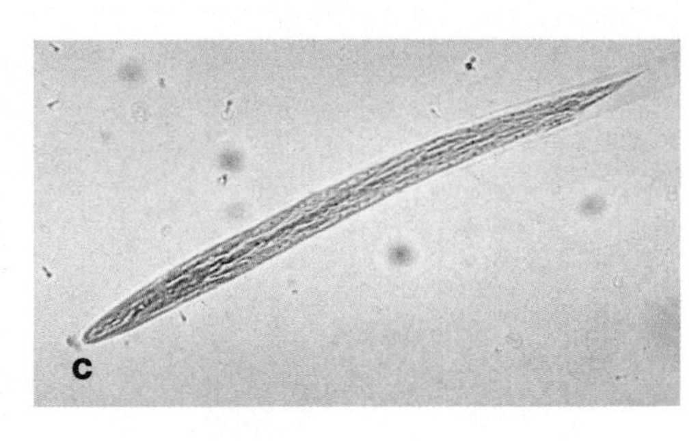


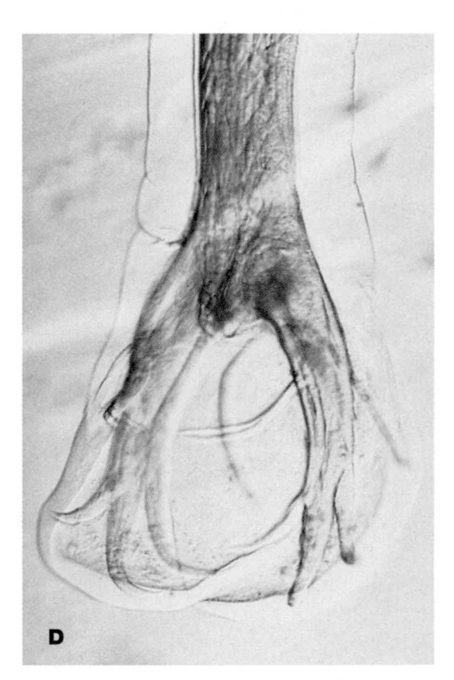


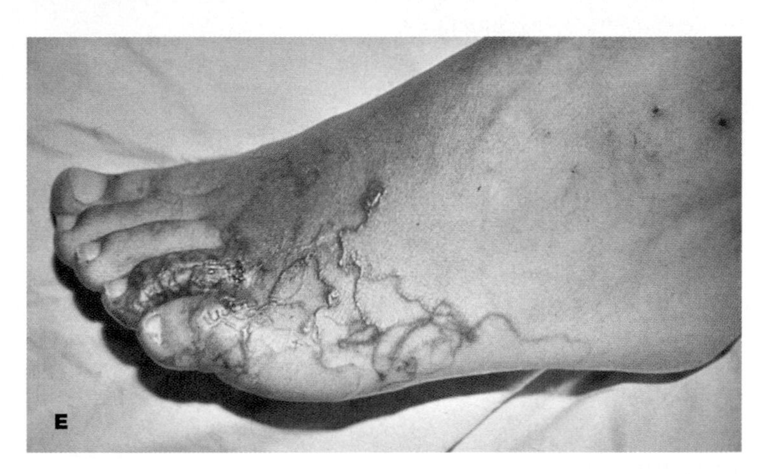


**Figure 304**
*Ancylostoma duodenale* and *Necator americanus* (hookworm). (a) *A. duodenale* ovum. (b) *N. americanus* ovum. (c) An infective (filariform) larva of *N. americanus.* (d) The bursa, or expanded posterior end, of the male worm. (e) A case of creeping eruption, also known as cutaneous larva migrans.

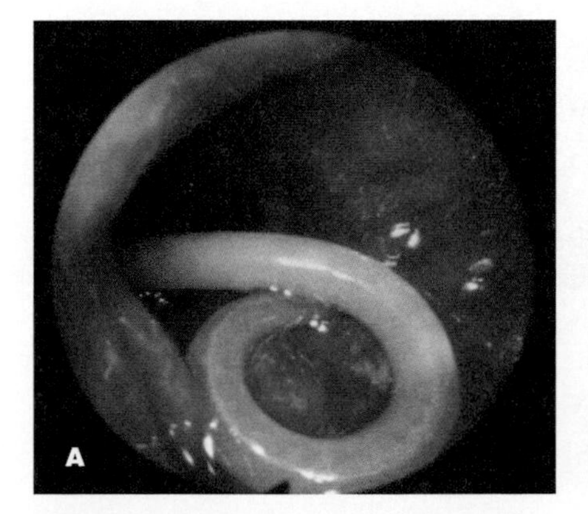


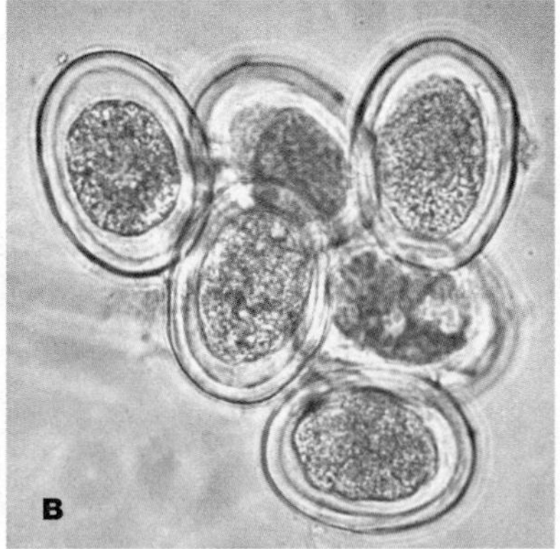


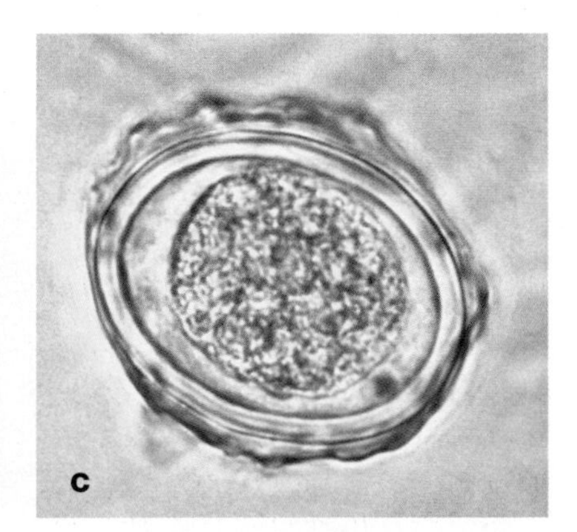


**Figure 305**
*Ascaris.* (a) A single adult *Ascaris* observed in the digestive tract. (From H.S. Fuessl *NEJM* 331(1994):301) (b) Unfertilized ova. (c) A fertilized ovum.

## Nematodes *(Continued)*

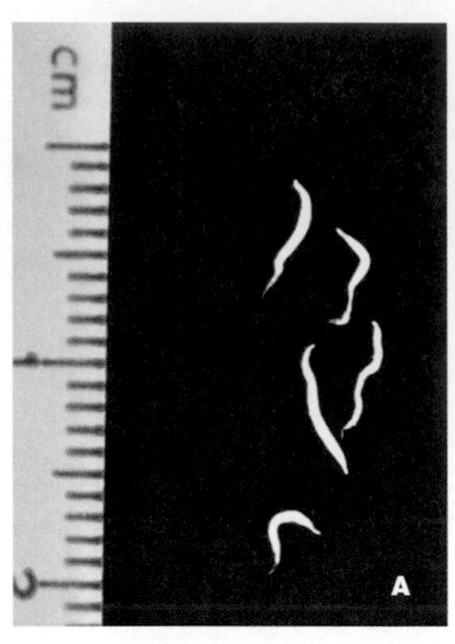


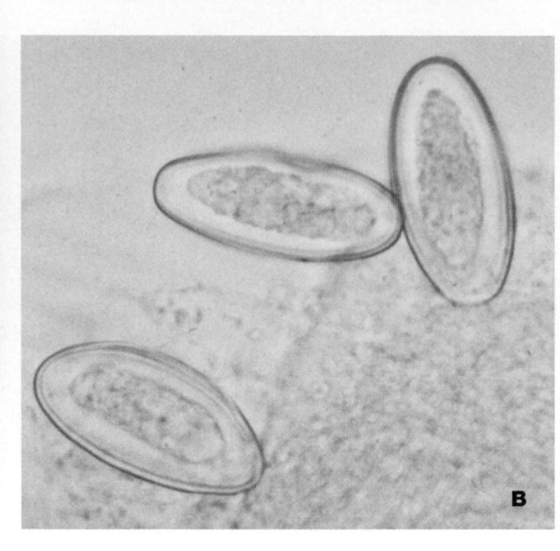


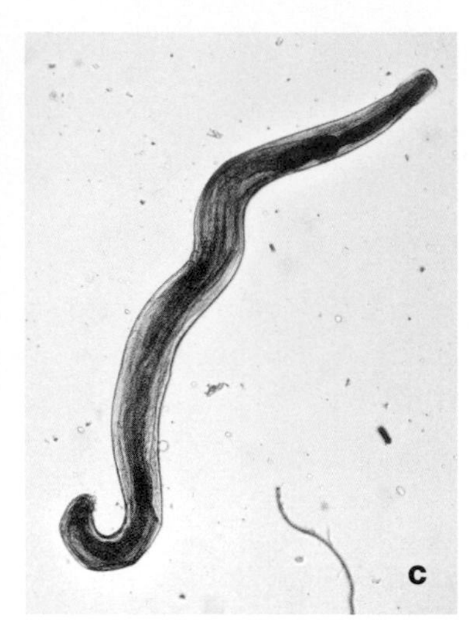


**Figure 306**
*Enterobius vermicularis* (pinworm). (a) Pinworms recovered from the perianal region of a young child. (Courtesy Dr. L. Alpert, Pathology Department, The Sir Mortimer B. David Jewish General Hospital.) (b) Ova on the surface of transparent tape. (c) An adult male worm stained.

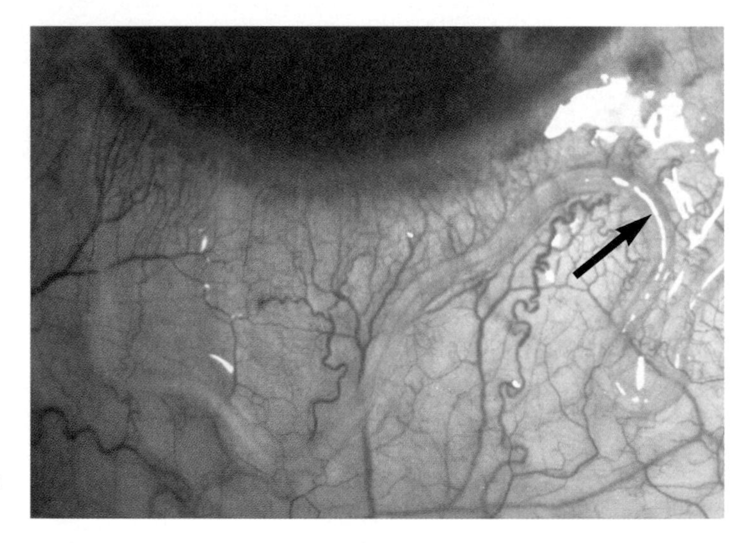

**Figure 307**
*Loa loa* (the eye worm). The worm (swollen area) can be seen near the surface of the eye.

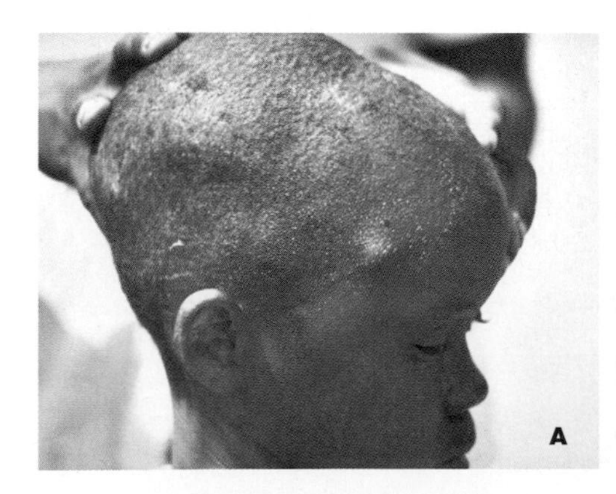


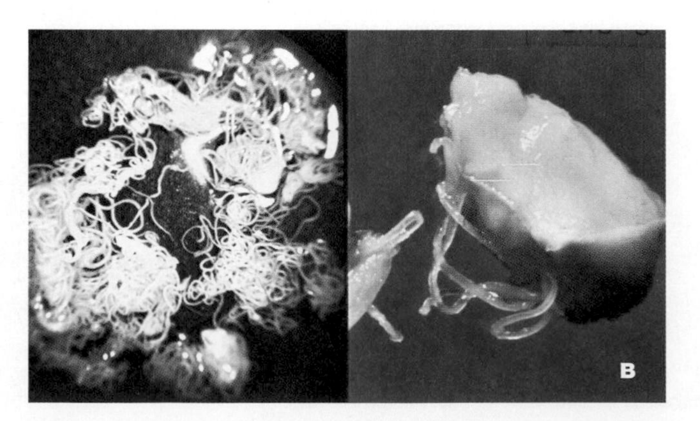


**Figure 308**
*Onchocerca volvulus,* the cause of onchocerciasis (river blindness). (a) Nodules on the head of an infected child. (b) A mass of live male and female *O. volvulus* (left) and a nodule with worms (right).

## Nematodes *(Continued)*

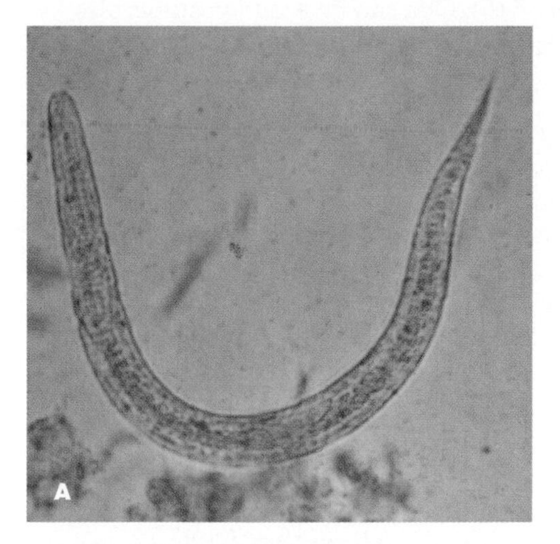


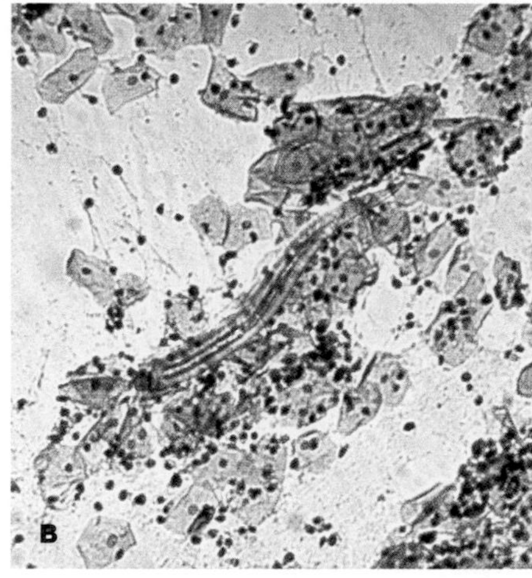


**Figure 309**
*Strongyloides stercoralis.* (a) The rhabditiform larva. (b) The presence of a long larval form of the roundworm in a vaginal smear. The worm is covered by a number of host cells (400×).

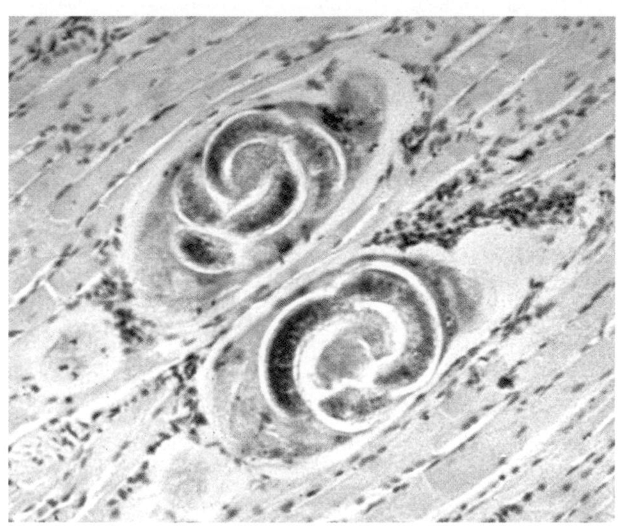

**Figure 310**
*Trichinella spiralis* (the pork roundworm) in muscle.

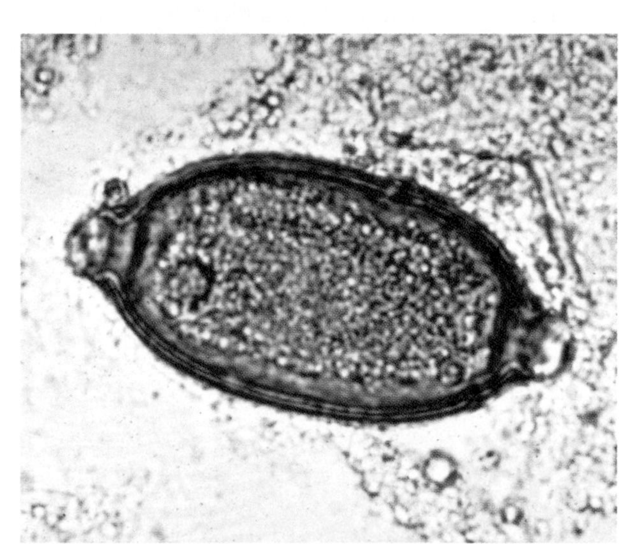

**Figure 311**
*Trichuris trichiura* (the whipworm) ovum.

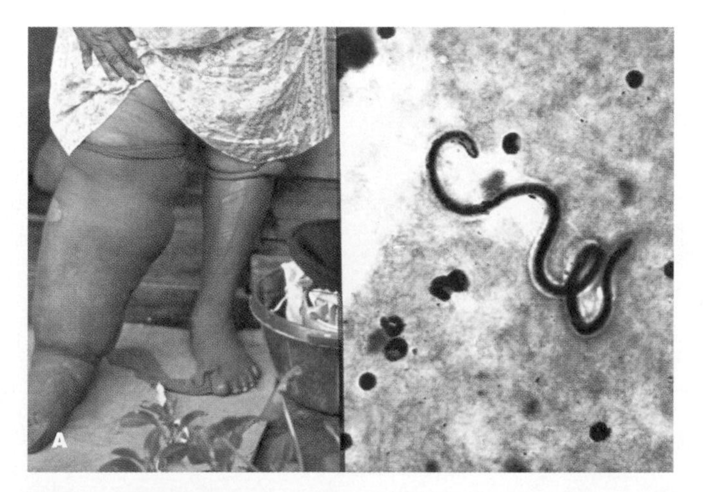


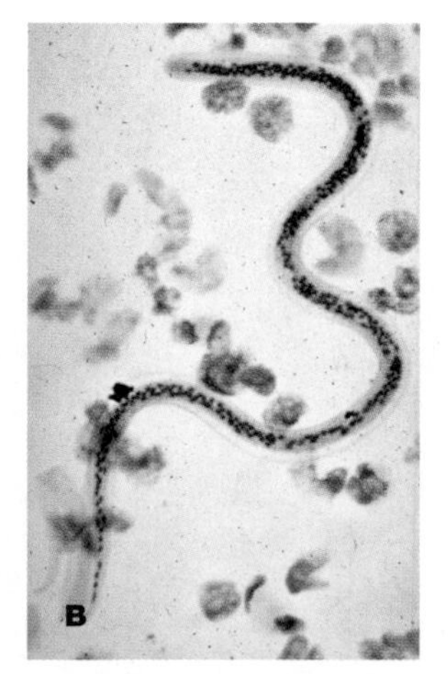


**Figure 312**
*Wuchereria bancrofti,* one cause of elephantiasis. (a) A case of the disease (left); the nematode from the leg (right). (b) *W. bancrofti* (a microfilaria) in a tissue smear, showing its typical parts.

### *Fasciolopsis buski*

*Fasciolopsis buski* causes fasciolpsiasis. Symptoms vary depending on the number of flukes. Abdominal pain and diarrhea are commonly experienced. The disease is found in various areas in Asia.

**Transmission and Diagnosis.** Infection is acquired by the ingestion of freshwater aquatic plants contaminated with metacercariae. Pigs, dogs, and rabbits are reservoirs of the fluke.

Diagnosis is made by finding ova in stool specimens (Figure 215).

### *Paragonimus westermani*

*Paragonimus westermani* and other species of *Paragonimus*, the lung flukes, cause paragonimiasis. As is the case with most other fluke infections, the signs and symptoms depend on the number of flukes. Cough, increased sputum production, and chest pain are typcial. Paragonimiasis is found predominantly in Asia and Africa.

**Transmission and Diagnosis.** Infection is acquired by the ingestion of improperly cooked freshwater crabs or crayfish containing metacercariae.

Diagnosis generally is made by finding ova in sputum (Figure 316). Ova can be found in stools also because they may be swallowed.

### *Schistosoma*

Various species of *Schistosoma,* blood flukes, cause schistosomiasis (Figure 317a). Unlike other flukes, schistosomes have separate sexes (Figure 317b,c). The effects of schistosomes result from the depositing of ova (Figures 317d–g) in the large intestine and urinary bladder and the reactions of blood vessels in the liver and lungs. Adult worm live in veins and venules of infected hosts. Schistosome species are found in different parts of the world: *S. haematobium* (Africa and the Middle East), *S. japonicum* (various regions of Asia), *S. mansoni* (tropical Africa, eastern part of South America and some Caribbean islands), and *S. intercalatum* (portions of Africa).

**Transmission and Diagnosis.** Schistosome infection is acquired when cercariae (Figure 317h) from infected snails penetrate intact skin upon contact in freshwater environments such as ponds and rivers.

Diagnosis includes demonstrating the characteristic ova of schistomes in stools. Serological tests such as enzyme immunoassays also are of value.

## Trematodes

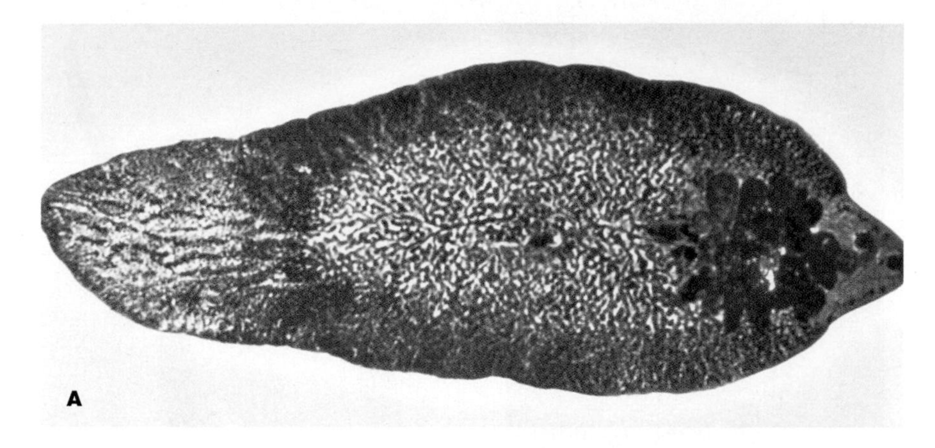


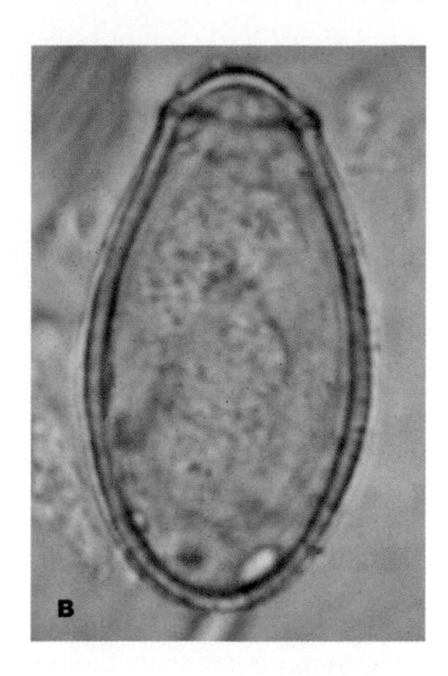


**Figure 313**
*Clonorchis (Opisthorchis) sinensis* (the Chinese liver fluke). (a) The adult fluke and its parts. (b) The ovum.

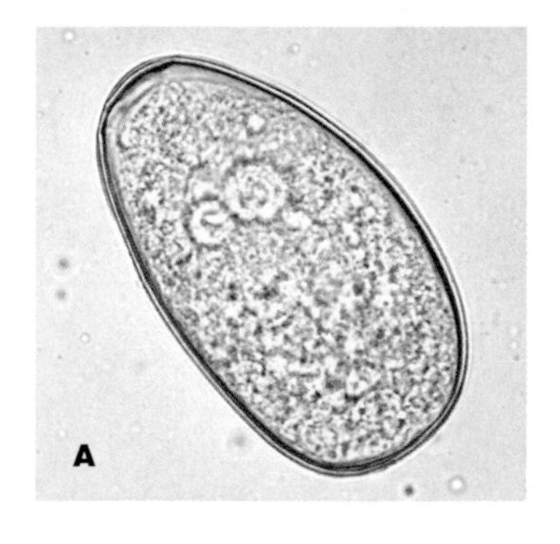


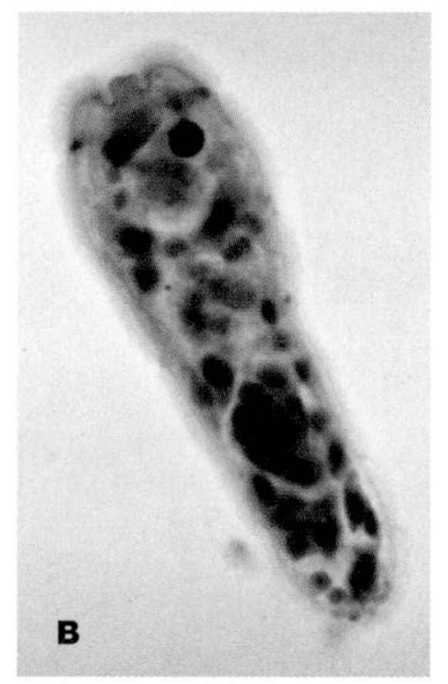


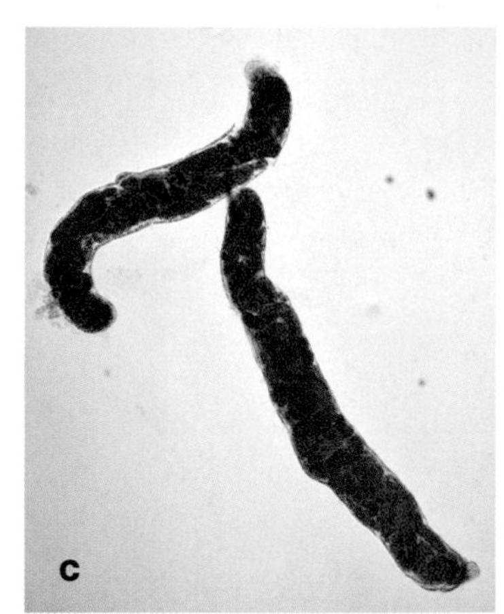


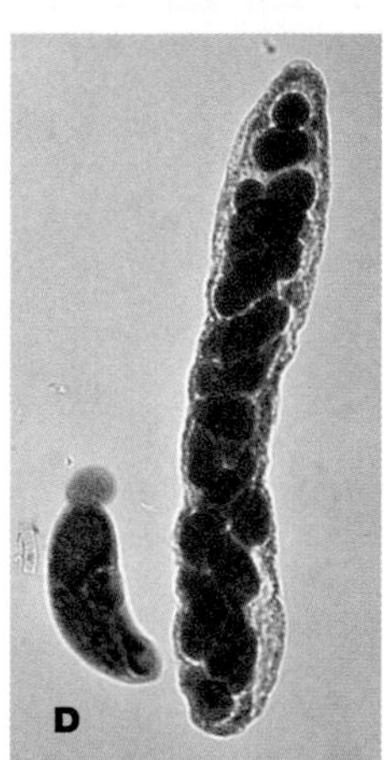


**Figure 314**
*Fasciola hepatica* (the sheep liver fluke). (a) The ovum. (b) A stained miracidium (2.5×). (c) Stained sporocysts (2.5×). (d) Stained rediae.

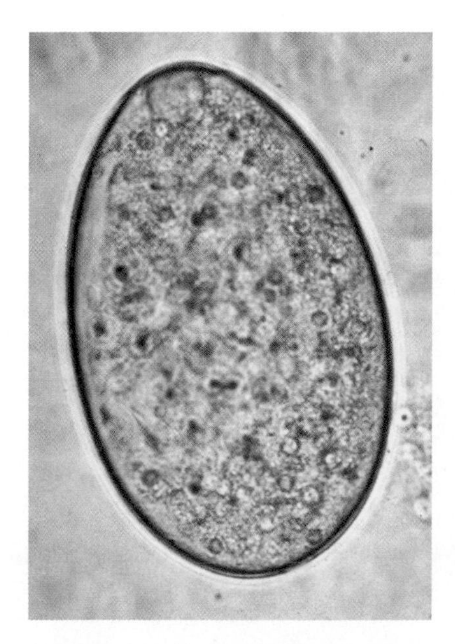

**Figure 315**
*Fasciolopsis buski* ovum.

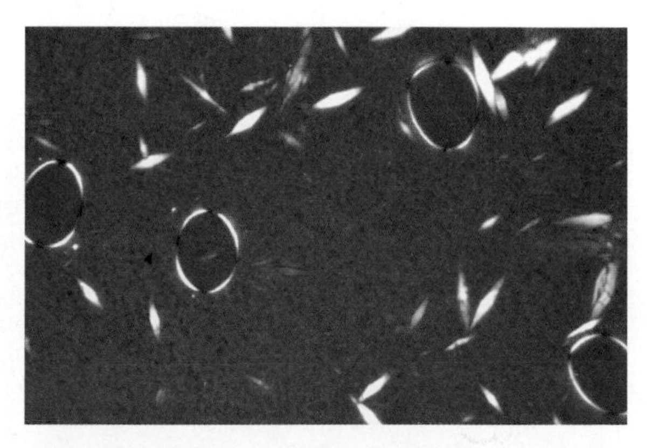

**Figure 316**
*Paragonimus westermani* (a lung fluke). Oval ova and Charcot-Leyden crystals are shown under polarized light. (Courtesy of Dr. L. Alpert, Pathology Department, The Sir Mortimer B. David Jewish General Hospital.)

## Trematodes *(Continued)*

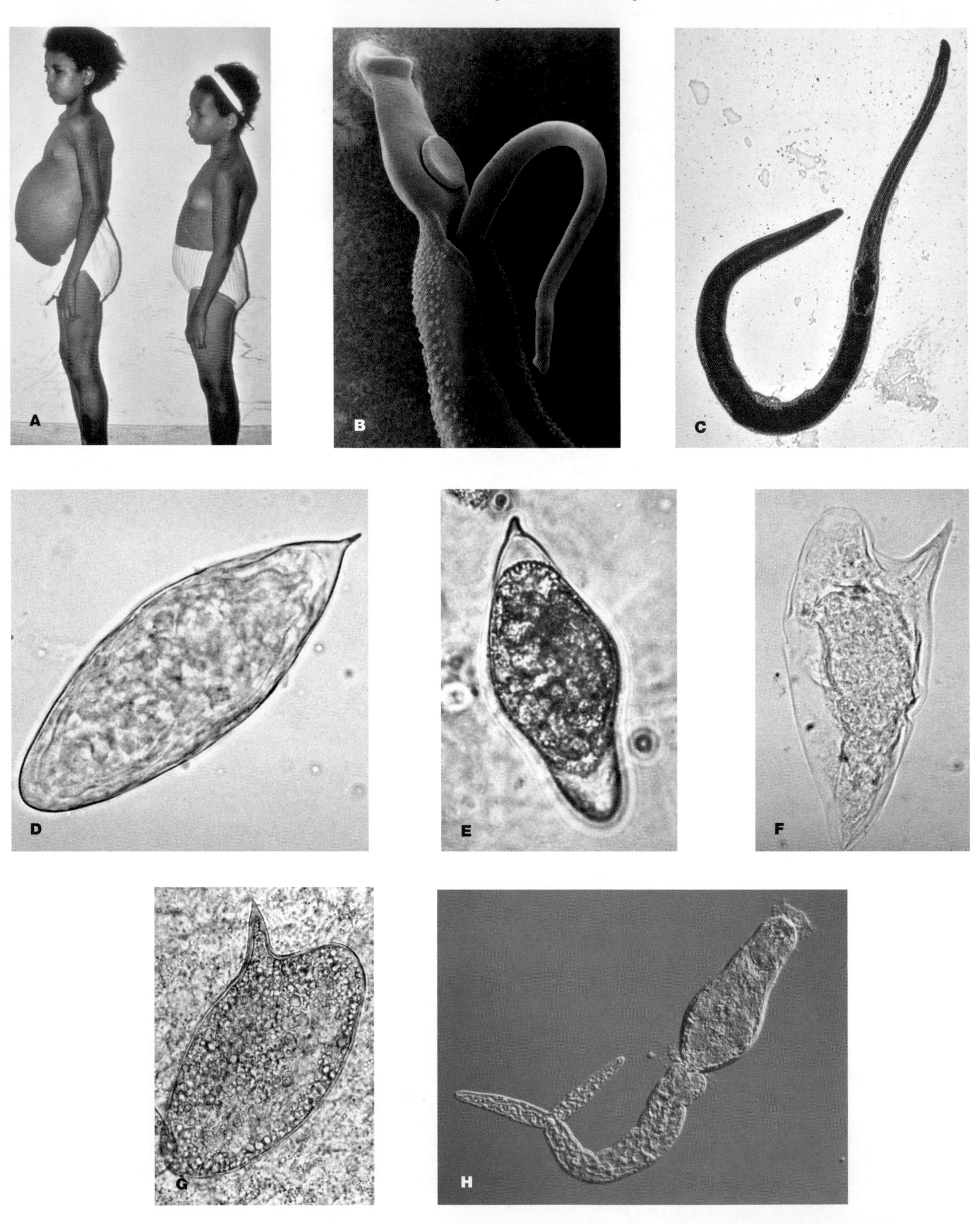


**Figure 317**

Schistosomes. (a) Victims of schistosomiasis. (Courtesy of the World Health Organization.) (b) A scanning electron micrograph of both the larger male and the smaller female schistomes (400×). (c) Adult female *Schistosoma mansoni* (2.5×). (d) *S. haematobium* ovum. (e) *S. intercalatum* ovum. (f) *S. mansoni* ovum. (g) *S. japonicum* ovum. (h) A live schistosome cercaria. Note the forked tail.

# SECTION 9 Arthropods and Disease

*So, Naturalists observe, a flea hath*
*smaller fleas that on him prey;*
*And these have smaller still to bite 'em;*
*And so proceed, ad infinitum.*

—*Johnathan Swift*

Several arthropod species are well known for the effects they cause by stinging or biting. Others, as earlier sections have described, serve to transmit certain microbial and helminthic infections. Still others cause human diseases and use human tissues in their development. This last group of arthropods includes *Pediculus humanus corporis,* the body louse (Figure 318); *Phthirus pubis,* the pubic or crab louse (Figure 319); and *Sarcoptes scabiei*, the itch mite, which causes scabies (Figure 320).

## Arthropods and Disease

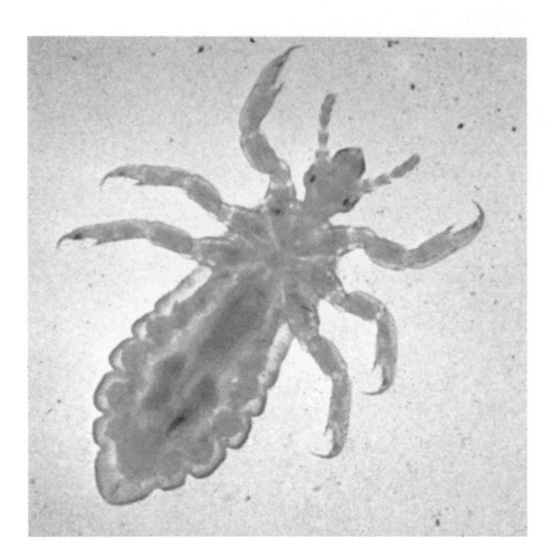

**Figure 318**
*Pediculus humanus corporis* (the body louse).

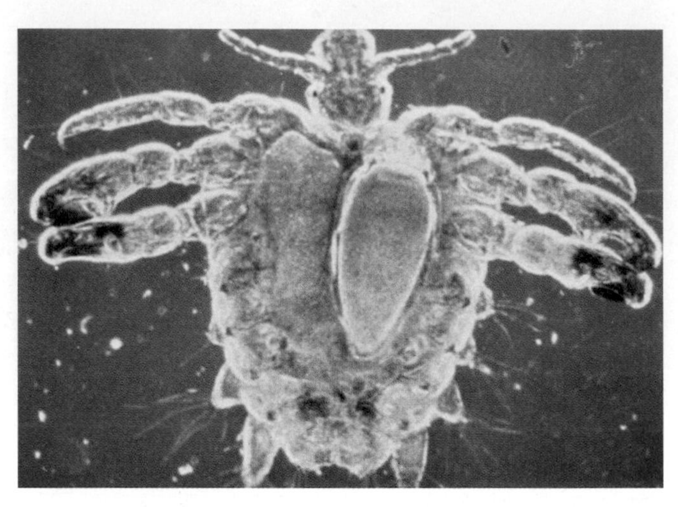

**Figure 319**
*Phthirus pubis* (the pubic or crab louse).

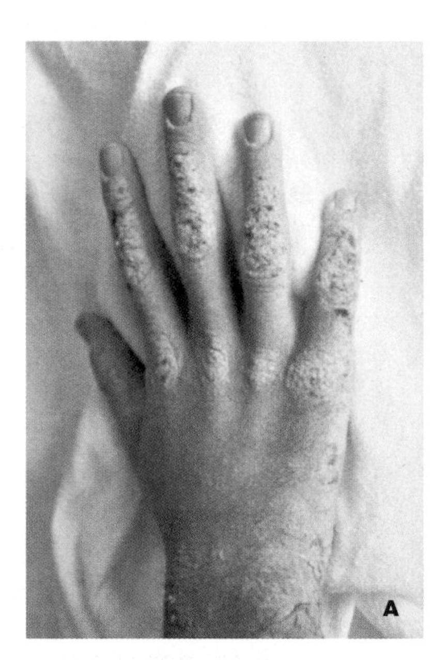


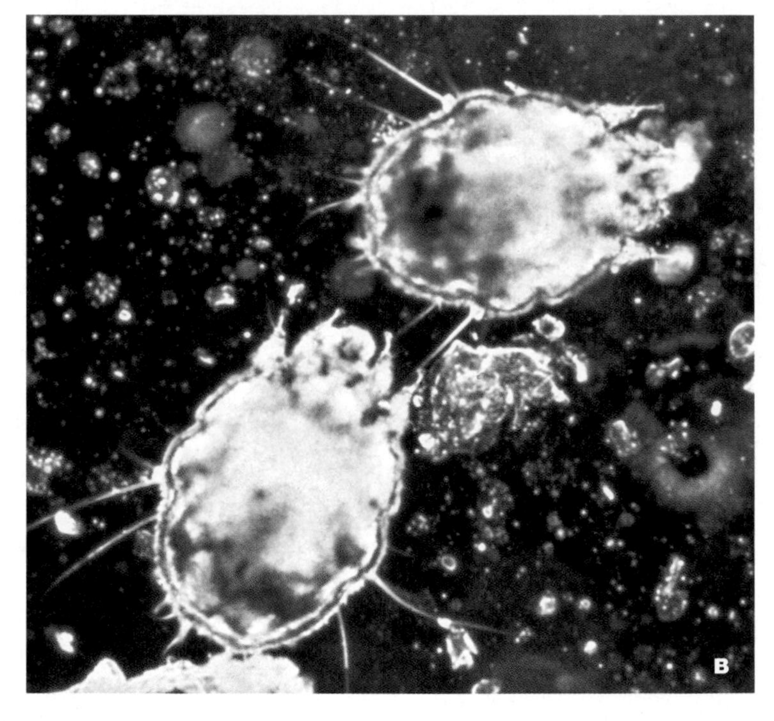


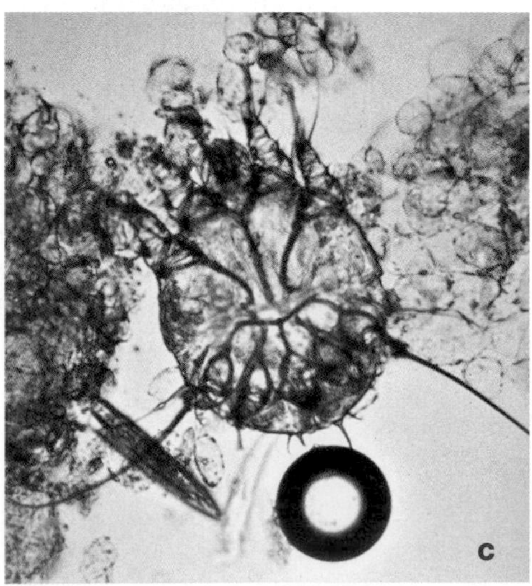


**Figure 320**
*Sarcoptes scabiei* (the itch mite), the cause of scabies. (a) A case of Norwegian scabies infestation in a patient with AIDS. (b) Live scabies mites in a skin scraping seen under dark field microscopy. (From T.L. Meinking et al. *NEJM* 333(1995):26–30) (c) The itch mite.

# SECTION 10 Immunology

*Science advances through tentative answers to a series of more and more subtle questions which reach deeper and deeper into the essence of natural phenomena.*

*—Louis Pasteur*

Humans and other vertebrates are protected in varying degrees from disease-causing microorganisms and cancer cells by a surveillance mechanism referred to as the **immune system.** Collectively, the various components of this system provide protection, or **immunity,** by imposing barriers to invasion by microorganisms or other disease agents or by selectively neutralizing or eliminating materials that they recognize as being foreign. Immunologic responses, which may be either **nonspecific** or **specific,** serve three functions: defense against invasion by microorganisms, maintenance of a stable internal environment (homeostasis), and surveillance, or recognition of abnormal and foreign cell types.

## Immune System Components

The various components of the immune system are combined in an exquisitely complex communications network that functions as an effective defense against foreign microorganisms and against body cells that have become abnormal (cancerous). Such components include the thymus gland, where T lymphocytes develop and mature (Figure 321); lymph nodes, where host immune responses are initiated (Figure 322); plasma cells, the immunoglobulin- (antibody-) producing cells of the body (Figure 323); and eosinophils, one type of granulocyte that defends against many pathogens and participants in common allergy reactions (Figure 323).

## Diagnostic Tests

A large number of immunodiagnostic tests are used to detect and monitor infectious diseases and to follow the recovery from disease. Several of these techniques are applicable not only to the areas of infectious and immunologic diseases and disorders but also to the entire spectrum of clinical medicine. Specific procedures are used for measuring levels of certain drugs, hormones, serum proteins, and tumor and transplantation antigens and for determining blood group incompatibilities. Table 20 briefly characterizes a number of tests, and Figures 324 through 334 show the results of such tests.

## Electron Microscopy

Immunoelectron microscopy is a relatively new analytical approach used to identify viruses, to demonstrate differences among microbial structures, and to detect minute amounts of immunoglobulins and related substances. Probes made of gold particles combined with specific immunoglobulins can be used to detect antigens and chemical components of cells (Figure 335).

## Commercial Devices

A number of commercial devices are available to either directly or indirectly diagnose diseases. These include agglutination tests (Figure 336) and applications of enzyme immunoassays (Figure 337 and Table 20).

**Table 20** Immunologic Procedures Used in Diagnosis and/or Microbial Identification

| Procedure | Principle Involved | Positive Test Result | Application(s) |
|---|---|---|---|
| Agglutination | Antibody clumps cells or other particulate antigen preparations (insoluble particles coated with antigens—e.g.,latex particles, *Staphylococcus* protein A, etc). | Aggregates (clumps) of antigens (Figure 325) | 1. Diagnosis of typhus, Rocky Mountain spotted fever (Weil-Felix test), typhoid fever (Widal test), and infectious mononucleosis<br>2. Identification of disease agents including *Haemophilus influenzae* type b, *Neisseria meningitidis,* and *Streptococcus pneumoniae.* |
| Complement fixation | Antigen-antibody complex of test system binds complement, which thereby becomes unavailable for binding by sheep red blood cells and hemolysin of the indicator system. | Cloudy red suspension (Figure 328) | Diagnosis of various bacterial, mycotic, protozoan, viral, and helminth (worm) diseases |
| Enzyme-linked immunoabsorbent assay (ELISA) | Antigen or antibody from specimens trapped by corresponding specific antibody or antigen coating a solid phase support combines with enzyme labeled specific antibody. The formed complex reacts with an added enzyme substrate in proportion to the amount of antigen or antibody first bound by the antibody or antigen coating. | Color changes occurring with the addition of an enzyme substrate are proportional to either antibody or antigen in specimens (Figure 327) | 1. Detection of IgM to rubella and influenza A<br>2. Identification and/or detection of herpes simplex viruses types 1 and 2, cytomegalovirus, measles, hepatitis B, and AIDS viruses<br>3. Detection of antibodies to bacterial antigens |
| Hemagglutination | Homologous antibody causes (hemagglutinin) aggregates of red blood cells to form[a]. | Aggregates of red blood cells (Figures 324 and 326) | Blood typing |
| Hemagglutination-inhibition (viral) | Antibody inhibits the agglutination of red blood cells by coating hemagglutinating virus. | Formation of a circle of unagglutinated cells (Figure 329) | 1. Determining the immune status toward rubella (German measles)<br>2. Virus identification |
| Immunodiffusion | Antibody and soluble antigen diffuse toward one another through an agar gel and react where homologous antibody is in proper proportion to homologous antigen. | Lines of precipitate form within the agar (Figure 327) | Antigen and/or antibody identification |
| Immunofluorescent microscopy | Antibody (usually) or antigen is labeled with a fluorescent dye, which fluoresces on exposure to ultraviolet or blue light. | Glowing on exposure to UV light (Figure 330) | 1. Detection of antigen or antibody<br>2. Identification of microbial pathogens of diseases such as rabies, syphilis, Legionnaires' disease, etc. |
| Immunoperoxidase | Antibody is labled (conjugated) with an enzyme, usually horseradish peroxidase, which is detected by a color reaction produced upon treatment with a peroxidase substrate. | Color changes occurring with the addition of a peroxidase substrate (Figure 331) | Detection and identification of several viruses, including cytomegaloviruses, rabies viruses, and herpes viruses |
| Precipitin | Antibody and soluble antigen react when they are in proper proportion to one another. | Lines of precipitate form (Figure 327) | 1. Diagnosis of microbial diseases<br>2. Detection of antigens |
| Western blot | Proteins of antigen are separated by electrophoresis, transferred to and immobilized on nitrocellulose strips, and then exposed to serum specimens. Antigen-antibody reactions are detected by an added enzyme-linked antihuman immunoglobulin reagent. | Formation of a black precipitate in the regions where the enzyme-immunoglobulin reagent is bound (Figure 334) | 1. Diagnosis of infectious diseases such as AIDS<br>2. Detection of antibody against different antigenic components |

[a]Hemagglutination reactions caused by certain viruses and bacteria generally do not involve antibody.

# Immunology

## Immune System Components

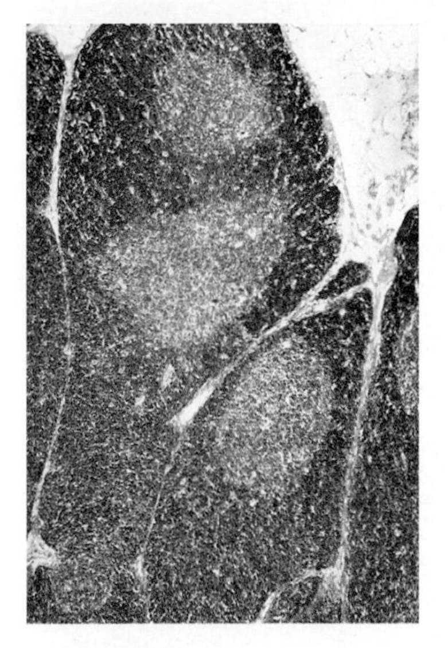

**Figure 321**
A section of the thymus gland. The outer cortex and the inner (lighter) medulla are shown.

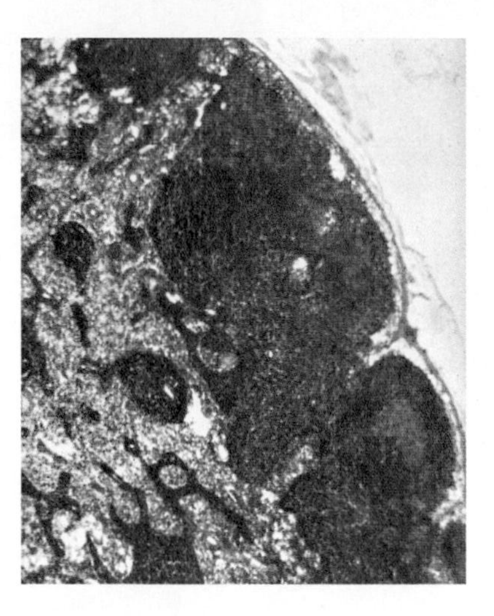

**Figure 322**
A view through a lymph node.

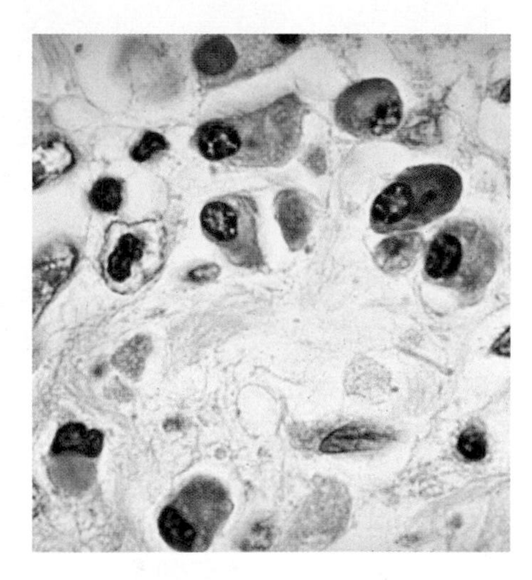

**Figure 323**
Plasma cells. Note the presence of eosinophils (cells with red-staining granules).

## Diagnostic Tests

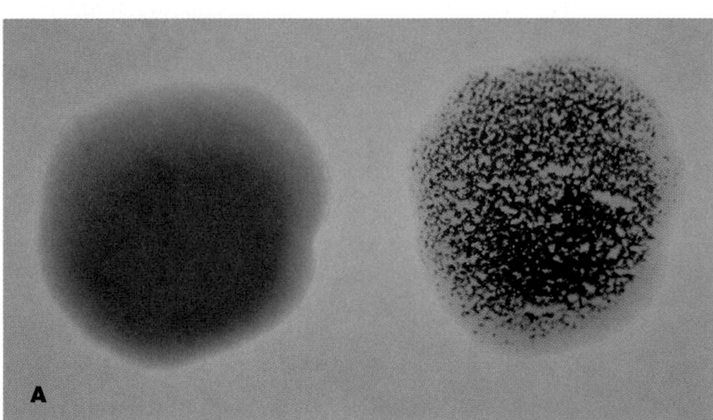


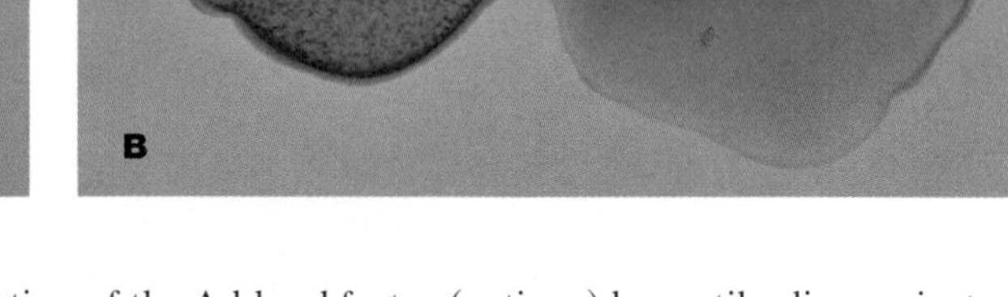


**Figure 324**
Blood typing. (a) An A+ (clumping) reaction showing agglutination of the A blood factor (antigen) by antibodies against the factor. (b) A positive (clumping) reaction for the Rho, or D, factor.

**Figure 325**
Bacterial agglutination (clumping) in a tube.

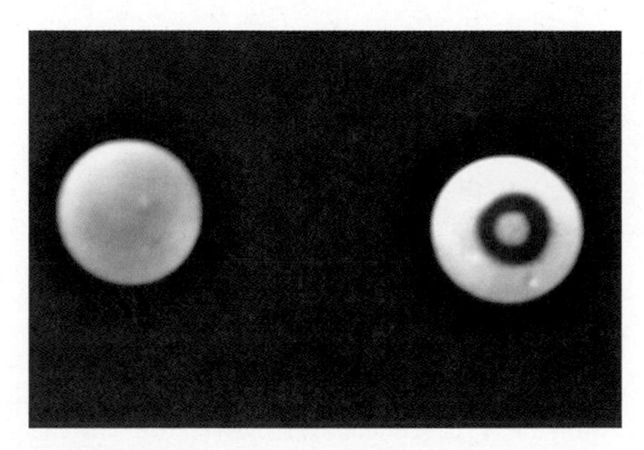

**Figure 326**
Hemagglutination reactions. The red doughnut reaction is a negative result.

## Diagnostic Tests *(Continued)*

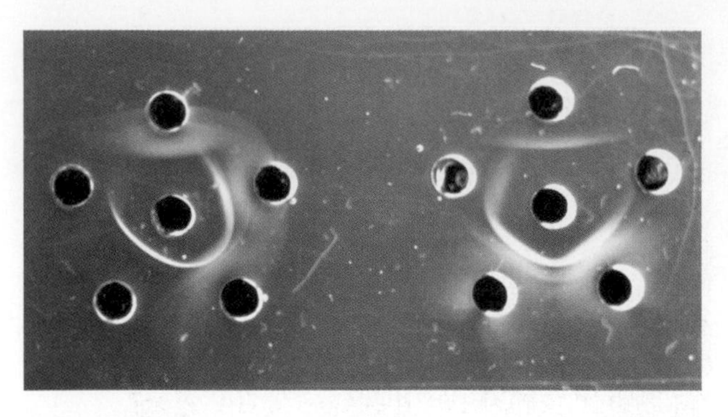

**Figure 327**
Immunodiffusion reactions (a variation of the precipitin reaction).

**Figure 328**
Complement fixation. A series of serum dilutions were used here. The cloudy solutions are positive.

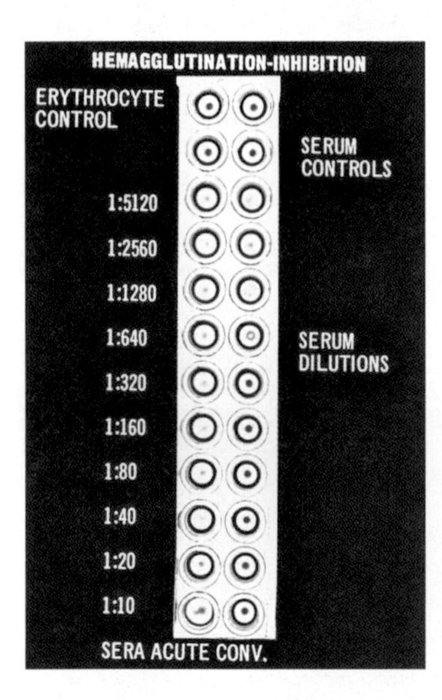


**Figure 329**
Hemagglutination-inhibition (HAI). A pattern of reactions shown with a series of serum dilutions from 1:10 to 1:5,120. The dotlike results are negative reactions. Erythrocyte and serum controls also are shown.

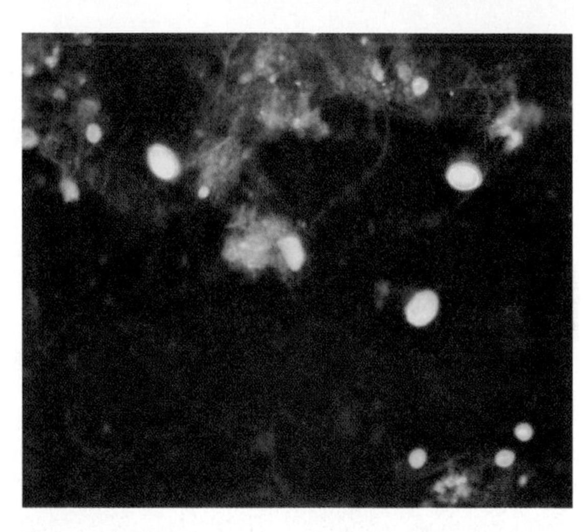

**Figure 330**
Immunofluorescence. The Meriflour procedure, a commercial system used to detect the presence of cysts of the protozoa *Cryptosporidium* (small green circular forms) and *Giardia* (yellow ovals). (Courtesy of Meridium Diagnostics.)

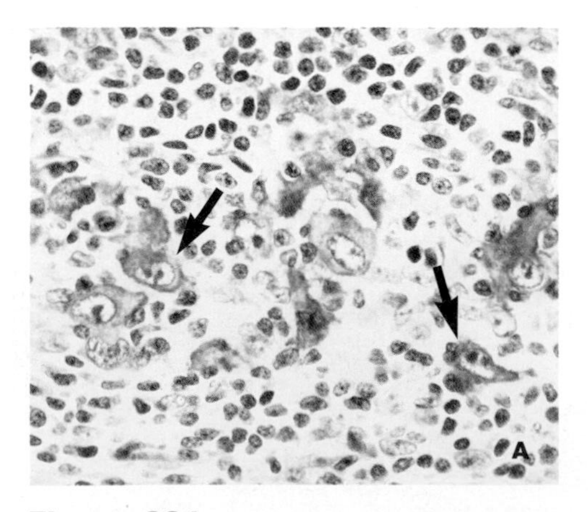


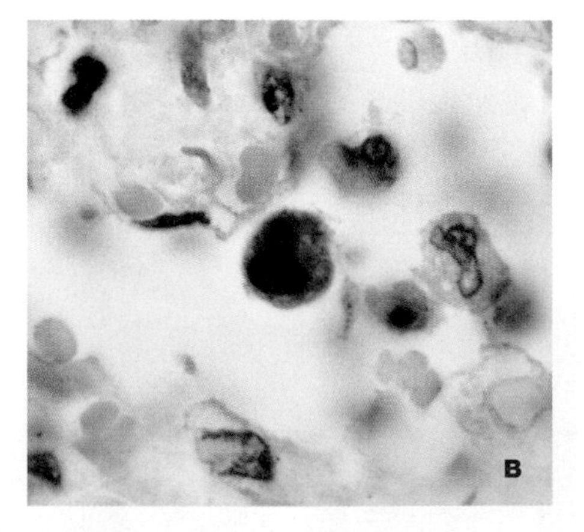


**Figure 331**
Immunoperoxidase reactions, examples of immunostaining. (a) A number of specific antigen "+" cancer cells (arrows) shown with the use of avidinbiotin-complex immunoperoxidase. (Courtesy of Dr. M. Miettinen, Jefferson Medical College of Thomas Jefferson University.) (b) Immunoperoxidase reaction demonstrating the presence of virus-infected cells. (From C. Sinzger et al. *J. Inf. Dis.* 173(1996):240–45.)

## Diagnostic Tests *(Continued)*

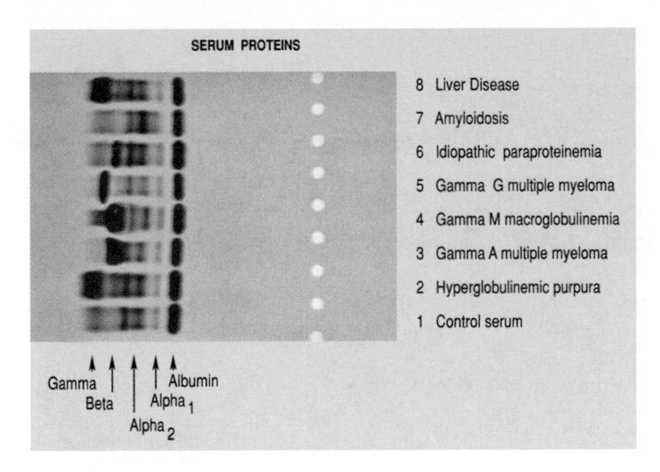


**Figure 332**
Electrophoretic patterns showing the concentration of serum proteins in normal and abnormal situations.

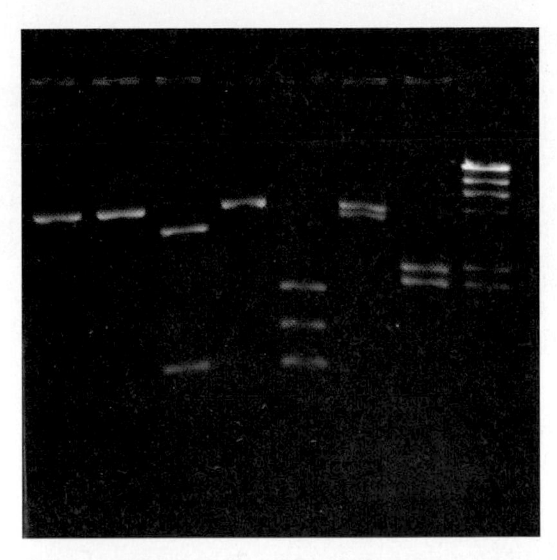

**Figure 333**
Electrophoretic patterns obtained with polymerase chain reaction (PCR) products. The violet background is provided by ultraviolet light, which was used to show the pattern.

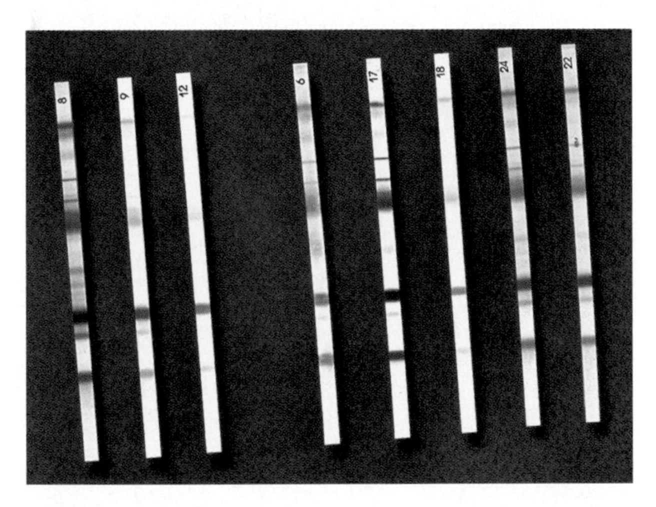

**Figure 334**
Results of the Western blot assay. This procedure is used to detect specific proteins.

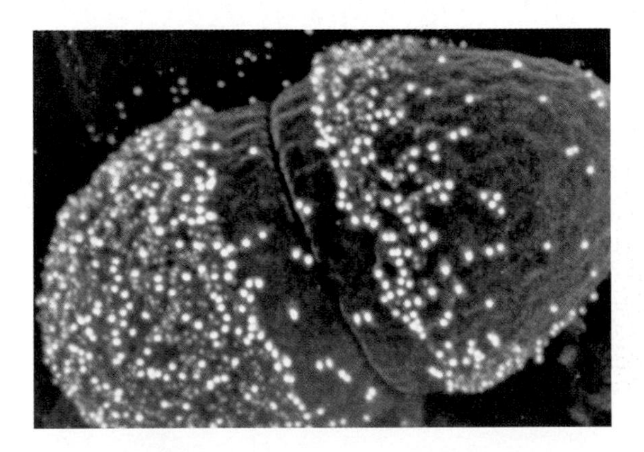

**Figure 335**
Electron microscopic techniques. Colloidal gold labeling of bacterial surfaces (dots). (From S.B. Olmsted, G.M. Dunny, S.L. Erlandsen, and C.L. Wells. *J. Inf. Dis.* 170(1994):1549–56.)

## Commercial Devices

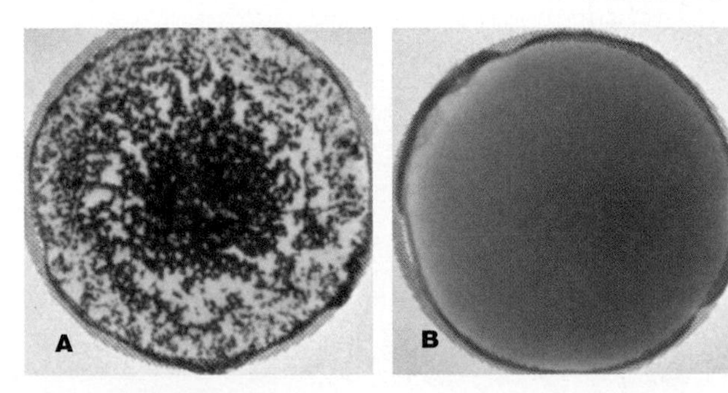


**Figure 336**
The Staphyloside, a rapid hemagglutination test to detect *Staphylococcus aureus* cell wall clumping factor. (a) A positive reaction. (b) A negative reaction. (Courtesy of Becton Dickinson Microbiology Systems.)

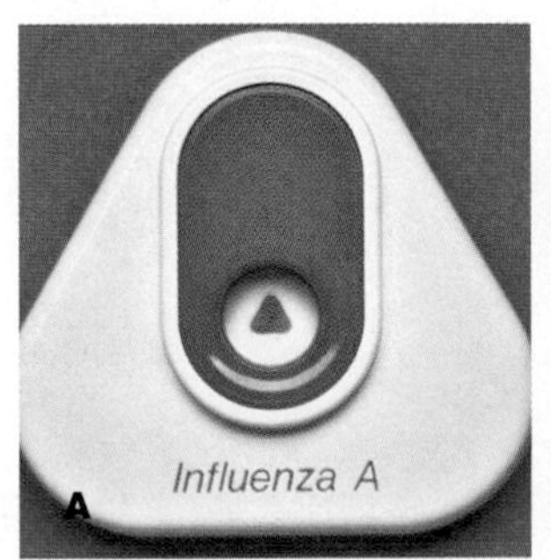


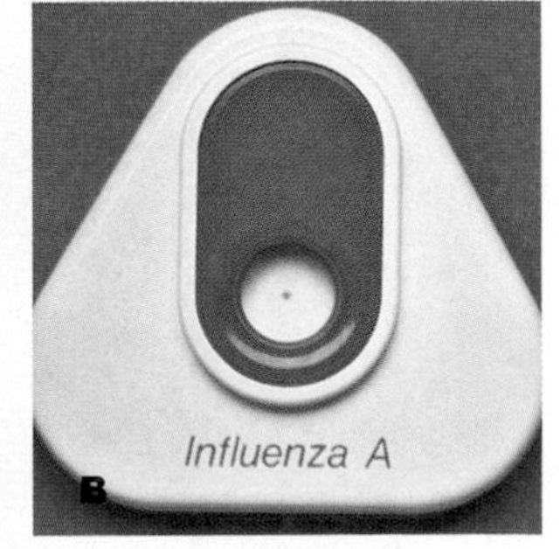


**Figure 337**
Enzyme immunoassay for influenza. (a) A positive reaction. (b) A negative reaction. (Courtesy of Becton Dickinson Microbiology Systems.)

# SECTION 11 APPENDIX FIGURES

The following figures are included in the *Color Atlas* to provide some perspective as to the body regions involved with microbial and helminthic diseases. Note that different disease agents (e.g., bacteria, fungi, protozoa, and viruses) are shown in different colors. **Red**: viruses; **Green**: bacteria; **Blue**: fungi; **Orange**: protozoa; **Black**: helminths.

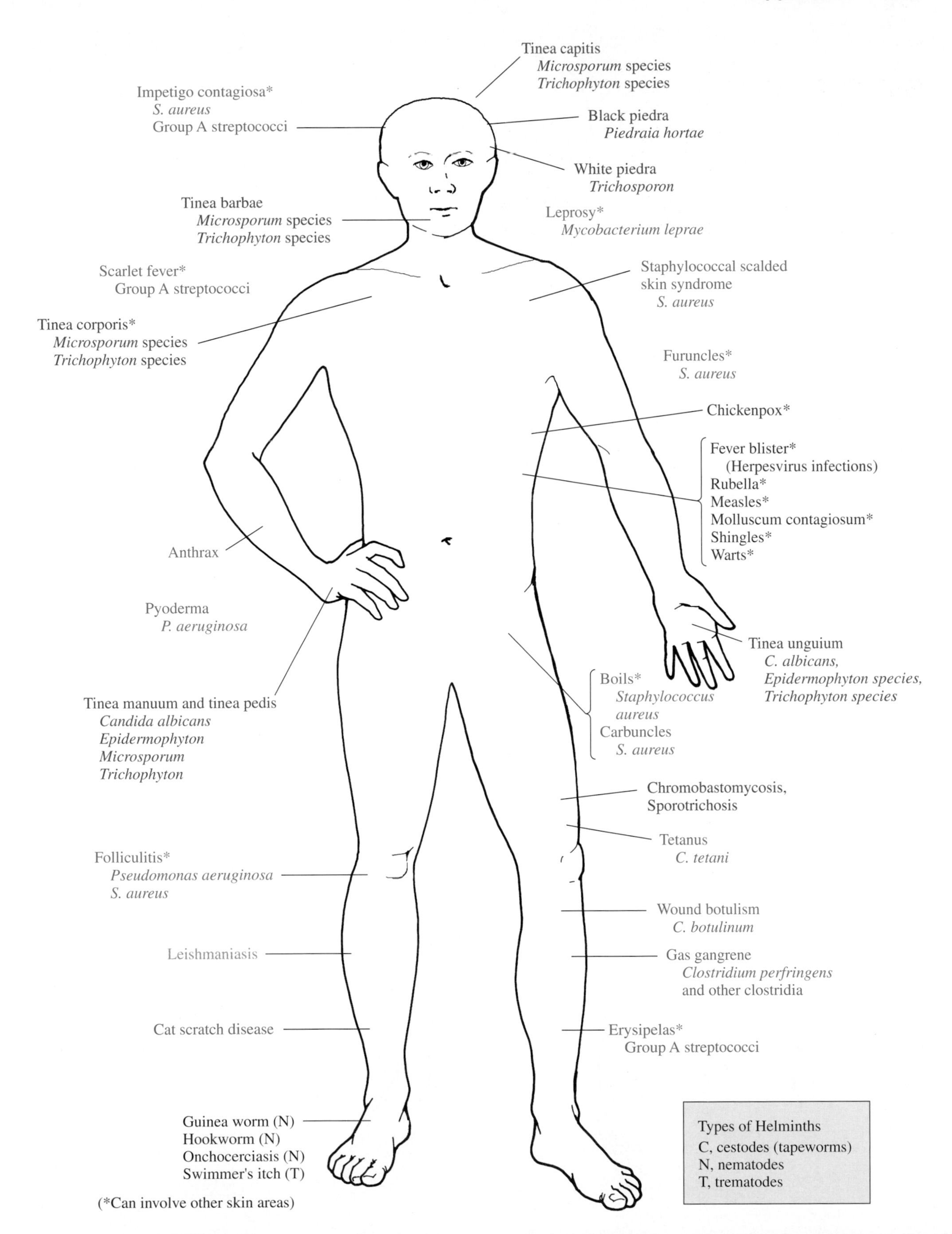


**Figure A-1**
Infectious Diseases and Disease Agents Affecting the skin, nails, and hair (Integumentary System).

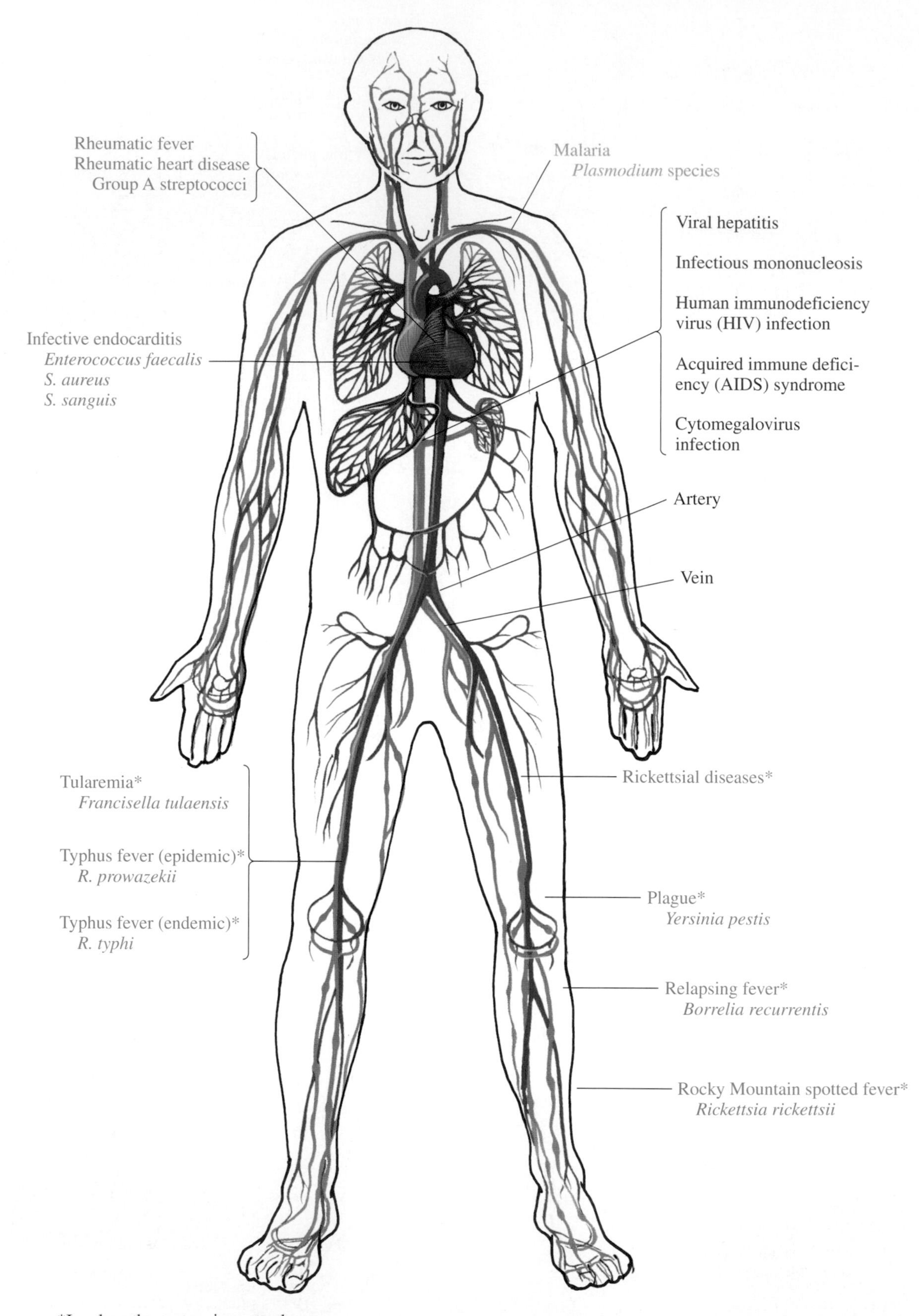


*Involves the system in general

**Figure A-2**
Infectious Diseases and Disease Agents Affecting the Cardiovascular System.

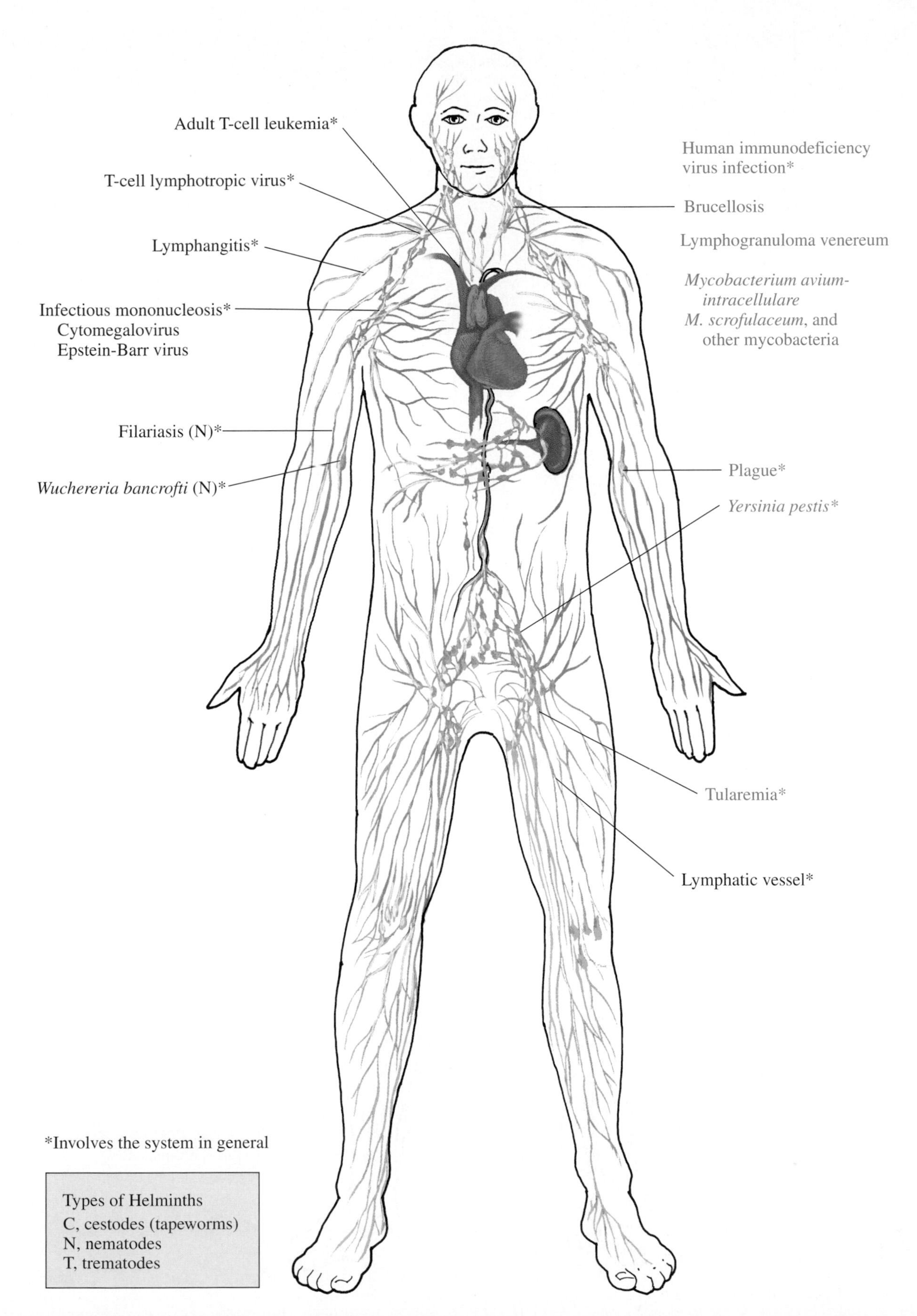


**Figure A-3**
Infectious Diseases and Disease Agents Affecting the Lymphatic System.

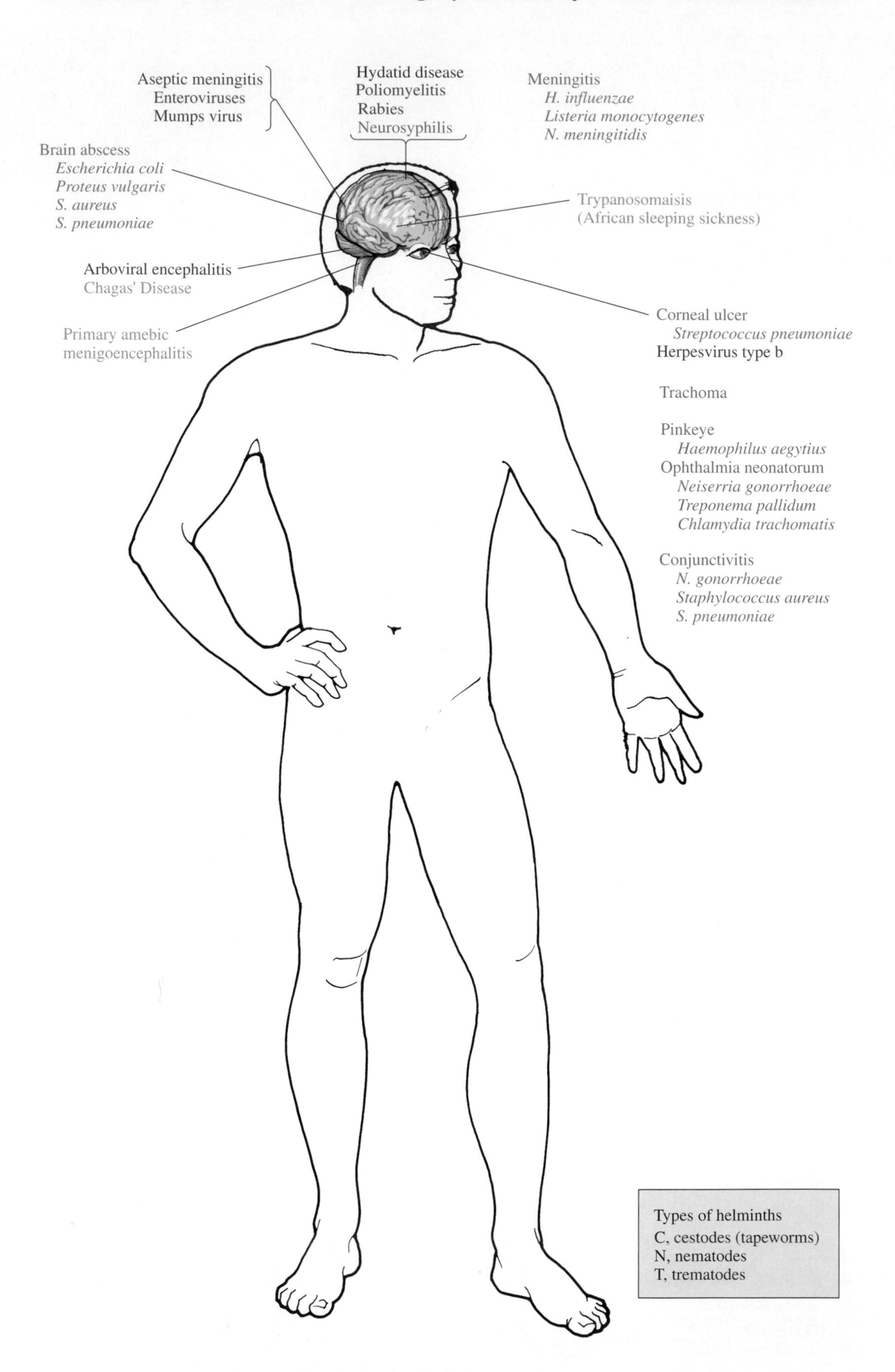


**Figure A-4**
Infectious Diseases and Disease Agents Affecting the Nervous System.

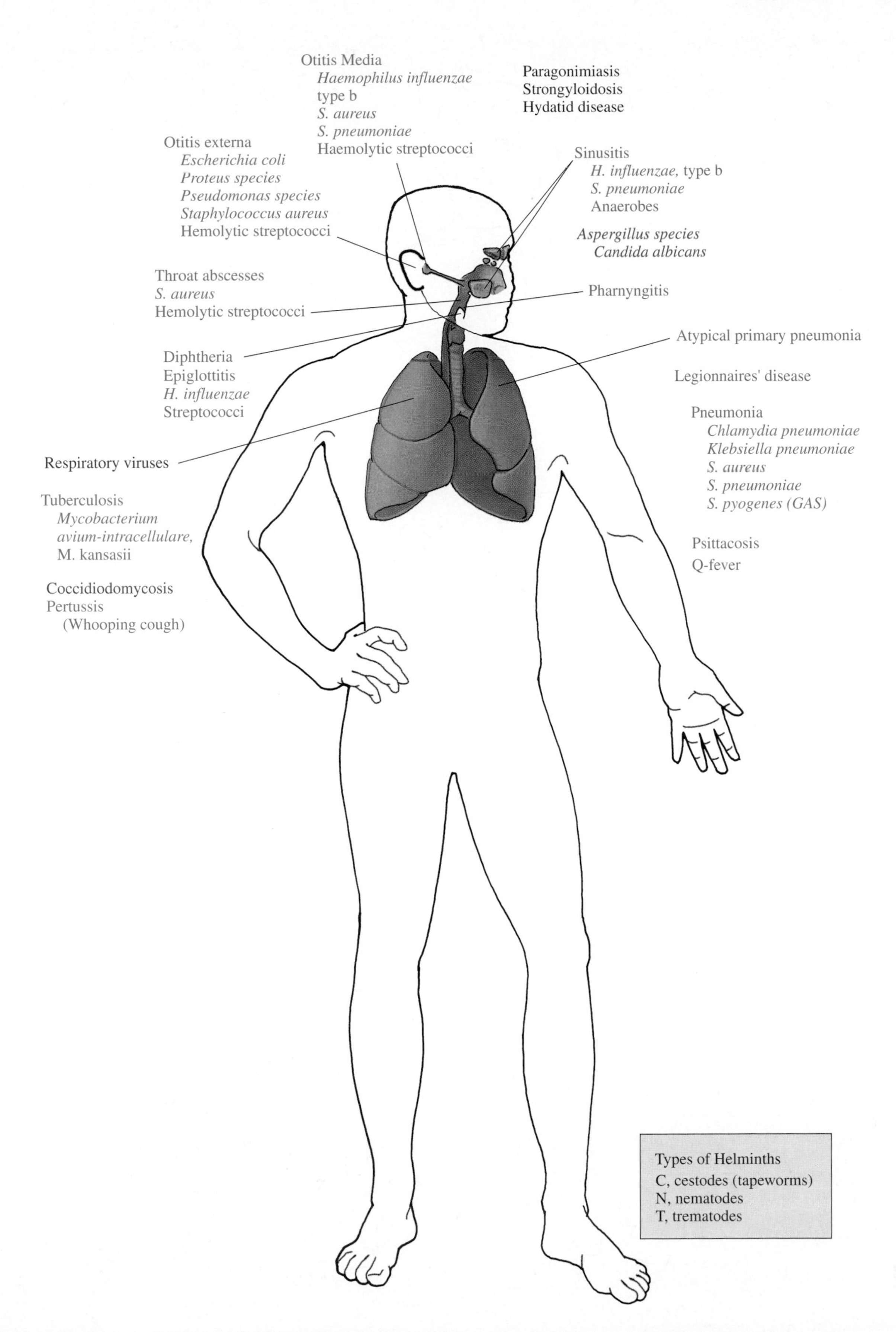


**Figure A-5**
Infectious Diseases and Disease Agents Affecting the Respiratory System.

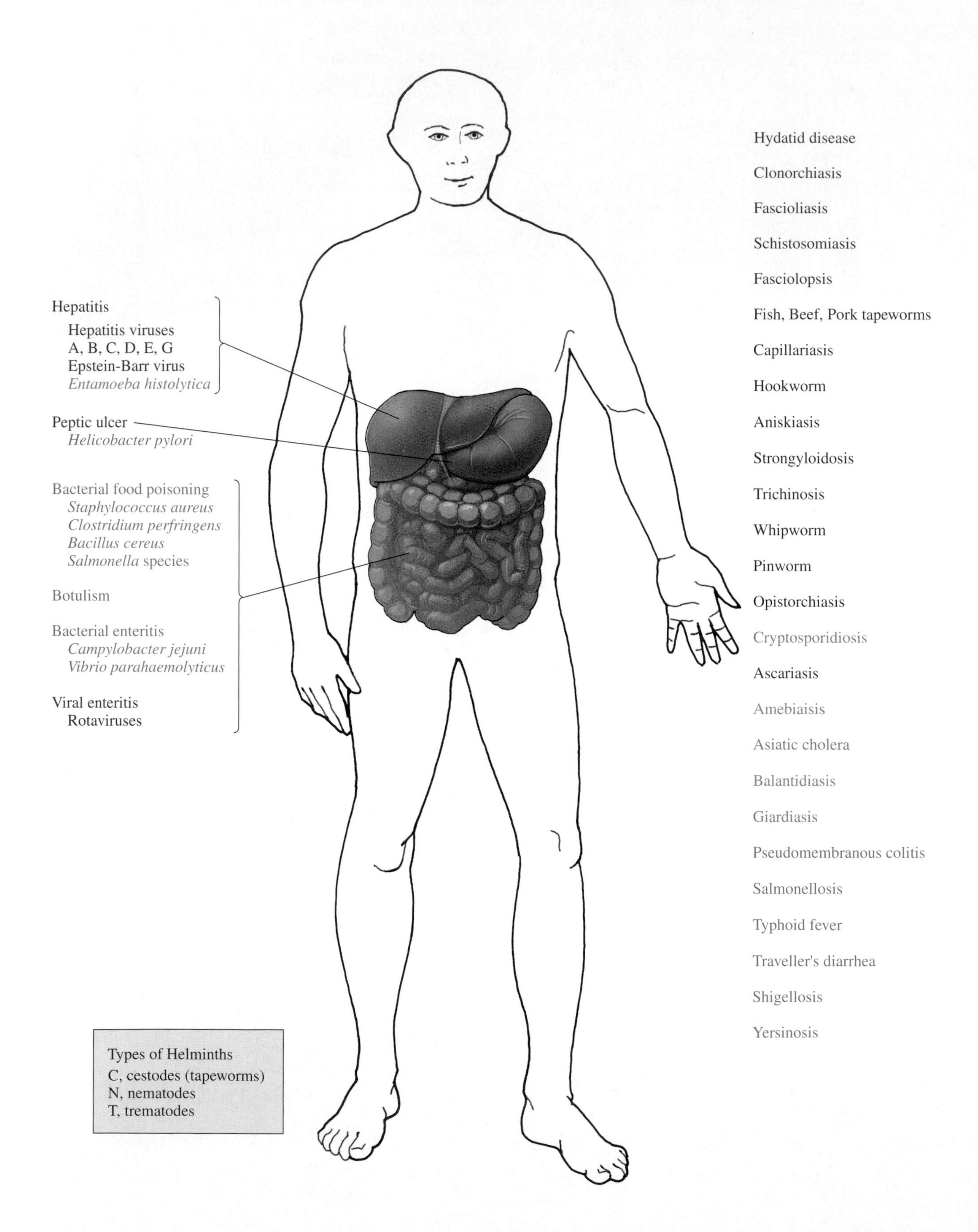


**Figure A-6**
Infectious Diseases and Disease Agents Affecting the Digestive System.

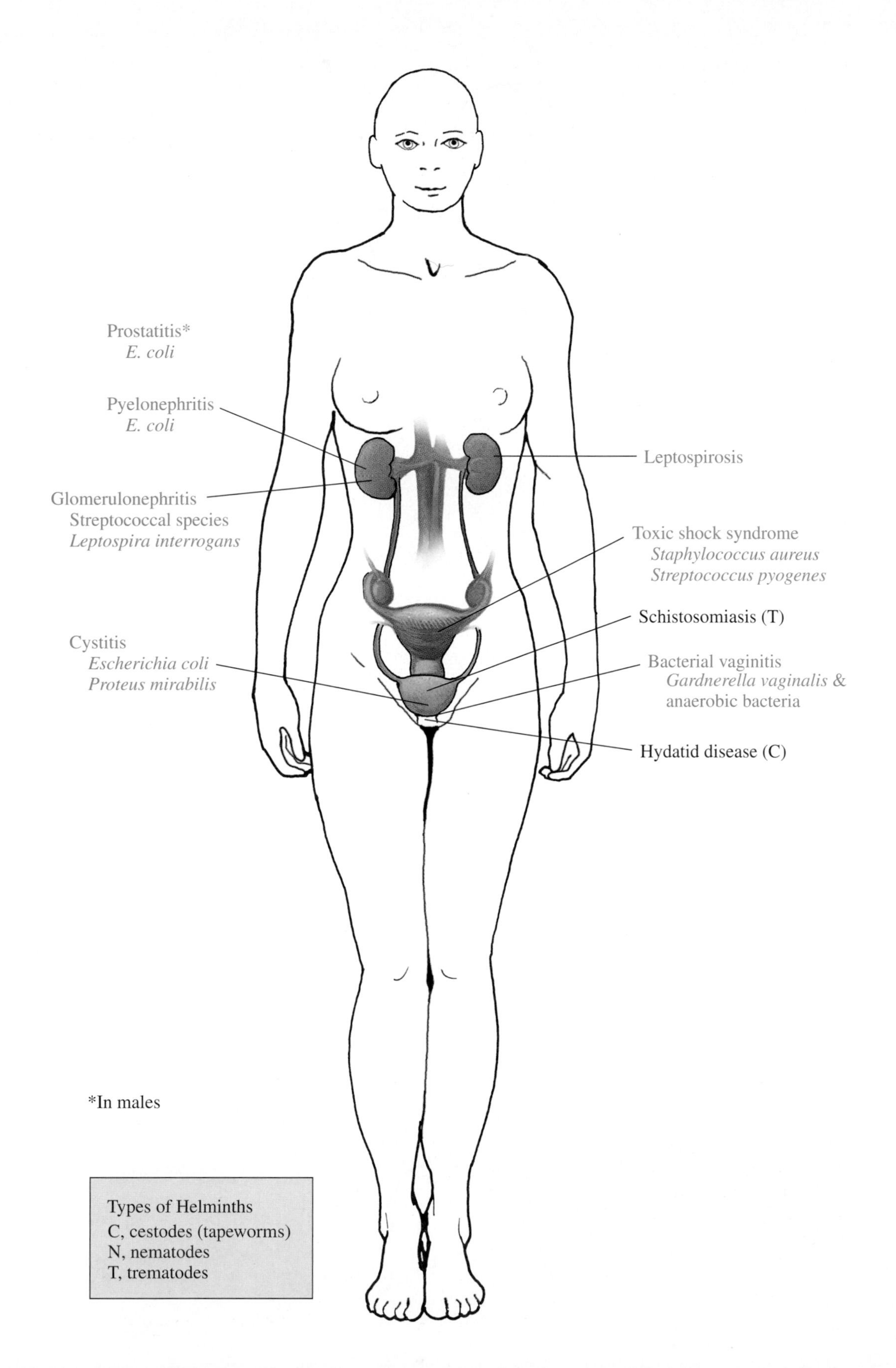


**Figure A-7**
Infectious Diseases and Disease Agents Affecting the Urinary System.

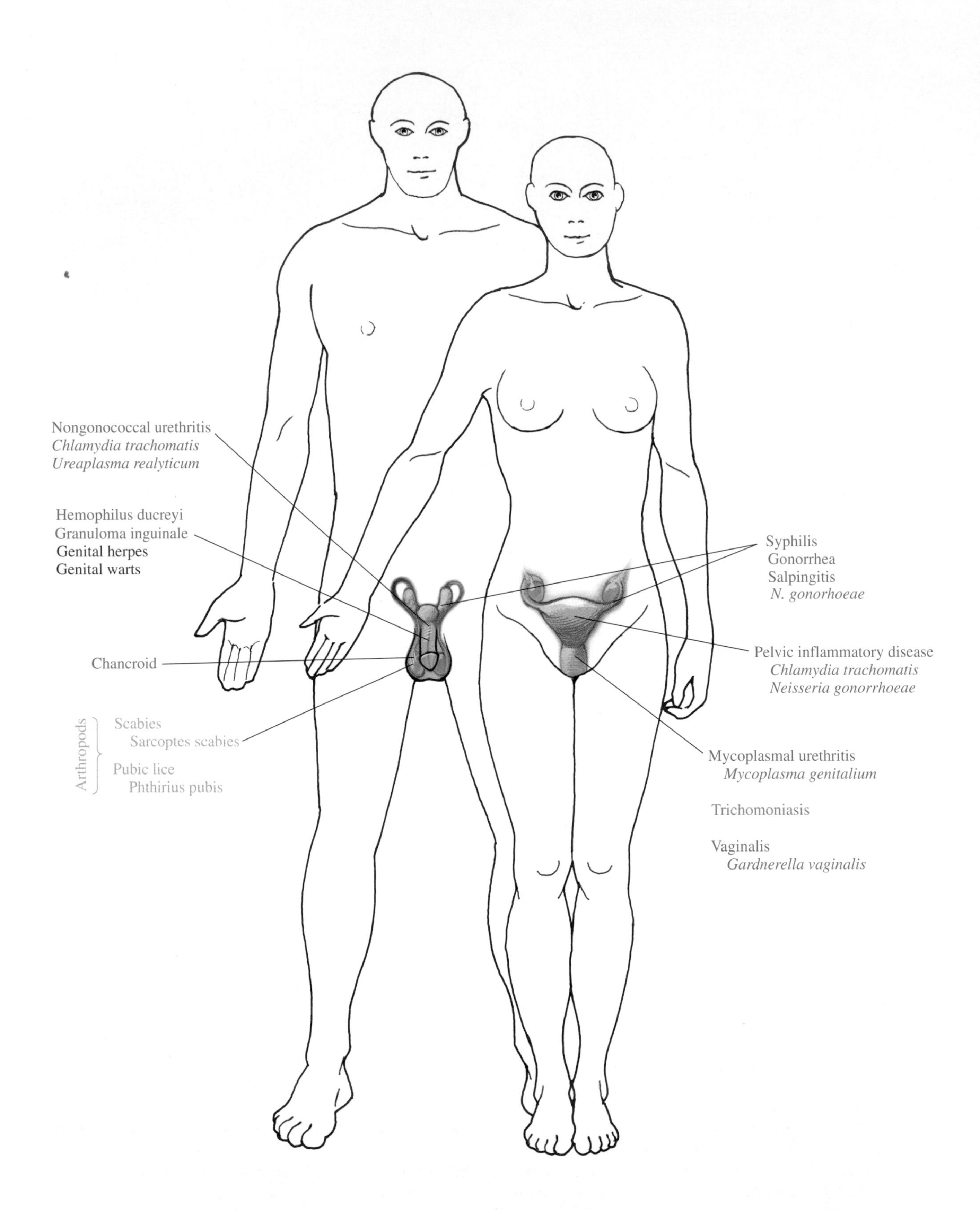


**Figure A-8**
Infectious Diseases and Disease Agents Affecting the Reproductive System.

# Glossary

**abscess** (AB-ses): an accumulation of pus in cavity hollowed out by tissue damage.

**aerobe**: a cell that is able to use oxygen as a terminal electron acceptor.

**acid-fast reaction**: a staining reaction in which organisms resist decolorization with acid-alcohol and retain the primary dye; acid-fast bacteria contain large amounts of long-chain fatty acids in their cell walls.

**aerial mycelium** (er-Ē-al): the portion of a mycelium that develops on surfaces.

**aerosol** (Er-ō-sol): atomized particles or droplets suspended in air.

**aerotolerant anaerobes**: bacteria that can survive in the presence of oxygen but do not use it in their metabolism

**agar** (Ā-gar): a dried polysaccharide extract of red algae used as a solidifying agent in microbiological media; agar generally melts at 99°–100°C, is solid at 42°C, and is attacked enzymatically by very few microbial species.

**agglutination** (a-gloo-ti-NĀ-shun): the clumping of cells.

**agranulocyte** (a-GRAN-yoo-lō-sīt): a leukocyte lacking cytoplasmic granules and having rounded nuclei.

**AIDS**: acquired immune deficiency syndrome; a disease caused by human immunodeficiency viruses (HIVs), resulting in the destruction of $T_4$ ($CD_4$) or T helper lymphocytes and other components of the immune system.

**alga** (pl, **algae**): any member of a group of eukaryotic, photosynthetic, and unicellular or multicellular forms of life; members belong to either the kingdom Plantae or Protista.

**alkaline** (AL-ka-līn): the condition caused by an abundance of hydroxyl ions ($OH^-$), resulting in a pH of greater than 7.0.

**alpha-hemolysis**: partial or incomplete breakdwon of hemoglobin by bacteria or other microorganisms growing on a blood agar medium; results in green zones surrounding bacterial colonies.

**amastigote**: the intracellular form of *Leishmania* species and *Trypanosoma cruzi* found in the human and reservoir hosts; the flagellum does not extend beyond the margin of the protozoan parasite.

**amphitrichous** (am-FĒ-tri-us): describes a cell that has a single flagellum at each end.

**amylase** (AM-i-lās): an enzyme that hydrolyzes starch.

**anaerobe** (an-ER-ob): an organism that cannot grow in an environment containing oxygen (air); a compound other than oxygen is the terminal electron acceptor; $O_2$ is toxic for the organism.

**annulus** (AN-ū-lus): a ring-shaped or collar structure found on the stak of some mushrooms.

**antibiotic** (an-tī-bī-OT-ik): a chemical produced by microorganisms and/or synthesized commercially that can inhibit the growth of or kill other microbes.

**antibody (immunoglobulin)**: a protein produced in response to an antigen that is capable of binding specifically to that antigen.

**antigen** (AN-ti-jen) (immunogen): a substance that stimulates the immune system to produce specific proteins known as antibodies (immunoglobulins) that react with the antigen and/or activate specific cells of the immune system.

**Apicomplexa**: intracellular protozoan parasites that have their organelles organized into an apical complex; such organelles are used to penetrate host cells.

**arthropod** (AR-thrō-pod): an invertebrate with a segmented body and jointed legs, such as a mosquito, tick, or related forms.

**arthrospore** (AR-thrō-spōr): an asexual spore formed by the fragmentation of specific fungal mycelial filaments; also called an arthroconidium.

**ascospore** (AS-kō-spōr): a sexual spore, characteristic of the Ascomycetes (Ascomycotina), formed in a saclike structure (*ascus*).

**asexual spores**: spores formed by a cell, without the fusion of nuclei of two different cells (sexual reproduction).

**bacterial colony**: a visible accumulation of bacteria on a solid culture medium.

**bacteriophage** (bak-TĒ-rē-ō-fāj): a bacterial virus; also referred to as *phage*.

**bacteriuria** (bak-tē-rē-yoo-rē-a): the presence of bacteria in urine.

**Basidiomycotina** (ba-sid-ē-ō-mī-KŌ-tē-na): filamentous fungi that form spores externally on a culb-shaped cell, the *basidium* (ba-sid-Ē-um).

**basidiospore** (ba-sid-ē-Ō-spor): a specialized asexual reproduction unit found with filamentous fungi in the Basidiomycotina.

**basophil** (BĀ-sō-fil): a class of leukocyte that is characterized by being stained with basic dyes producing blue-colored granules and lobed nuclei.

**beta (β) hemolysis**: a complete breakdown of hemoglobin, resulting in sharply defined clar zones surrounding colonies on blood agar.

**binomial system of nomenclature**: the scientific method of classifying and naming organisms, using a *Genus* and *species* designation.

**blackwater fever**: a complication of falciparum malaria consisting of massive hemolysis resulting in bloody urine.

**blood agar**: a differential and enriched medium used to identify microorganisms that cause hemolysis.

**blood group antigens**: genetically determined antigens (immunogens) found in with the membranes of red blood cells and realted tissues.

**bloom**: a heavy growth of algae or cyanobacteria on a water surface.

**bradyzoite**: the stage of *Toxoplasma gondii* that develops within tissue cysts.

**brood capsule**: the cystlike structure that originates from the germinal membrane of hydatid cysts associated with the worm *Echinococcus* species; the brood capsule contains primitive worm scolexes.

**bubo** (BOO-bō): a swollen lymph node, especially in the groin and armpit; a characteristic lesion in bubonic plague, chancroid, and certain other infectious diseases.

**budding**: a form of asexual reproduction in which a new cell is formed as an outgrowth from a parent cell; typcial of several yeast species.

**capsid**: the protein shell of a virus particle.

**capsule**: a polysaccharide or protein-containing structure surrounding and external to the cell walls of certain microorganisms.

**catalase**: an enzyme that catalyzes the breakdown of hydrogen perioxide ($H_2O_2$) to water and oxygen.

**cell wall**: a structure of unique chemical composition that lies close to and external to the plasma membrane; it confers rigidity and shape to bacteria, algae, and fungi.

**cercaria** (ser-KA-rē-a): the larval stage of trematodes (flukes) that after its development leaves the snail intermediate host and invades a final host.

**chancre** (SHANG-ker): the initial lesion in syphilis; a hard, painless, nondischarging lesion.

**chemically defined media**: media for the exact chemical composition is known.

**chlamydospore** (kla-MID-ō-spōr): a thick-walled, fungal resistant spore formed by the direct differentiation of mycelia cells; also referred to as a chlamydoconidium (kla-MID-ō-kōn-id-ē-um).

**chromatodial body**: deeply staining bundles of crystalline RNA found in young cysts of the genus *Entamoeba*.

**coagulase** (kō-AG-yoo-lās): an enzyme produced by pathogenic staphylococci that coagulates human blood plasma.

**cocco-bacillus**: an oval bacterium that is intermediate between the coccus and bacillus shapes.

**coenocytic** (SĒ-nō-sit-ik): a condition in which crosswalls are absent, thus allowing cellular material to flow uninterrupted in certain hyphae.

**colony**: a visible accumulation of microbial growth on the surface of a solid culture medium.

**complement fixation**: the binding (fixing) of complement to an antigen-antibody complex so that the complement is unavailable for a subsequent reaction.

**conidiophore** (kon-ID-ē-ō-for): a hypha that holds (bears) condiospores.

**conidium** (kō-NID-ē-um): an asexual fungal spore that may be one- or multi-celled and of various sizes and shapes; also called conidiospore.

**crutose** (krus-TŌS): a flat, crustlike growth of lichens.

**cyanobacteria** (sīn-an'-ō-bak-TĒ-rē-a): photosynthesizing prokaryotes that contain chlorophyll and phycocyainin.

**cyst** (sist): with bacteria, these are spherical, thick-walled resting cells; with protozoa, thick-walled environmentally resistant structures formed for reproduction purposes.

**cysticercus** (sis-ti-SER-kus): the larval encysted stage of the tapeworms *Taenia solum* and *T. saginata* containing fluid and a single head, or scolex.

**cytopathic effect**: the visible destructive effects frequently seen with virus-infected tissue culture cells.

**dark-field microscopy**: a form of microscopy in which organisms appear white against a dark background; spirochetes that stain poorly or not at all with the usual dyes are best observed in the living state by this method.

**defined medium**: a preparation that contains known specific kinds and concentrations of chemical substances.

**definitive** (de-FIN-i-tiv) **host**: the final host.

**denitrification**: the process by which nitrates ($NO_3$) are reduced to nitrous oxide (NO) or nitrogen gas ($N_2$).

**dermatophytes** (der-ma'-tō-FĪTS): fungi that invade the skin, nails, and hair; ringworm.

**Deuteromycotina**: a group of fungi in which no sexual stage occurs; also known as the Fungi Imperfecti.

**diatoms** (DĪ-a-toms): plantlike protists that are characterized by their glasslike outer shell.

**differential medium**: a culture medium capable of distinguishing bacterial species from one another by differences in colonial appearances or by changes produced in the medium.

**differential stains**: a procedure that uses more than one stain to distinguish parts of a cell from one another.

**dimorphic**: having two morphological forms, one displayed in the tissue of a host, the other in nature and commonly in cultures in the laboratory.

**ectoparasite**: a form of life that lives on the surface of another organism.

**ectothrix**: a sheath of fungal spores on the outside of a hair shaft.

**electrophoresis** (e-lek-trō-fōr-Ē-sis): a laboratory technique used to separate proteins and certain other large molecules by passing an electrical current through a specimen on a gel.

**elephantiasis**: an enlargement of the legs and/or scrotum in chronic lymphatic filariasis, a nematode infection.

**ELISA** (enzyme-linked immunoabsorbent assay): a highly specific immunologic diagnostic procedure that involves the linking of soluble antigens or antibodies to an insoluble solid surface to retain the reactivity of the antigen or antibody.

**endothrix**: a cluster of arthroconidia (arthrospores) within a hair shaft.

**euglenoid** (yoo-glen-oid): an alga containing chlorophyll *a* and *b* and belonging to the taxonomic division Euglenophyta.

**eukaryote**: a cell with a well-defined nucleus surrounded by a nuclear membranous envelope and having other membrane-bound organelles such as mitochondria.

**exospore**: a spore resistant to heat and drying that is formed external to the vegetative cell by budding.

**facultative anaerobe**: an organism that grows well under both aerobic and anaerobic conditions and for which oxygen is not toxic.

**fermentation**: the metabolic process in which the final electron acceptor is an organic compound.

**fluorescence** (floo-RES-ents) **microscopy**: a type of microscopy in which cells, their parts, or related structures are stained with a fluorescent dye and when exposed to ultraviolet light appear as glowing objects.

**fluke** (trematode): with the exception of the schistosomes, flukes are flat, leaf-shaped worms having both sexes (hermaphroditic) and oral and ventral suckers.

**foliose**: leaflike.

**fruticose** (FRU-ti-cōs): shrublike.

**gametocyte**: either the male or female form of the malaria and related parasites (sex cell).

**gamma hemolysis**: the absence of an enzymatic breakdown of hemoglobin; red blood cells in a blood agar medium remain intact.

**gelatinase**: the exoenzyme (extracellular) that degrades the protein gelatin.

**Gram stain**: the differential staining procedure developed by C. Gram to distinguish two groups of bacterial cells from one another, gram-positive and gram-negative.

**granulocytes**: one group of white blood cells having differently staining granules within their respective cytoplasms; neutrophils, eosinophils, and basophils.

**gumma**: granulomatous lesion found in the skin, bone, and liver during the tertiary stage of syphilis.

**helminth**: a worm.

**hemolysis** (hē-MOL-i-sis): disruption of red blood cells.

**hermaphroditism** (her-MAF-rō-dit-izm): the presence of both female and male organs in the same individual.

**hydatid cyst**: the larval stage of the sheep tapeworm *Echinococcus granulosus* consisting of a fluid-filled cystic (bladderlike) structure.

**hypha** (HĪ-fa): the structural vegetative unit of a fungal mycelium.

**indicator system**: one of the components of the complement fixation test; consists of hemolysin (antisheep red blood cells) and sheep red blood cells.

**immunogen** (im-MYOO-nō-jen): a substance that stimulates the formation of immunoglobulins (antibodies).

**immunoglobulin** (im-mu-nō-GLOB-yoo-lin): refers to protein molecules produced in response to immunogens (antigens).

**intermediate** (in-ter-MĒ-dē-at): a form of life used for development of larval stages of helminths.

***in vitro*** (VĒ-trō): refers to procedures and/or tests performed outside of the body, usually in test tubes, microtiter plates, etc.

**larva** (lar-VAH): developing stage of an insect or a worm.

**lesion**: an area of injury or a circumscribed pathological tissue change.

**lichen**: a plantlike form of life consisting of a symbiotic relationship between a fungus and a photosynthetic alga or cyanobacterium.

**litmus**: plant extract used as a pH indicator; also an oxidation-reduction indicator; turns blue when alkaline and red when acid in reaction.

**lymphocyte**: a type of leukocyte arising in lymphoid tissues and the most important cell in specific immunity.

**lysis**: disruption, or breaking apart, of cells.

**macroconidium**: the larger of two types of conidia produced by a single species of fungus.

**macrogametocyte**: the female form of the malaria and related parasites usually found within red blood cells.

**medium**: a nutrient preparation used to grow microorganisms; it may be a liquid or a solid.

**merozoite**: the form of the malaria parasite that results from asexual multiplication within the RBC; the form that is released from the mature schizont to infect other RBCs.

**metabolism**: the total chemical activities of an organism; consists of anabolism and catabolism.

**metacercaria**: an encysted larval trematode stage; it is found in second intermediate hosts and is infectious for the final (definitive) host.

**metachromatic** (met-a-krō-MAT-ik) **granules**: a reservoir of inorganic phosphate within a bacterium that is stainable by basic dyes.

**microaerophilic** (mī-krō-AR-ō-fil-ik): refers to aerobes that require environments with small amounts of oxygen or less than atmospheric oxygen levels for growth.

**microconidium**: the smaller of two types of conidia produced by a single species of fungus.

**microfilaria**: the embryonic roundworm produced by adult female filarial parasites including *Onchocerca volvulus*.

**microgametocyte**: the male form of malaria and related parasites within the RBC.

**micrometer**: unit of measurement for microorganisms and other forms of life; 1 micrometer (μm) equals 0.001 mm or 1/25,400 in. (*micrometer* replaces the older term *micron*.)

**microsporidia**: protozoa that are characterized by the production of spores containing coiled polar tubes.

**miracidium** (meh-ra-SID-ē-um): a ciliated free-swimming development form of a fluke that hatches from a helminth egg.

**mitosis**: nuclear division that follows duplication of chromosomes and results in daughter nuclei with chromosomes identical to the parent cell.

**molds**: filamentous fungi.

**monoclonal antibodies**: antibodies of a specific type produced by cells arising from a single clone of antibody-producing cells.

**monocyte**: actively phagocytic, monoculear white cell found in the bloodstream; as monocytes mature, they emigrate into tissues and differentiate into macrophages.

**monotrichous**: having a single flagellum at one end of the cell.

**mordant**: a substance that fixes (precipitates) a stain; e.g., iodine in the Gram stain.

**negative stain**: a staining procedure in which the background surrounding a specimen is stained, but the specimen is not.

**nematode** (NEM-a-tōd): a roundworm.

**nitrate reduction**: reduction of nitrates to nitrites or ammonia.

**nitrification** (nī-tri-fi-KĀ-shun): the conversion of ammonia to nitrate.

**nonseptate** (non-SEP-tāt): having no dividing walls (septa) in hyphae (filaments).

**nosocomial** (nōs-ō-KŌ-mē-al) **infection**: an infection acquired in a hospital or other health care facility.

**obligate aerobes**: organisms that must use oxygen as their final electron acceptor.

**obligate anaerobes**: organisms unable to grow in the presence of oxygen.

**oncosphere**: the embryo in the eggs of tapeworms; it has six hooks.

**oocyst**: a stage of development in the life cycle of apicomplexans. The mature oocyst contains infectious sporozoites; it is the diagnostic form found in feces in infections with *Cryptosporidium*, *Cyclospora*, and *Isospora*.

**ookinete**: a motile zygote formed by sporozoans such as the malarial parasite (*Plasmodium* species).

**operculum**: the lidlike portion of the eggshell of most trematodes and the fish tapeworm *Diphyllobothrium latum*; the larva hatches from the egg through the opening made by detachment of the operculum.

**opportunistic infection**: infectious disease caused by a microorganism without major virulence factors that takes advantage of a host's immunosuppression.

**ovum** (Ō-vum): the female reproductive cell; egg.

**oxidase test**: a test to distinguish colonies of *Neisseria* from other bacteria; does not differentiate among *Neisseria* species.

**plaque** (viral): a clear area in the confluent growth of a bacterial or cell culture due to lysis by a phage or virus.

**pock**: a local blisterlike formation caused by virus infection of the chorioallantoic membrane.

**pour** (PŌR) **plate**: a basic technique used in the culturing and/or isolation of bacteria in which melted, yet sufficiently cooled, medium is inoculated with a bacterial culture, introduced into a sterile Petri dish, and allowed to harden; the individual bacteria trapped within the medium grow and eventually form colonies.

**precipitin** (prē-SIP-ē-tin) **reaction**: the basis of an immunologic reaction in which antigens and antibodies diffuse toward one another, resulting in a precipitate in which these reactants are in equivalent proportions.

**proglottid**: an individual segment of a tapeworm; a mature proglottid contains both male and female sex organs, and a gravid proglottid contains eggs (ova).

**prokaryote**: a cell lacking a membrane-bound nucleus and membrane-bound organelles.

**promastigote**: the form of *Leishmania* species found in the sand fly vector and in culture; it is elongate and has a prominent anterior flagellum.

**Rh (factor) blood group**: a blood group agglutinogen discovered on the red blood cells of Rhesus monkeys and given the designation *Rh;* the factors are found to a variable degree in human blood cells.

**rhizoids**: rootlike structures characteristic of some molds.

**schizont**: a stage in the prerythrocytic reproductive cycle of the malaria parasite.

**scolex** (SKŌ-leks): the head of a tapeworm.

**selective and differential medium**: a preparation that incorporates the features of both selective and differential media.

**serology**: systematic study of blood serum; e.g., reactions between antibodies and antigens.

**spore print**: a technique used to determine the properties of spores and lamellae.

**sporocyst** (SPŌR-ō-sist): a developmental stage of a fluke formed from a miracidium containing reproductive cells and found in snail (intermediate host) tissues.

**sporozoite**: form of malaria parasite injected into the bloodstream by an infected mosquito.

**streak plate**: basic technique used in the culturing and/or isolation of bacteria in which an inoculum is spread over the surface of a medium by means of an inoculating loop or needle; the isolated bacteria in the inoculum grow and form colonies on the surface of the medium.

**substrate**: substance acted on by an enzyme.

**tachyzoite**: the rapidly dividing, crescent-shaped form of *Toxoplasma gondii* and related parasites.

**tapeworm**: a ribbonlike helminth (cestode) that has an attachment organ called a scolex and developing segments called proglottids.

**tinea**: the medical term for the ringworms.

**trophozoite**: the motile, feeding form of protozoa.

**trypomastigote**: also called trypanosome; this form of the genus *Trypanosoma* is found in the human host; the organism is elongate and has an undulating membrane that runs the length of the body and extends as a flagellum from the anterior end.

**vector**: any agent that carries a disease from one host to another; may be animate, as insects, or may be inanimate (nonliving), as soil.

**Voges-Proskauer reaction**: test for acetylmethylcarbinol (also called acetoin) production; used to differentiate *Escherichia* and *Enterobacter* of the coliform group.

**Ziehl-Neelsen** (zēl-NĒL-sen) **method**: a differential staining procedure used to distinguish species of *Mycobacteria* and *Nocardia* and certain pathogenic protozoa; these microorganisms are acid-fast.

**zoonosis** (zō-NŌ-sis): a disease communicable from lower animals to humans under natural conditions.

**zygospore**: sexual spore that results from the fusion of like gametes; characteristic of Zygomycetes (Zygomycotina).

# Index

Page numbers followed by a *t* indicate a table; numbers followed by an *f* indicates a figures.

## A



## B

## T

# Additional Photo Credit

| | |
|---|---|
| Figure 7b | Lesley C. Alpert, M.D., Ph.D. |
| Figure 8 | Richard M. Gore, M.D. Dept of Radiology Northwestern University Medical School Chicago IL 60611 |
| Figure 9b | University of Rochester, Department of Radiology |
| Figure 10 | Philips Ultrasound, Inc. |
| Figure 33 | Photo courtesy of Becton Dickinson Microbiology Systems. Gas Pak and Gas Pak Plus are trademarks of Becton, Dickinson and Company. |
| Figure 75 | Centers for Disease control, Atlanta, GA 30333 |
| Figure 90 | Centers for Disease control, Atlanta, GA 30333 |
| Figure 92 | Centers for Disease control, Atlanta, GA 30333 |
| Figure 98a,c | U.S. Public Health Science |
| Figure 99 | Centers for Disease control, Atlanta, GA 30333 |
| Figure 119 | Centers for Disease control, Atlanta, GA 30333 |
| Figure 140 | Centers for Disease control, Atlanta, GA 30333 |
| Figure 178a–c | Department of Health & Human Services |
| Figure 191 | Centers for Disease control, Atlanta, GA 30333 |
| Figure 194 | Centers for Disease control, Atlanta, GA 30333 |
| Figure 198 | Centers for Disease control, Atlanta, GA 30333 |
| Figure 203 | Centers for Disease control, Atlanta, GA 30333 |
| Figure 237a–g | Department of Health, education, and Welfare. Public Health Service |
| Figure 264 | Department of Health, education, and Welfare. Public Health Service |
| Figure 265a–j | Department of Health, education, and Welfare. Public Health Service |
| Figure 265a–j | Department of Health, education, and Welfare. Public Health Service |